AF725158

Disrupt Science

Mihai Nadin

Disrupt Science

The Future Matters

Springer

Mihai Nadin
The University of Texas at Dallas
antÉ—Institute for Research
in Anticipatory Systems
Richardson, TX, USA

ISBN 978-3-031-43956-8 ISBN 978-3-031-43957-5 (eBook)
https://doi.org/10.1007/978-3-031-43957-5

© The Editor(s) (if applicable) and The Author(s), under exclusive license to Springer Nature Switzerland AG 2023

This work is subject to copyright. All rights are solely and exclusively licensed by the Publisher, whether the whole or part of the material is concerned, specifically the rights of translation, reprinting, reuse of illustrations, recitation, broadcasting, reproduction on microfilms or in any other physical way, and transmission or information storage and retrieval, electronic adaptation, computer software, or by similar or dissimilar methodology now known or hereafter developed.
The use of general descriptive names, registered names, trademarks, service marks, etc. in this publication does not imply, even in the absence of a specific statement, that such names are exempt from the relevant protective laws and regulations and therefore free for general use.
The publisher, the authors, and the editors are safe to assume that the advice and information in this book are believed to be true and accurate at the date of publication. Neither the publisher nor the authors or the editors give a warranty, expressed or implied, with respect to the material contained herein or for any errors or omissions that may have been made. The publisher remains neutral with regard to jurisdictional claims in published maps and institutional affiliations.

This Springer imprint is published by the registered company Springer Nature Switzerland AG
The registered company address is: Gewerbestrasse 11, 6330 Cham, Switzerland

Paper in this product is recyclable.

To make public my love for her would be an indiscretion.
To make my respect known, a moral obligation.

Through their patient reviews, A. H. Louie, George Păun, and Sanda Reinheimer-Râpeanu helped me clarify the text. I am grateful to Ioan-Aurel Pop, President of the Romanian Academy, and Marius Andruh, Vice-President, for making possible an in-depth discussion of the book by eminent scholars from Romania who are members of the Academy. Dr. Thomas Ditzinger and his team at Springer Nature helped me cope with the publication effort. Many others (requesting anonymity) deserve credit for serving as sounding boards, often bringing me back from the abstraction of thinking to the concreteness of life.

Contents

A Woke Moment: The Time for a Second Revolution in Science is Now

A lifeless corpuscle brought to its knees a civilization that takes credit for—when not boasting of—stellar scientific and technological achievements. *Perseverence* landed on Mars on February 18, 2021. During the same time period, lockdowns and a slew of arbitrary measures were tried for containing what became the COVID-19 pandemic. As the Atlas V-541 rocket (launched July 3, 2020) made it to the Red Planet, vaccines, genetically concocted days after the virus was sequenced, were fast-tracked with the hope of accelerating artificially induced herd immunity. During this time, an additional 10 million people got infected and over a million more died. Quite a spectacular image in contrasts. Physics, chemistry, computer science at their best in dealing with the lifeless universe—but notwithstanding in search of the origins of life—in contrast to the rather primitive understanding of what is alive, and how lives can be saved. All of this under circumstances in which nothing other than sustainability—i.e., the future—is at stake. The James Webb Space Telescope was launched (December 25, 2021) while a virus variant defied the inadequate science meant to tame it. The huge eye might peer far back in time—1000 years into the past, or maybe even to the beginning of the universe, millions of years ago. But here on planet Earth, the future, above and beyond climate change, cannot be taken for granted. It appears as though the much acclaimed and generously supported science—over 2.5 trillion dollars annually—blind to the future, might actually undermine life.

The COVID-19 pandemic—almost seven hundred million infected worldwide, almost seven million dead—is part of a larger breakdown: CoVID, i.e., the Crisis of Vision[1] into which humankind drove itself, not realizing that the opportunities afforded by reductionist-determinism pale in comparison to the self-inflicted vulnerabilities of a skewed perspective of reality. When these became apparent, it felt as though somebody pulled the plug on the machine of never-ending progress. The magic of continuous progress was suspended; the hour of anguish began. It was a

[1] Crisis (C) of (o) Vision (VID, from the Latin for "seeing").

© The Author(s), under exclusive license to Springer Nature Switzerland AG 2023

M. Nadin, *Disrupt Science*, https://doi.org/10.1007/978-3-031-43957-5_1

revolution whose time has passed. Robbed of its sovereignty, society got sucked into the whirlwind of woke-isms of all kinds: rage—justifiable or not, accumulated over time—against real and imaginary expressions of privilege, including that of science. Cancel culture and the obsequious acceptance of the tyranny of the few revived the dictatorship of political correctness handed down from bygone dictatorships. Newly concocted prejudices were added to those inherited from the justifiably maligned past. Reaction, to which our now global civilization is conditioned, metastasized. Within CoVID, social media (which are anything but social), in cahoots with those who wield power, made it all but impossible to distinguish the real from the opportunistic fake. "React to the past" did not translate into the willingness to "Act for a different future."

Concretely, the reaction to the pandemic means almost one hundred trillion dollars lost worldwide. Worse yet: lives ended or made miserable. Social, political, and moral breakdowns piled up. Over two years were wasted in sub-mediocre online educational improvisations. Remote work has done away with any sense of solidarity—an unintended consequence of technological progress. Inherent liabilities, resulting from sacrificing the future for the sake of immediate gratification, revealed themselves through the reality of pain and loss. This could have been prevented. In the name of the science that actually led to the breakdown, the pandemic became the never-ending tale of costly and often ineffective reactive measures. Like so many reactions preceding this crisis—blind, expensive, futile. Under the pressure of government interventions, COVID-19 mutated into the fastest and largest transfer of wealth, mostly through increased systemic corruption. "Scientists are going to kill us all!" became part of the pandemic demagoguery. Many wondered: After the Cartesian Revolution replaced religion, wasn't science supposed to save us? Heroic efforts of physicians notwithstanding, medicine—ever more a business than a sacred, or at least moral, vocation—failed society. In the embrace of a science adequate for guiding rockets but not for healing, medical science ended up scoring for physics and chemistry, but not for life. Ironically, publicly funded failures claim the right to be celebrated—a grotesque expression of entitlement.

CoVID is a self-inflicted wound. It originates in the oblique perspective of the Cartesian Revolution that has governed science since the eighteenth century. It advances the rationality of a limited and limiting understanding of causality:

(1) The effect is contingent on past causes (determinism).
(2) You can understand the whole if you understand its parts (reductionism).

The ill-informed determinism and misleading reductionism of the Cartesian Method are embodied in a mechanistic view of reality: Everything, including the human being is a machine. The Industrial Revolution legitimized the rationality of unlimited expectations of progress at any cost. If anything goes wrong—pandemic, economic crisis, social breakdown—reaction, i.e., fixing the machine, is the way to go. The price is charged to coming generations. Within the broader context of CoVID, the pandemic was approached in full disregard of the fact that reaction is more expensive than anticipatory action. And in the long run, it is unsustainable.

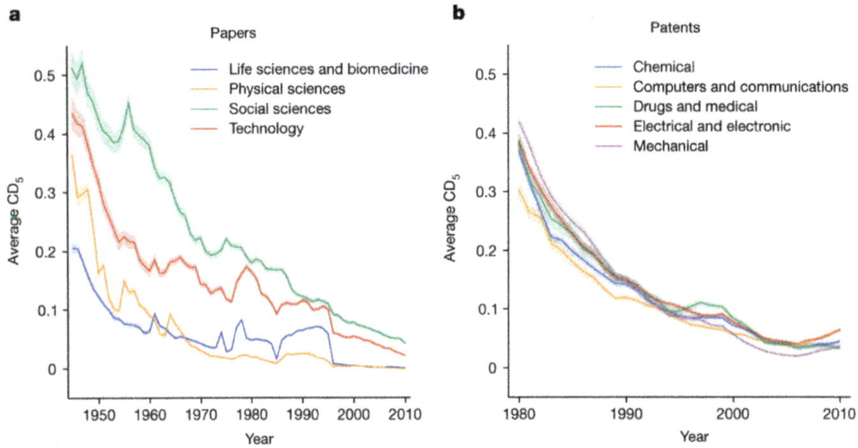

Fig. 1 While publications are gaining in number of references, they are less and less opening new directions of research. The magnitude of decline ranges from 91.9% (social sciences) to 100% (physical sciences). For patents, lines correspond to National Bureau of Economic Research (NBER) technology categories; from 1980 to 2010 the magnitude of decline ranges from 93.5% (computers and communications) to 96.4% (drugs and medical). Over 25 million papers and 3.9 million patents from the web of science (WoS) were analyzed. A citation-based measure (pertinent to 390 million citations) was applied in order to identify new directions

Science and technology grounded in the Cartesian Method entail risks of an order of magnitude larger than that of the indisputable progress made possible. The Upside-Downside ratio—visualized as the inverted yield curve—flipped. In keeping with the language of the Upside/Downside ratio, let us consider all that it takes to achieve a goal, compared to what it takes to deal with its consequences. For the sake of illustration: making available fossil fuels as an energy source (for heating, for cars, trucks, ships, etc.,) vs. dealing with the consequences of becoming dependent on them. Likewise, for the use of antibiotics and pain killers. The underpinning of the modern technology of wars, of nuclear weapons, of genetically engineered viruses is the same as that upon which the prosperity of "ever more and ever cheaper" is based. In short: living at the expense of the future. Addicted to the illusion, citizens increasingly become dependent on governments controlling their choices. In the ongoing Crisis of Vision (CoVID), the long-term dependencies of the industrial and post-industrial ages are accepted as the price for progress. They are as dangerous to the future of humanity as is the engineered human being (and by extension, the engineered nature), treated as a machine after being forced to behave like one.

Science is expected to guide humankind into the future. Surrendering itself into the arms of an incomplete view of reality, the science of the last hundred years has been goading society towards a dead end. Indeed, regurgitated in millions of publications and patents of dwindling impact (Fig. 1), it brought humanity to the verge of a chasm.

As the headlines testifying to the sad state of science have it, "No one knows why" (Kozlov 2023). *Disrupt Science*, arguing for a change of perspective, is the answer.

The impressive accomplishments of science, as well as its abject failures, are testimony to the impact it has. However, an inherently benign condition, sometimes morphing into the malign, renders progress at the cost of the future questionable. The argument set forth in this book is straightforward: The past is relevant, but so is the possible future—the unavoidable vector of life. The living, in its open-ended variety of embodiments, does not only react to change. It makes the future in an anticipatory manner, corresponding to its purposefulness.

Awareness of the fact that the human condition transcends that of the matter in which it is embodied explains, and indeed justifies, the call to *Disrupt Science* in its current state. The dynamics of living matter includes reaction, quite well described in deterministic and reductionist science, but it is not limited to it. Determinism informs reactive actions—progress at the cost of depleting resources. Reductionism does away with complexity, in which life is couched. The living is by necessity anticipatory. Preserving life is not in reaction to, but in anticipation of change. Living entities, from the simplest bacterium to plants, and insects, to the complex human being, are adaptive, goal-oriented, and capable of self-repair. Anticipatory actions involve non-deterministic processes and engage the wholeness of the living, including its environment. The adaptive nature of the living and the goal-driven dynamics of life are expressed in actions that are anticipatory in nature. Awareness of consequences informs actions that reflect the creative nature of human beings.

Redefining science—and implicitly, medicine—is not a negation of its past, but rather an affirmation of trust in its capacity to renew itself. The moment of truth can no longer be postponed. At stake is the future of humankind, and even of life on planet Earth. The record of breakdowns (including so-called natural disasters), by now global in scale, is part of the empirical premise for the call for completing the *Cartesian Revolution.* A "Second Revolution in Science" could unleash humanity's remaking: free of surrendering to want, and finding the opportunity not only to measure everything—life included—but also to realize its meaning. The future matters.

Reference

Kozlov, M.: 'Disruptive' science has declined — and no one knows why, Nature, January 4 (2023) DOI: https://doi.org/10.1038/d41586-022-04577-5

CoVID—A Confluence of Breakdowns

The knowledge perspective of Descartes's Method was revolutionary, but rather skewed. The Cartesian view of reality explains change exclusively in reductionist-deterministic terms: the whole is the sum of its parts; the past (cause) determines the present (effect). Left out was the realization that life, in all its embodiments (from the simplest mono-cell to the complex human being) is goal driven. In contradistinction to non-living matter, life is not locked in a predefined way of functioning, like machines are. Anticipatory action, complementary to reaction, prepares the living for possible changes. *Disrupt Science* expresses the urgent need for science to reinvent itself. By integrating determinism and reductionism (to which science owes spectacular achievements) and anticipation, with its implicit holistic view of change, it will better reflect the dynamics of reality. Sustainability is the outcome of the realization that *The Future Matters.*

Forty years ago—but still ongoing—the HIV/AIDS global epidemic took center stage. Almost 40 million individuals have died so far, and close to 100 million got infected. A multi-pronged attempt at treatment (with antiretroviral therapy) eased the world into accommodating the individuals affected by this life-threatening condition. There is no cure, there is no vaccine, only maintenance. The costs incurred in a variety of reactive activities—fix what can be fixed—amounts to an initial 500 billion dollars in treatment, cash and housing assistance, research funding, and prevention. To state it loud and clear: only 3%—three percent—of the total was earmarked for prevention. The value of lives lost or otherwise affected cannot be captured in a dollar-indexed ledger.

Around that time (1979–1982), the USA experienced two recessions. Unemployment reached 10.8%; the Savings and Loan banks lost over 400 billion dollars in book value. The bank bailout—not the first, and not the last—cost 293 billion dollars. Such numbers seem modest today, but the USA budget was in the low hundreds of billions (the Carter budget was 304 billion dollars). At that time, the GDP (the total monetary value of finished goods and services produced in a nation usually within a year) was around three trillion dollars. Then came *Black Monday* (October 19, 1987), and the losses—to which the world reacted as though a dam had burst—added up to 1.7 trillion dollars. For each of those episodes, the value of preventive measures was below 0.1 percent of the volume transacted. A partial list of what could have been done

© The Author(s), under exclusive license to Springer Nature Switzerland AG 2023
M. Nadin, *Disrupt Science*, https://doi.org/10.1007/978-3-031-43957-5_2

to prevent similar situations includes mitigating risk, deploying intelligent analysis, educating individual investors and investment professionals, enacting regulation that connects compensation to performance in a manner equitable to all involved.

These events were followed by the "dot-com" bubble (2001)—1.755 trillion dollars vanished; the severe acute respiratory syndrome/SARS (2002–2003)—over 200,000 people dead; the "Great Recession" (2008–2009)—costing over $70,000 *per capita* in income (in the USA), or, expressed otherwise, over two trillion dollars in global economic growth. The unfortunate war in Afghanistan (20 years old when it was called off, in defeat, in 2021) adds up to more than that. It might be that many social and political forces were triggered in what amounts to a confluence of polarizing moments that preceded the housing crisis leading to the Great Recession. The economic damage cannot be effectively separated from the social and political consequences that extend to the current global breakdown of the system of values. The numbers used to describe the cost are of a different order of magnitude from 25 years ago. The subsequent H1N1 "swine flu" (2009), over half a million dead; the Ebola scare (2014–2016, and again in 2018), the MERS epidemic (2015), and the indelible Zika virus (2019) are like seismographic warnings for what followed. Covid-19 (since December 2019) became the ongoing earthshaking event. Nobody can yet put an exact number to what is still advancing, wave after wave, all over the world.

If this retrospective is not sobering enough, consider natural disasters, with their horrendous consequences. Volcanic eruptions, droughts, torrential rains, landslides, floods, fires, earthquakes, and weather and temperatures extremes lead to death, injuries, displacement, disease, economic loss, and disastrous ecological damage. Those who are poor and less capable to prepare for facing such disruptions take the brunt of the suffering. Between 1980 and our day, the number of events qualifying as natural disasters almost quadrupled (according to Our World in Data, based on the International Disaster Database maintained by the Centre for Research on the Epidemiology of Disasters, Ritchie and Roser 2021). The graphs, documenting the heavy toll of the various aspects of such events are, illustrate not some curse, but rather inadequacies in facing the challenges of extreme events.

A Confluence of Many Tributaries

There is a confluence of breakdowns: pandemic—for which Covid-19 (with the implicit time reference in the name) stands—is only part and parcel of a broader crisis (Fig. 1). For this broader, open-ended crisis—economic breakdown, financial meltdown, social and political turmoil, education disruption, healthcare failure, global

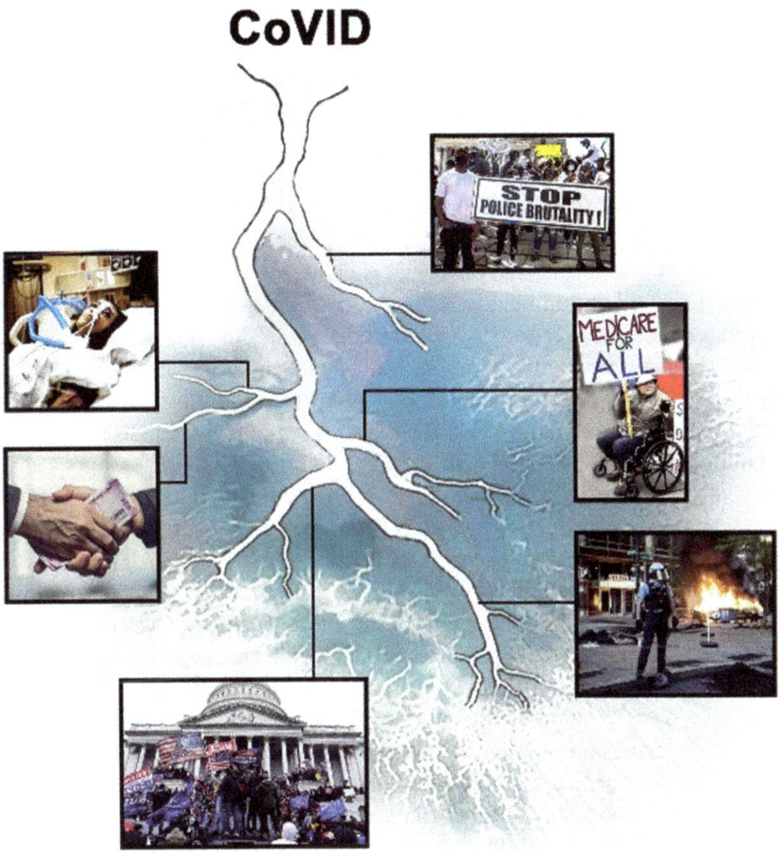

Fig. 1 Suggestive snapshot for the crisis of vision—CoVID. The catharsis of disease and unbearable legacies, as well as of the consequences of natural disasters, fused into a state of despair, moral depravity, and confused revolt against the lack of the sense of future

instability, etc.—the moniker CoVID (without a number) will be used henceforth. It stands for Crisis (C) of (o) Vision (VID, from the Latin for "seeing).[1]

The numbers, very much in flux, add up to beyond 100 trillion-dollar mark, a large portion of the global GDP. Only the blow to the global economy was estimated at over 50 trillion dollars (according to the International Monetary Fund 2021). At least 200 to 300 million people were forced into a new wave of refuge seekers. Even more significant, they add up to a realization of being powerless in a world that claimed otherwise. Imagine someone, like in a Charlie Chaplin comedy, reacting to

[1] The Latin root *vid* (as well as *vis*) means *see*. It is present in a good number of words in English—e.g., video, provide, evidence, providence, visual, invisible, view—and in Indo-European languages. It indicates that the way we see things informs our actions.

the buckling of walls in his house: his bare hands pushing against the walls, running from one falling brick to another, stacking furniture against the wall to support it, taping over cracks. Of course, the comedy would make the viewers laugh as the futile reactive actions fail. But when your own house is falling, laughter gives way to a cry of desperation. So much could have been done to prevent the collapse. An illustrative example: "Removing one million homes from flood zones could save one trillion dollars" (Frank 2020). Building a house on sand," instead of on a rock-solid foundation, is costlier in the long run.

It is too early to even try to find out what, if anything, was, or could have been, done to prevent the still unfolding crises. Vigilance? Containment? A better medical system for everyone? A fairer society? Better education? Just one detail here: The corona virus, in a different configuration, has been around for at least 20 years; and a vaccine for a possible flu outbreak was conceived more than ten years ago (Hixenbaugh 2020). As we analyze CoVID, this observation will help us understand the shortcomings of the obsession with the here-and-now of fixing as opposed to steady prevention activities. In hindsight, it is probably justified to state that only a pitiful sum was spent on pro-active measures. The three percent of the entire cost of reacting to the AIDS/HIV epidemic, and the 0.1 percent of the volume of transactions in the financial breakdowns qualified as prevention are hypocritical alibis for covering failure. Prevention does not figure in the profit machine driving the world today. To develop and test a vaccine before a pandemic occurs would qualify as an investment in prevention. It is different from reacting to a pandemic with vaccines rushed through Emergency Use Authorization (with billions of vaccines discarded because their effectiveness expired). Natural disasters, more frequent and more damaging, cannot be prevented. But preparation for such events is possible.

Against the background of more and better science and technology, reactions to breakdowns reveal a pattern: the cost of reacting to one crisis after another, taking place ever more often, irreversibly increases in every respect. As well-intended as reactions are, the success—i.e., return on investment—diminishes. The limiting determinism of cause-and-effect—in short, reaction—as opposed to prevention proves inadequate for coping with the many aspects of the inexorable change the entire world is going through. It has no moral dimension. The high price and low rate of success are the result of opportunistically focusing on particular aspects—the reductionism of easy targets—to the detriment of seeing the whole. More important, failure to understand the relation among the many components makes it impossible to prevent the impairment of highly interconnected processes, such as political and economic activities.

What the world has experienced since the end of 2019 is indicative of what it means not only to reap the benefits of globalization, but also to face danger on a global scale. Around the time of the HIV/AIDS crisis, which is indexing CoVID, the scale tipped. It became obvious that opportunity and risk cannot be disassociated. When damage (immediate and potential) exceeds benefits—the Upside/Downside Ratio—a new context arises. The SARS-CoV-2 virus that triggered the pandemic and the associated, politically exploited, anxiety pandemic are entangled. It takes very little to bring down an interconnected unstable system—which the global economy became.

If opportunity scales up, so does risk—at a higher rate. Interconnectedness, as the highly negative messages of social media reveal, can become a liability. Evidently, Covid-19 took place in the broader context of many conflicts: military, economic, political. Some are still unfolding—all preventable.

Just to illustrate the thought: the Chernobyl nuclear power plant explosion. The benefits of electricity provided to millions of households turned out to be lower than the damage to all affected: people, animals, the environment (local and distant). The chasm between those who govern and the governed widened and deepened. Ever larger and more powerful bureaucracies, associated with an ever more vertical power structure, robbed citizens of what was left of their sovereignty. Organizations (mostly self-serving), governments, political parties, lobbies, the military- and medical-industrial complex, academia, etc. empowered themselves to the disservice of members of society. They embody the mechanics of determinism, to the detriment of allowing for individual choice: "Don't worry. Relax. We'll take care of everything. We know better what's good for you." In reaction to the pandemic, governments all over the world placed their citizens in various lockdown modes. As the economy went into a nosedive, and education was replaced by wasting time online, panic set in. The drastic impact on the lives of the people affected cannot be fully accounted for. A second breakdown shadowed the so-called second wave of the pandemic. And after that, more of the same, wave after wave, leading to a reality of never-ending warnings. Ill-conceived social distancing, misunderstood effectiveness of masking, unjustified obsessive disinfection, temperature checking, and testing became the new gamble: "Do it and we'll beat the pandemic." All reflected in the grotesque image of the blind leading the blind. The third wave, associated with a mutated virus, prompted new vocabulary and yet another promise: "Get vaccinated and we'll attain herd immunity." The fourth followed. And with omicron, the fifth, the sixth, the seventh. There is no end in sight, but rather a muddled perspective: we will have to learn to live with this. The talk about yearly jabs—booster shots—like with the flu, became a new policy. The model advanced sounds more like the razor blade story: Get a cheap razor with easily replaceable blades; the repeat business generates profits. It is more a business model than science. This applies to vaccines, rapid test kits, and antiviral pills. It provides the heavily subsidized pharmaceutical companies (which insisted on legal immunity and protection from competition) with an endless revenue stream. With no provisions regarding the obligation to provide the scientific evidence for the method they promote, society was forced to invest, and private companies are cynically cashing in.

Health and the health of societies stand in strong correlation. Therefore it is surprising that the political discourse in America about Covid-19 never extended to medical care for all. The fact that the disastrous performance of the most advanced economy in the world reflects the lack of an affordable healthcare system is simply pushed aside: "That's socialism!" The ensuing social unrest in the USA and around the world reflected "sins" of the past and refocused attention on the contemporary and historical shortcomings of all kinds: unfair policies (protecting business over society), unequal opportunity, racist attitudes, inadequate medical care, etc. Sectarian violence increased; the power-hungry political class cynically turned corruption into another "freedom of expression" issue. The record of protests and social activism proves that they are not incidental, but congenital. For those who were and still are miserable within the walls of their abode—elegant house or rudimentary apartment, prison cell or room in a nursing home, workers' dormitory—science, explaining how a virus interacts with the body, does not put anyone's mind or heart at ease. The same applies to the insecurity triggered by all kinds of expressions of revolt. "When can I escape this state of being a prisoner to fear or to some government edict?" was and still is on the minds of millions around the world. "Why do I have to wear a mask?" became a declaration of resistance more than a question only hinting at it. "My body, my choice!" took on a new meaning as the cry of the right to abortion.

The next question loomed large, especially since, as a consequence of the pandemic, science has lost its credibility: "Why accept vaccination—and which one—as long as we still don't know how effective (or harmful) it might be?" Benefits of the vaccine notwithstanding, the risk of blood clots and heart problems does not suggest that medical science is fully aware of possible consequences (short- and long-term). Vaccine escape—the breakthrough cases, even after boosters—became another source of worry. Another threat: the post-treatment rebound (the Paxlovid story). Even if the risk is low, it is still a risk, not taken lightly by millions of people. It does not help that in keeping up with a pattern of being less than honest with their consumers, the pharmaceutical companies have failed by not disclosing whatever information accumulated as they raced to win a greater share of the market. Their contracts with the government are convoluted to the extreme of making the government an accomplice.

Becoming a number added to the total of those who didn't make it through the pandemic is nobody's voluntary choice (except for those—not few—who committed suicide—a pretty damning proof of societal shortcomings). Facing the consequences of economic collapse is a disheartening prospect: putting food on the table, paying the rent, medical insurance, education are concrete expressions confronting the majority, especially those not benefitting from some social programs (as in Europe, New Zealand, Singapore). But more than that, the uncertain future—which the millennials and Generation Z have good reasons to fear. The here and now of instantaneous gratification to which the public is conditioned through the promise of unlimited progress override the sense of future. The expectation of more and better, and of course cheaper—the model of progress currently pursued—and of sustainability are ultimately incompatible. How long will abuse of the ecology continue before it can no longer sustain life?

As usual during times of crisis, experts—some legitimate, of course—multiply as fast as the virus. More precisely: while the physical presence of the actual virus confuses the immune system and involuntary host cells multiply the invaders, opportunists of all kind hijack the communication channels. Freedom of expression—invoked to the extent of justifying corruption—is a shield for political hacks masquerading as experts. It is also misused by scientists impervious to the damage they cause through confusing and (often) partisan pronouncements. The tireless "medical officials" initially assured the public that Covid-19 is of minimal concern (Moreno 2020). After that, they spoke to TV cameras about the uselessness of masks

("There is no reason to be walking around with a mask," MacCandless Farmer 2020). They argued for the need to scrub every inch of people's homes, doorknobs included. A whole book was dedicated to misleading pronouncements (Deace and Erzen 2021). Authority, however compromised, took over the lives of those with no choice but to trust them. Official science, often justified for reasons of politics, systematically eliminated any challenge detrimental to deterministic dogma. Posing as experts or as prophets of a new order that cancels the past, they confuse the public with ill-informed and contradictory messages. Incompetence sprouts from the fertile ground of a crisis. Lying, not reserved only to politicians, became the rule and no longer the exception. And if the pronouncements—repeated and justified as "It's science!"—are not lies, they testify to incompetence—more dangerous than lies.

Dubito Ergo Sum

The last 40 years—to limit revisiting the past to only one generation—were memorable in so many ways. This timeframe can be associated with *mondialisation*—in reference to the dissolution of national identities and the abolishment of borders inside the world network of economic exchanges (Peyroux 2020). However, the most significant change is in the fact that the cost of fixing the damage brought about by reductionist-deterministic science and technology is greater than the benefits these afford. Consider not only the exceptional explosion of a nuclear power plant, but the more common medical treatment that sometimes saves lives at the price of a future lifelong dependency on drugs and medical equipment. Accomplishments associated with the Cartesian action-reaction perspective of thinking reflect the view according to which thinking is proof of one's existence. An open-ended list of the "iconic" results is quite inspiring: the Solar Maximum Mission (Solar Max) launched by a Delta rocket (1980); identification of the early-onset Alzheimer's gene and the Duchenne muscular dystrophy gene (1987); the first detection of Space–Time ripples (2016). To the list belong also the first human CRISP gene editing trial and the DeepMind algorithm for detecting breast cancer (2020). The claim that protein folding can be predicted, as impressive as it is misleading, scores very high among the successes of "Deep Learning" (Service 2020; Marx 2022). It got all the recognition one can hope for. Editing genes, the Cassini spacecraft, the ancient genome sequencing, vaccine and treatment for Ebola, and detection of the Higgs Boson at CERN are success stories impossible to downplay. The mRNA vaccine method, of over 20 years ago, also belongs to the list of victories. Such attainments (only partially illustrated through the examples mentioned) are recalled not as justification for Cartesian-grounded science; rather, these are exemplary steps on the ever faster moving treadmill of progress by now taken for granted. Many enjoy the ride: prosperity, another outcome of the scientific-technological revolution, is addictive. So is glory (deserved or not).

The prophetic *dubito* (doubt) amendment that Descartes articulated is rarely, if ever, systematically pursued. Let us recall it: *Dubito, ergo cogito, ergo sum* (I doubt, therefore I think, therefore I am). In the absence of self-questioning, perspective is lost. Funding of the spectacular science and technology premised on reductionist-determinism is usually seen as an investment in the future. Around the world, almost nine million individuals are employed in research. The number increase more rapidly than the economy. Research and development (R&D) is an investment that has reached 2.5 trillion dollars. Rarely, if ever, do the long-term consequences of generously funded, ever larger, damage (unintentional, of course) justify blind trust in science. The civilization inspired by the Cartesian Revolution, affirming rationality over faith-based explanations of reality, brought about a world of progress and plenty (not equally or equitably distributed). Not surprisingly, in the absence of doubt, i.e., of self-questioning, extreme vulnerability became unavoidable. The disadvantaged suffer economically, educationally, health-wise, or are victims of discrimination. They experienced the embedded vulnerability more than those who are in positions of privilege (class, culture, race, gender).

Giving credit where it is due should go hand-in-hand with a critical assessment of the limits of the Cartesian-inspired understanding of causality divorced from doubt. In particular, CoVID manifested in the "Woke" cancel culture (but not only) summons the highlighting of the annoying intolerance towards alternative views. Descartes advanced an alternative view—in opposition to the deeply ingrained conceptions of his time. Given the ever-higher price civilization pays in trying to survive crisis after crisis, *it becomes necessary to question the limited model of causality anchored in knowledge of the past, as well as the disputable assumption of the sufficiency of inferring from parts to the whole.* Acknowledging that the crisis of vision is congenital with the machine model underlying the global economy could help in a better understanding of sustainability.

Dynamics and Sustainability

The dynamics of physical phenomena is convincingly represented along the time continuum between the past, where cause resides, and the present, where effects are noticeable or expected. The change in the living involves, in addition, the future into which life inevitably extends. The conceptual premise of this study is the understanding that in the living, anticipatory actions complement reactions.

Action-reaction underlies the dynamics of matter. The dynamics of the non-living can be fully described in terms of action-reaction. Anticipatory processes underlie the dynamics of life. Action-reaction descriptions of the dynamics of the living are incomplete. In the absence of awareness of the future, such descriptions guide an inadequate, mechanistic course of action that undermines life. These observations, to be properly detailed, are the logical premise for the realization that the conception and practice of reaction and the conception of pro-active activity (prevention, for instance) could and should be understood in their unity. Implicit in this understanding is the acceptance of the distinction between what is alive and what is not. What this means in practical terms is essential for gaining effective knowledge about the variety of forms through which the living interacts with the world it belongs to and depends upon. This is not a matter of the rhetoric of environmental awareness, but rather about the concrete actions through which this awareness contributes to sustainability. Indeed, environmental awareness is a matter of survival, not of a political discourse of convenience, or of inconsequential philosophy.

Confluence, as a process easy to identify in nature (confluence of rivers, for example), as well as in the functioning of the organism (confluence of physiological processes, such as digestion, for example), is an intuitive way to characterize behavioral aspects, i.e., qualitative distinctions between various states (de Kleer and Brown 1984). It offers a non-mechanistic view of how processes that seem independent of each other lead to situations impossible to predict within a reductionist-deterministic method.

In Fig. 1, the confluence of Black Lives Matter, #MeToo, Antifa, etc. is telling about how various forms of social activism (some begun over 30 years ago in the context of the "Great Recession," others newly emerged) influence each other, without being reducible to each other. Their respective programmatic documents reveal what they might have in common, but also what distinguishes them. For instance, the call to dismantle the Western monogamous family, already in self-destructive mode, extends into the practice of polygamy and polyamorous relations and efforts to legitimize them. In association with affirming transgender identity and sex fluidity, the questioning of family (including the legitimization of the homosexual family) leads to a rather confused political message. On university campuses, the New Left of the 1960s seems suddenly resuscitated in extreme forms. Civil rights movements, individual freedom initiatives, and divergent calls to address injustice (many other causes can be added to the list) entered the confluence. New are the belligerence of intolerance and the associated tyranny of the few. Sixty years ago, Students for a Democratic Society (SDS) had a broad, ambitious perspective. They denounced the fact that higher education became a machine for reproducing all society's structural failures. At institutes of higher education, the current confluence of many causes (against White prejudice, for inclusion, for equitable representation, for opportunity) does not, however, protest that the university campus caters to their students' addiction to consumption. One movement hoists the flag of self-righteousness—"This campus is on land stolen from the Lenape natives"—over the spa and fitness center where, incidentally, credits are disbursed for reasons quite remote from educational goals. Rigor is denounced as a practice of White supremacy, and those who are only *suspected* of affirming rigor end up "cancelled." Diplomas are becoming a matter of entitlement. Confusion became endemic. Polarization is extreme to the extent that left and right (and in-between, wherever in the world there is still an in-between) expect viewpoints to be imposed, with no attempt at critical evaluation: "Put up or shut up!" In the words of Ortega y Gasset (1930): "the reason of unreason" took over. Within CoVID, there is only guilt, but never responsibility; "self-responsibility" was erased from the vocabulary.

Confluence is a matter of compatibility. White Supremacy (in its various expressions) and the Black Lives Matter movement are probably as different as Covid-19 and the seasonal flu. However, in an extremely fractured social reality, they capture the extremes of a conflictual situation that might lead to the breaking of the dam represented by the social contract currently in place in the USA. This applies as well to other self-styled progressive countries around the world. Demonstrations in the rich neighborhoods of Hamburg (Germany) and Paris (France), and what seemed as the never-ending protests in Portland, Oregon (back in 2020) are different in nature. But they were fueled by indignation, as well as by a state of desperation: When will this end? Covid-19 pulled down the curtains behind which illusion after illusion were concocted by the power-greedy who are running the world, whether in democracies or any other form of governing. The Hong Kong protest movement and the Chinese government's attempts to squelch it are part of the same narration. And so is *el estallido* (the social explosion, 2020–2022) in Santiago de Chile. Many movements fizzled out: from Occupy Wall Street to the LGBTQ Day of Silence, to the #BlackTransLivesMatter, to Reopen (protesting Coronavirus lockdowns in India, Lebanon, and Iraq, for instance), to #RunWithMaus. These illustrate aspects of a symptomatic *état d'esprit* (frame of mind). In France, generals found solidarity in glorifying the nation and its past glory (never mind its putrid colonialism) in the age of the global economy, or at least of the European Community (from which their *République* benefits more than others). A hashtag—so easy to acquire in the age of social media—is by itself not a source (of ideas, revindications, reforms) in the vast landscape of confluence. It is not where streams (the source of confluence dynamics) start, but rather where demagoguery originates. The *"When will this end?"* of the Covid-19 is as impotent as the rhetoric of the question itself. What should end: the pandemic? entitlements? sectarian violence? racism? intolerance? self-righteousness? sexual domination? The list can go on. The more important question never comes up: When will the wholesale disregard of the future end?

The qualitative aspects captured in the dynamics of confluence are useful in understanding how unrelated events, of divergent ideologies, can, under certain circumstances, contribute to the instability of the economic/social/political system. The breakdown (rather, implosion, Nadin 2013) of the Soviet Union exemplifies a rich confluence of events (some related, some not). It was just extended in the Russia-Ukraine crisis of vision. They are, as we understand them in retrospect, testimony to the system's inability and refusal to adapt. Anticipation was "written" large on the flag of the 1917 Revolution (as it was written on the flags of the American Revolution). It ascertained visionary ideals, eventually sacrificed to reaction. All totalitarian regimes, regardless of their underlying ideologies, work in the machine modus. The integrated economy of today's world is but a machine tuned to maximize the return on investment. An implosion at a global scale—i.e., the breakdown of the machine—cannot be excluded. Something like a pandemic or a natural disaster could turn the metaphor of the flapping of a butterfly's wings leading to a major storm into a new reality. The mathematics of dynamic systems undergirds this possibility.

The most recent crisis prompts the need to reassess the understanding of change, including inevitable breakdowns. A broad view, meant not to debase the Cartesian perspective and its expression in science and technology, implies acceptance of the anticipatory perspective. This could be the concrete method for integrating multicausality and the holistic view in the science that guides society. Ignoring or militantly rejecting the anticipatory perspective is more than a matter of intellectual integrity. Locked in its Cartesian foundation, from which empirically grounded doubt was excluded, science ended up undermining its own legitimacy. The crises of the recent past suggest a malign condition: progress at the price of irresponsibly borrowing from the future.

CoVID (confluence of breakdowns, some associated with Covid-19, some not) is the outcome of the dogma that has dominated science since the Cartesian Revolution. To disrupt science is to reaffirm confidence in science, not to question its legitimacy, provided that scientists themselves are willing to revolutionize it. Without awareness of the limits of the views that scientists hold, they become captive to them, and regurgitate them instead of questioning, even against evidence. The limits inherent in the premise of the science that brought humankind to the condition in which it now is—more successful than ever, but also more vulnerable than ever—need to be overcome.

Let's Start with a Conclusion

Sustainability Thesis (ST): Reaction is more expensive, by many orders of magnitude, than anticipatory actions. In the long run, it is unsustainable.

This thesis might qualify as a fancier way of saying "An ounce of prevention is worth a pound of cure,"[2] if indeed prevention and healing, including treatment, are seen in their unity. An overview of the costs associated with reaction to disasters makes apparent that the costs are increasingly higher, and unsustainable in the long run. Also significant: Examination of the timeline of breakdowns—including natural disasters (some anthropocentric in nature)—reveals how limited is the effort in prevention and preparedness. Empirical evidence—the numbers spelled out in this chapter's introductory lines—justifies the ST in its generalizing formulation. The ST is an effective expression in the sense that it informs action regarding how society meets the challenges of change. It is formulated against the background of a new systemic condition: the short- and long-term benefits of deploying reductionist-deterministic-based science and technology are increasingly outweighed by the short-and long-term unavoidable, unintended, and unpredictable consequences. The Upside/Downside ratio (of benefits vs. undesired costs) started flipping around the time of the HIV/AIDS pandemic. Unlimited progress and entitlement became the goals. To achieve them, society made the fundamental choice of pursuing, under the guidance of science and with the assistance of technology, the reactive path. The thought of questioning deterministic reductions, or better yet, complementing them with an anticipatory course of action, was eliminated.

---

[2] Various languages have their own version: Praevenire melins est quam curare (Latin). Mieux vaut prévenir que guérir (French). Más vale prevenir que curar (Spanish). Vorsorge ist besser als Nachsorge (German). " 預防勝於治療." The literal translation is "prevention is better than cure." "болезнь легче предупредить, чем вылечить". "The disease is easier to prevent than to cure".

In proving the ST, the need for a revolution in knowledge and action informed by sustainability considerations will become evident. In particular, we shall subject the ST to examination based on evidence from the most recent large-scale, undesired experiment—if the open-ended CoVID crisis can be considered worthy of the label. For the sake of establishing a basis of shared knowledge, definitions are provided. They are meant to translate this shared knowledge into an alternative course of action. The purpose is not to add more pages to the literature of critically examining the Cartesian foundation of modern science and questioning why the doubt component was abandoned. Rather, the goal is to advance a non-reductionist understanding of science. Within such a new foundation, the spectacular accomplishments of physics and chemistry are complemented with an appropriate view of what life is.

The reductionist-deterministic *Method* advanced by Descartes (1637) is at the core of the Cartesian Revolution—and explains its successes. However, the original systematic questioning—doubt—was replaced by the acceptance of the certitude of numbers meant to explain life in the same terms that led to the landing on Mars. Without entering into the rich elaborations on determinism and reductionism, and without spelling out the *Rules for the Direction of the Mind* (1701), we will refer to them within the Cartesian understanding of reality according to the following definitions:

- *Determinism*: All events are determined by previous causes.
- *Reductionism*: The properties of the whole can be subsumed as the properties of the parts into which it can be reduced.

Examples can guide in understanding the contradistinction between two different knowledge domains.

(1) A stone rolling down a hill: the forces at work can be easily identified, and the process described in the mathematical expression of the laws of physics.
(2) A moving snail—as a living entity embodied in matter—is subject to the same laws of physics. Its behavior, consequence of interactions with the world, is self-initiated. The gliding muscular foot, covered with epithelial cilia, and lubricated by what seems a kind of mucus (made by the snail and secreted), is the evolutionary outcome of purposeful actions.

For the description of movement of a non-living entity, it suffices to measure quantities: the geometry of the hill, the stone's weight, the characteristics of matter. For the description of the locomotion of gastropods, in addition to quantity, context is important. Survival, acquiring energy, or avoiding predators is a matter of meaning. The living, at all levels of life, interprets the world. For an observer, the organism acts on the meaning of the data describing it, not on the numbers making up the data. A limpet sticks to a rock, and if the rock is acted upon by a wave, the limpet moves together with the rock. At high tide, it crawls away from the rock in search of food. This is its survival method.

Quantitative descriptions, obtained by humans from measuring whatever changes, are sufficient for explaining how inert objects move—the stone rolling down a hill is the example given. In its material embodiment, the living does not escape physics, but has control over its direction, and even over the choice of moving or not from one position to another. The dynamics of the parts and the dynamics of the whole (e.g., a stone falling, a car in motion, a rocket) can be separately defined (and controlled). This does not hold true for the living entity, be it a snail, some animal, a human being. An observer of a living process cannot infer from a cell or collections of cells to the whole entity. Indeed, the 20,000 neurons of a limpet, or for that matter the 75 million neurons of a mouse, are part of a larger configuration in which the whole and the parts depend on one another. The state of the living changes in a manner different from that of a non-living entity. (For an evolutionary report on the process through which minds emerge, see Goas and Gaddam 2022).

An Overdue Revolution

"A Second Cartesian Revolution is Long Overdue" could be appended to this book's title. The meaning of this declaration goes beyond an improved explanation of the movement of a stone as fundamentally different from that of a snail. Examples do not make a theory. But they can suggest that alternative perspectives might explain what contradicts the premises of established views. For an alternative to be effective, its ascertainments must be actionable upon. The legitimacy of the anticipatory perspective, complementary to deterministic causality and to reductionism derives from the consequential nature of the distinction between purposeful living, based on the ability to learn, and the inert physical existence of matter, unable to learn and devoid of purpose. The default state of the non-living—subject to interactions described by physics—is inertia. Stones do not move on their own. The living—subject to a dynamic that is the subject of biology—is in a default state of continuous activity (Longo and Montévil 2012), from its inception (birth) to its death. The living does not need a physical cause to be active. Its interactions with the world it belongs to are in preparation for change—from low tide to high tide for the limpet—in anticipation, not in reaction to it.

The pandemic, initially described as the 2019 novel corona virus, and even as the 2019-nCoV (same meaning) is now labeled Covid-19 (short for corona virus disease time-stamped at 2019). It is a collective name for a rather diverse pathology, made even more variegated by virus mutations. CoVID, as already suggested, is the broader *Crisis of Vision* affecting society driven into the future under the guidance of an incomplete rear-view mirror take of the world. It is a confluence of breakdowns originating years before the new virus: three pandemics (in the last 20 years), natural disasters, failure of science and technology (in particular in medicine), economic and financial fluctuations, disparities (e.g., race, class, gender), social and political turmoil, and global instability. So much hiding behind the shiny veneer of apparent prosperity and progress in science and technology; so much to fix, if not to discard. In retrospect, it all boils down to a still rejected realization: *the failure of the reductionist-deterministic view to address the multi-causality characteristic of the dynamics of human existence from a holistic understanding of life informed by awareness of sustainability.*

"This too shall pass" is what everyone aligned with the Cartesian-based practice of science and technology can hope for. But at what price? Lives lost, well-being spoiled for good, hopes shattered. And the renewed illusion of a future of progress that cannot be sustained. When—and if—a return to "normality" will happen, nobody but demagogues, and especially those profiting from the breakdown, could call it a victory. The renewed wake-up call that CoVID prompted all over the world should inspire a necessary revolution in science. Its goal: the integration of the complementary dimension of anticipation in our understanding of causality. In the context of increased focus on data, there is a need for a new understanding of knowledge itself: how it is acquired, disseminated, and validated. The premise for this new understanding: knowledge acquisition cannot be reduced to measuring. The data that measurements generate incompletely represent what is measured. Moreover, science limited to considering change only as a quantitative process fails to understand the meaning—exactly the reason why measuring is carried out. To disrupt science does not mean to get rid of it. Rather, it means *to make it meaningful.* In other words, to reference it to life, on account of which it is performed, and for the benefit of which it is deployed. A sense of responsibility extending beyond successes at the price of humanity's future—this is the meaning of "Disrupt Science." Otherwise, we could miss yet another (if not one of the last) chance.

Celebrating victories—such as vaccines and antiviral drugs, as successful as they are (or we wish them to be), but also as controversial as they will remain—is justifiable. But way too often, the celebration is either premature or undeserved. Those who celebrated progress in combating racism—e.g., the Emancipation Proclamation, Constitutional Amendments 13 and 15, the Supreme Court ruling in the landmark civil rights case Brown vs. the Board of Education, the Fair Housing Act (Title VIII of the Civil rights Act of 1968 as amended)—lost sight of the fact that symptoms are not the disease. Almost everything done was in reaction to disturbing manifestations. The remedies were bromides, not attempts to heal and empower the sufferers. The same holds true in respect to acknowledging contributions made by women, and by handicapped individuals. The Great Recession (2007–2008) and the "too big to fail" model of reaction left behind not only accumulated injustice and unfairness, but also a polarized world taken over by zealots. Victory laps occasioned by other incidental events are no substitute for what, within CoVID, became legitimate calls for fundamental change. Those affected by Covid-19 know that the medical system is deficient to the extent of celebrating one or another drug, but without any clue as to a convincing individualized treatment. Therefore, to recreate the reality of an incomplete view of the world that dominated before the breakdown (and definitely led to it)—in other words, to return to the *status quo ante*—would be a self-inflicted curse. Sentencing a police officer accused of the death of a Black man in his custody is no answer to the larger problem of law and order, corrupted in many ways, in the USA and around the world. Defunding the police—a reactive measure—quickly resulted in more crime. The anticipation view would have produced a solution reflecting the dynamics of society, especially regarding criminality. The crew that made it through a devastating storm even though it charted the wrong course celebrates survival, but the compass still points in the wrong direction. This aptly describes the political system (regardless of party) in America, which claims credit for yet another expensive reaction. But it is not prepared to change course! Actually, the political class, including by now the plutocrats and the media, is change-adverse. Nobody is willing to give up privileges.

Life in the universe of our existence, including each individual's life, can be saved and made meaningful if what in the final analysis defines it would no longer be reduced to dead matter—and manipulated as such. This entails a different understanding of the role of society, not only in respect to medical care—where it all seemingly started—but also to its own health and viability. Virchow (the originator of cellular pathology) was convinced that "Medicine is a social science and politics is nothing else but medicine on a large scale" (Ashton 2006). This "medicine," captive to politics, proved quite lethal (Cole 2021). Gender, race, sexuality, ecology, civil rights, generational relations, privilege, opportunity, justice, education, and especially ecological awareness are issues of shared concern—even though their confluence became capricious as time went by.

Beyond discussion is the fact that, generally speaking, biological characteristics, inseparable from the larger dynamic of society, supervene on the matter in which life is embodied. But beyond this generality, in the spirit of Descartes' rationalism, lies the concreteness of life, in particular its non-deterministic and creative nature. Doubt, of which he was aware, is part of it. The living is infinitely diverse—a lesson that society still must become aware of. Particular properties of biological matter, expressed in unlimited diversity, supervene on the dynamics of life, in particular on societal change. To know the living cannot succeed in the absence of acknowledging the uniqueness of each of its embodiments. This epistemological premise deserves to be examined.

"Hijacked" by a Virus

A grotesque situation: The corpuscle (0.1 μm in size, of one millifluid of an ounce in volume, one femtogram in mass, Bar-On et al. 2020) called the SAR-CoV-2 virus enters the body through mucosal surfaces—i.e., columnar epithelia (such as the gut, lung, endocervix) or squamous epithelia (nose, extocervix, eyes, vagina). Eventually, the oral cavity was identified as a site for infection (Huang et al. 2021); and the eyes (Dawood 2021). The virus—an entity ensconced in the fuzzy border between living and non-living—"hijacks" the human organism's amazing anticipatory ability to continuously remake itself. And turns it against itself. For the sake of clarity: viruses do not have agency; they don't *do* anything. Their presence, as foreign bodies, leads living cells to interact with them. The body itself—more precisely, the cells the virus attaches to through physio-chemical interactions—reproduces the invader. Sometimes the copy is almost like the original, other times it can be different. The so-called mutations, or variants, originate in infected bodies. Contrary to incompetent views disseminated not just for the public at large, but also in publications labeling themselves "scientific," nothing in the virus is of an evolutionary nature, and even less of an adaptive mechanism. Treatments, for example, anti-viral or otherwise (for those affected by comorbidities such as chronic lung disease, obesity, diabetes, cardiac anomalies, etc.), explain the emergence of strains more difficult to treat (Koyama, Platt, Parida 2020). From the initial (called Ancestral) D614G to the British b117 variant, to the California B1.429, New York B1.526, Brazil P.1, and to B1.351, referenced to South Africa, to, most recently, BA.5 or BQ11, immune evasiveness changed. This means variation in infectivity and, sometimes, in lethality. That so-called experts, taking ownership of all possible communication channels, fail to understand the process is not surprising. Their becoming the "celebrity of the hour" could be ignored if the damage caused to the credibility of science, of medicine in particular, were not as widespread as it is. They rely on primitive deterministic explanations similar to those characteristics of the larger failure of science in addressing the pandemic (WHO 2020).

It is probably less than helpful to start the argument for anticipatory-grounded medicine with an example of an anticipatory process that went wrong. The virus's genome (its genetic makeup, i.e., its chemistry) becomes the "program" that guides the invaded living cell to create the virions, i.e., copies produced in specific parts of the body—from head to toe! Those who are captive to the explanations of life in terms of physics and chemistry, but blind to the anticipatory characteristics of living processes, identify a machine at work in the virus's multiplication. Therefore, it is quite appropriate to examine the process in some detail (Fig. 2).

The virus's SARS-Cov-2 spike protein physically attaches to the host cells, in particular the ACE2 enzyme as receptor-binding domain (Futerman 2020). The anticipation that underlies the continuous remaking of the living is now driven by a misinterpreted "poison" (poison is the original meaning of the word *virus*). The virus attached to the living is reproduced by the host under the false assumption that this multiplication capability—an anticipatory capability—is beneficial. Those who understand how anticipation is expressed in the process we call *living* will not be surprised. Anticipatory processes ensure the continuous re-creation of the cells making up the organism. This is a form of maintaining life under continuously changing circumstances. But since cells are not identical, it is not surprising that the reproduction results in variants—some more infective than others. One person treated with drugs—chemistry at work—for an unrelated medical condition is obviously a host different from another who enjoys good health. In reality, virions are not identical copies of the virus. The cell is not a printing machine.

Life is an Expression of Anticipation

Anticipatory processes are non-deterministic: the same cause can have different effects. This is not the physics of the inanimate. It is a living process, constrained physically and chemically, but rather ambiguous biologically. Anticipatory processes

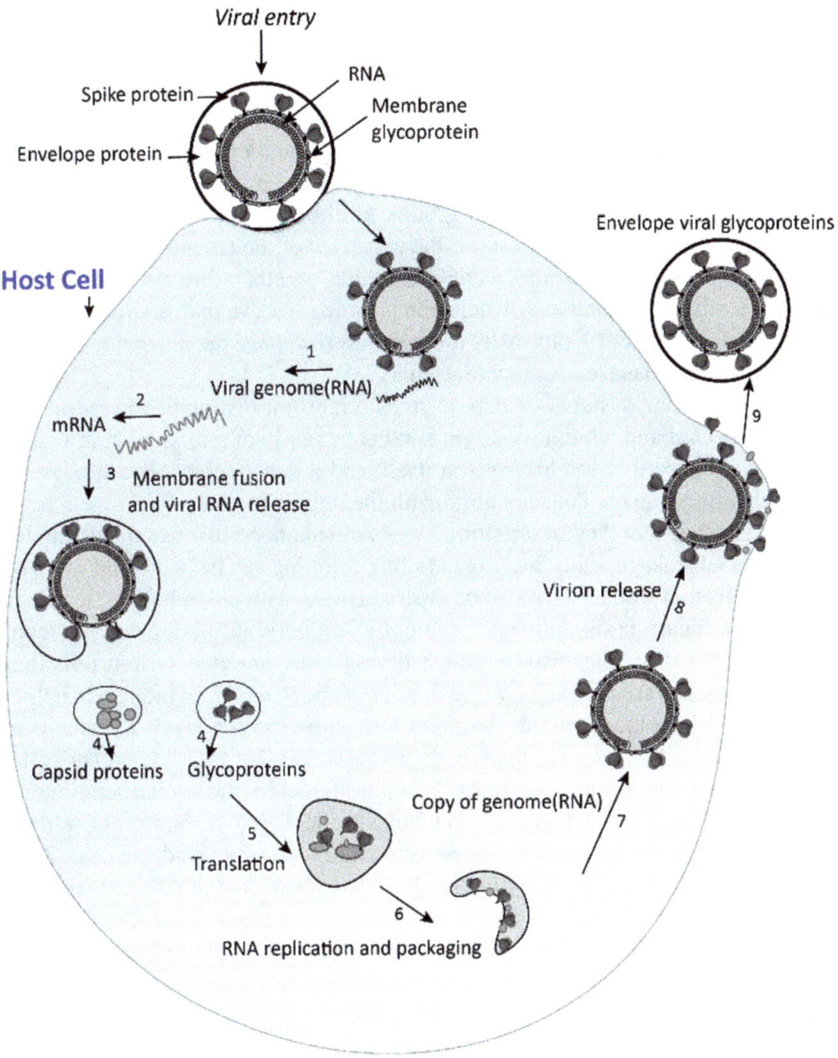

Fig. 2 The genetics of viral multiplication. Biochemistry explains how the spike attaches to the host cell. The beneficial genetic process of cell re-creation (remaking of the cells in the human body) becomes the multiplication process for the virus

are related not only to past states, but also to possible future states. They can be beneficial—successful motoric expression (such as in swimming or in avoiding injury in downhill skiing) is one example. They also can go wrong, as is the case with the rapid multiplication of the coronavirus in ACE2 cells. If we adopt this understanding of living processes, we can better grasp CoVID's social component: Socially legitimate ideas can be echoed in social actions aiming for desired outcomes. Let us recall the actions that led to the Civil Rights Act (1964) or to the "Affirmative Action" Executive Order (1965). Or rather destructive: The rollout of www.Healthcare.gov (October 2013), a healthcare exchange for the Affordable Care Act is one example among many botched government projects initiated with best intentions. Or for securing public support even for ideas not necessarily well thought out. Ethanol in gasoline is yet another fitting example here: it appeases corn farmers, albeit the expected pollution reduction entails increased ground-level ozone and smog, and moreover damages automobile engines. In short: the negatives of the ethanol mandate exceed the benefits. These consequences were foreseeable, and therefore avoidable through pursuit of a different plan, but self-delusion is more infective than a virus. Activism, triggered before Covid-19, but vastly more widespread once the pandemic took hold of society, is not a panacea for incompetence.

The autoimmune system—which at times might not distinguish between what has to be rejected and what not—is yet another example of anticipation at work. It does not ask for permission to manifest itself, and it is not subject to a peer-review process that rejects what does not align with the dogma in place. The over-reaction of the immune system ("cytokine storm," as it is metaphorically described) can lead to blood vessel leakage, clots, and organ failure. Clotting can be beneficial when one is bleeding from a wound; or it can be destructive, as with an embolism, leading to organ failure (heart, brain, kidneys). The out-of-proportion inflammatory response to the virus is yet another way in which the hijacked immune system fails those who are affected. Vascular biology (the dilation blood vessels), of interest in helping patients with high blood pressure, explains low blood-oxygen levels in patients not yet gasping for oxygen. Quite often, the heart fails. In other cases, the renal system abdicates. Physicians trained within the "machine model" of the human being address the low level of oxygen mechanically. Machines are deployed: respirators to help with taking in oxygen, or hemodialyzers to prevent kidney damage (Wadman et al. 2020). The reactive path is costly: many human lives have been lost due to a meaningless reactive protocol based on faulty medical science.

In addition to motoric expression (e.g., movement), anticipatory processes guide sexuality, metabolism, and, most important, self-repair. Anticipation is expressed in the workings of the autonomic system: blood pressure is maintained regardless of whether the body is standing or prone. (*Autonomic* means "on its own.") The stability of the living is the expression of autonomic processes. The organism, from its simplest form (e.g., the mono-cell) to the most elaborate—the complex human being—owes its adaptivity to anticipation.

Reaction—the approach to Covid-19, as well as to the main forms in which society has reacted to the failure of the system in place—is deterministic: do something (e.g., apply the respirator or prescribe a drug, or, for society, make money available as a quick fix) on the assumption that that will work as it did before. In determinism, the same cause leads to the same effect. Dictatorships embody the reductionist-determinist view: top-down actions (from limiting freedom of expression to the extremes of violently extinguishing expression of revolt). Paradoxically, within CoVID, the vision is the same: Instead of anticipatory action for a future in which systemic shortcomings (privilege, racism, intolerance, etc.) are addressed holistically, a piece-meal approach is adopted. The various movements, justified in many cases, limited themselves to reactions. Scapegoating, and the "Cancel Culture" reminiscent of the Cultural Revolution in Mao's communist China, have further undermined the health of a society dangerously fragmented:

> If we extend unlimited tolerance even to those who are intolerant, if we are not prepared to defend a tolerant society against the onslaught of the intolerant, then the tolerant will be destroyed, and tolerance with them. … We should therefore claim, in the name of tolerance, the right not to tolerate the intolerant (Popper 1945).

In respect to Covid-19 (but not unlike previous pandemics or epidemics), reaction was justified by the urgency of the task of saving lives. Biology shaped by the physio-chemical reductionism leads to a medicine of cause-and-effect directed towards *fixing* whatever impedes life functions. No doubt, reaction is important to maintaining life in the world in which the living emerged and changed through evolution. Nevertheless, reaction, characteristic of the physics of the universe, is not a defining characteristic of life. In its normal condition, the organism is prepared for change *before* it occurs. Anticipation underlies evolution. Preparation for possible changes increases the possibilities for adaptation, and thus for survival. When practiced in a reductionist-deterministic form as *the physics and chemistry of the living*, biology ignores anticipatory processes, or at most takes note of them without realizing their consequential nature. In its deterministic form, it can at most guide those who are focused on the mechanics of life, while ignoring the meaning of life. Reaction is short-lived. Anticipatory action is, by necessity, of the same rhythm as life itself. It takes time.

The non-living is embodied in matter. Its characteristics—that is, its form or its change over time—can be described through measurements of its quantitative aspects. The non-living neither becomes aware of nor experiences the world in which it exists. It interacts with it in a manner that reflects the characteristics of the matter it is made of. Biological matter entails meaning: it can be described through integrating both measurements of its quantitative aspects and interpretations of interactions with the world in a given context. Life is matter with acquired awareness, expressed and perfected in the evolution process (Goas and Gaddam 2022). The living experiences the world in which it comes into being. Its active nature is affected by it. Its interaction with the world is driven by purpose. As opposed to the stone rolling down a hill, the snail or the limpet, or anything that is alive, actively pursues a goal. Their state, at each instance, reflects the state of the world in which they move. The conception that maintaining life is like keeping a machine in working order—expressed through the *homeostasis* model (Cannon 1926)—is undermined by the premise informing it. Machines settle into stable steady states. They do not function based on their own choice: their states are predefined. In the living, at each scale of life (mono-cell or snail, or vertebrate, or human being), the state changes continuously. Nothing is predefined. *Allostasis* (Sterling and Eyer 1988), as a preparation for the future, is anticipatory. The allostasis process is driven by meaning. Stress can result in increased blood pressure, for example, but also in creative output: same numbers–different meanings. Stress is not reducible to a quantitative qualifier (description), but rather to one of quality, of confluence. In a given context, music can be stressful (triggering particular memories, for instance) or relaxing. In a broader sense, where there is no awareness of meaning, there is no life.

The argument for a science that integrates reaction and anticipation can be further specified, especially in acknowledging the role of doubt, which was consubstantial with Descartes thinking. Homeostasis and allostasis, understood in their reciprocal relation, could free medicine from blind determinism, and thus from sometimes harming persons hoping for healing. An intubated patient (or a patient attached to any machine) is deprived of allostasis, to the detriment of maintaining autonomous life. Undoubtedly, in many instances, machines have saved lives during the pandemic (and in other situations). But most of the time, machines did away with life as the experience of awareness. Through machines, the anticipatory nature of the living was sacrificed for the disputable benefit of a survival as matter void of consciousness. *Disrupt Science* implies the disruption of medicine—a goal of extreme urgency given the fact that deterministic medicine is a major cause of lethality. Covid-19 made this reality painfully clear.

The physics of cause-and-effect, in which forces and energy contribute to the success of an action, offers guidance for what happens when more and cheaper, regardless of meaning, becomes the target. When half of the food produced in the USA is thrown away, when the same takes place in Europe, high productivity (at any price, including damage to the ecology) can no longer be cause for the celebration it still gives rise to. In adopting reductionist determinism, a choice was made regarding how to shape human activity—from fulfilling needs (for survival) to satisfying ever-higher expectations, which are increasingly deemed as entitlements. Within the perspective chosen, this process is qualified as progress. Western civilization, which the world adopted, became obsessed with never-ending progress, setting ever higher goals in terms of higher production and consumption. New records in scientific and technological achievements, in prosperity, and even in re-engineering nature are celebrated. This understanding of progress is premised on the *limited* (and limiting) perspective of action-reaction, embodied in the machine of the Industrial Age, and on reductionism: explain life in terms of matter particles (electrons, protons, atoms, etc.) or genetic code (which means *chemistry*). Natural needs and natural forms of satisfying them are fundamentally different from artificially defined needs and artificially produced means and methods of satisfying them. Industrial production of food, clothing, housing, and means of mobility are possible on account of intense exploitation and despoliation of natural resources. The Cartesian Revolution destitute of its congenital doubt component paved the way to a state of mind, adopted by society, according to which all that counts is success, regardless of the cost.

It is not just a matter of more people living better, but also living longer and expecting more—even believing themselves entitled to a guaranteed minimum income, whether they work or already have other income sources. On account of the industrial model, population dynamics, once affected by limited natural resources—in the main, renewable—is practically unlimited. Today the world counts almost eight billion people dependent on resources—mostly non-renewable—"borrowed" from future generations. Those generations will inherit not only a depleted environment—unsafe air and water quality, infertile soil, shrinking forested areas, decrease in number and variety of flora and fauna—but also a new genetic make-up of extreme vulnerability. Their "natural disasters" will be more and more the making of their ancestors, who were unwilling to consider the consequences of their choices in terms of ecological impact. It makes no sense to contrast hunting and fishing to "fabricating" meat and fish filets (involving 3D printing, of course) in industrial agriculture "factories," using the romantic vocabulary of an idyllic past (when "everything tasted better") vs. a cynical present. The physics and chemistry involved in going beyond the natural to arrive at the artificial is based on quantitative descriptions of forces at work in processing materials, making machines, and eventually automating them. The path that humankind currently treads is actually one from evolution—as an encompassing process pertinent to everything that is alive—to engineered progress, and sometimes to degeneration. Everything that humans, with their inherent limitations, have performed within natural cycles is more and more entrusted to machines. A spectacular increase in productivity has allowed for population growth, as well as for lifestyles that border on the absurd. All this, however, is not only the return on higher creativity—the miracles of science and technology—but also on the higher use of energy, and the literal exhaustion of many resources. It takes fewer people than ever to feed a rapidly increasing population conditioned to expect even more. The Upside/Downside ratio changed dramatically. The world as a miracle machine ceased being an evolutionary environment.

Naming the Culprit

The knowledge domain called physics, sometimes expressed as "laws of nature," does not distinguish between the non-living and the living. Physicists assume that everything material has the same dynamics. Indeed, gravity—one of the forces described in Newton's "mechanics" (how the "machine of the universe" works)—affects the behavior of living bodies as much as it affects the interactions of non-living matter. At first, it seems to make no difference whether a person or a stone is falling, or whether a stone is rolling or a snail is moving. Upon further observation, it is clear that the stone has no active role in the process: its mass and position in space are all that count in the exercise of the gravitational force. The "history" of the stone (provided by tectonic reconstruction), its composition (what kind of silicates), or its current condition (wet, moss-covered, cracked, or homogenously compact) plays no role in the movement (the fall, in this case). The stone falls the same way, day and night, observed or not (Fig. 3).

The history of any and every living entity is its record of learning. Falling is not only an inevitable gravitational event, but also an expression of accumulated experience in preventing harm to life. The living (insect, animal, human being) most of the time "knows:" how to move, how to fall, how to jump, etc. A stone thrown into water falls to the bottom. A living organism (insect, animal) seeks ways to avoid

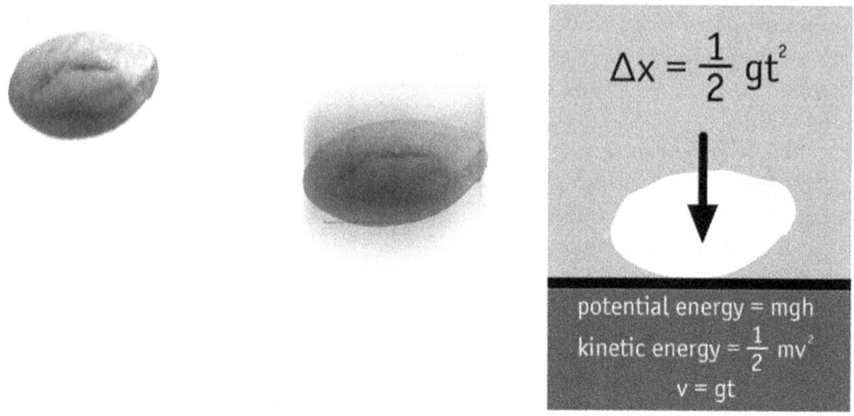

Fig. 3 The physics of stone falling (in simplified formulation)

drowning. Age plays an important role, since perception (in particular, vision and hearing) affects performance. Motoric abilities are also at work. Context is important, too: day (light) or night, noise, the presence of observers, for example. Behavior must be taken into consideration, as well as purpose: a gymnast avoiding a tumble; a pedestrian avoiding a slip on ice. Neither physics nor chemistry have a place for behavior or purpose in their conceptual toolbox. Neither do they have a place for Descartes' own doubt—supposed to trigger thinking, not to eliminate it.

The interaction of non-living entities is always only a reaction ("*Actio est reactio*"): there is a cause and there is an effect—one determines the other. There is no learning. There is only sameness. Physical interaction was quantified and experimentally confirmed: Galileo, Newton, Einstein, quantum physics—each on a different knowledge horizon. Living interactions are not free of the laws of physics. In addition to the limited past-present vector of purpose-free physical causality, living entities experience the future as a domain of possibilities. They make choices to preserve life. Doubt—sometimes deemed as hesitancy—is a guardian of such choices. Life, expressed in biological matter, instantiates a possible future. It is the outcome of a continuous holistic anticipatory process aimed at preserving the dynamics of biological matter. Given the non-deterministic condition of anticipatory processes—they can succeed or fail—life can come to an end, either on account of aging, or on account of failed anticipation (e.g., disease, accident).

Empirical evidence, such as the very long record—the entire evolution timeline—of various interactions involving living entities, shows that such interactions are, by necessity, both reactive and anticipatory (Kauffman 1993; Damasio 1994). The outcome is the change of the living over time—including the end of life, which neither physics nor chemistry can accommodate in their explanations. Physicists have no place in their vocabulary for death. However, physicists appropriated the notion from the living and use it metaphorically, as in the astrophysics of the "death" of stars. The meaning of dying from Covid-19 is, of course, quite different from the "death" of stars. Dying of old age (*senescence* includes the deterioration of anticipatory capabilities, Nadin 2004), or through a serious accident has a meaning different from becoming the victim of a condition which medicine was not prepared to address.

Based on considerations derived from statistics, the medical community suggested that the lifespan of the elderly affected by the pandemic will be shortened by eleven years (Hanlon et al. 2021). They also took note of a decrease (by one year) in longevity (Arias et al. 2021). Their conclusion was based on the model of physics and speculative statistics. Bach (2021) demonstrated that probably five days is closer to reality, but the demagoguery of "nation's worst public health crisis in 100 years" makes for better headlines and more expansive reactions. The meaningless "quantification" of life belongs to the same mechanistic view that determines the "lifespan" of any machine (e.g., car engine, computer, washing machine). Obsolescence—often built-in—is a feature, not an accident, since the economic machine must keep going. During breakdowns (the crises succeeding ever more rapidly), society is asked to pay for the replacement, borrowing more and more from the future. CoVID accelerated the rhythm and scale of borrowing; the debt grew without the public worrying about it. The cause-and-effect machine of capitalism cannot escape the law of maximizing profits through the mechanism of the ever-faster movement of capital.

Our progressive civilization, in all its glory, but with all its flaws, results from a blindly assumed deterministic view of life and reductionist way of doing things. Therefore—and this is a harsh inference, eventually to be made even more explicit—the outcome of the pandemic could not have been different. For those married to physics and faithful to its perspective, it makes no difference if death comes from a pandemic or from being hit by an asteroid. Therefore, an inference invites our attention: If there is a culprit for the lost lives and the economic breakdown, as well as for the turmoil integrating calls for fairness and equity as loud as calls for intolerance, it is the science and technology grounded in the castrated Cartesian Revolution. It once again forced the entire world into an absurd effort of reaction, to the exclusion of anticipatory action. The opioid crisis of recent years—caused by the machine-driven model of medicine (Nadin 2019)—extended into the pandemic; so did the abject failure of medical care—even when it is available—and the deepening crisis of social and political values. Why even try to distinguish between a Covid-19 case and some other medical condition when the reactive mechanism reimburses a hospital at a much higher rate for the victims of the scary pandemic than for other patients? There is no morality in the reductionist-deterministic view that has informed medical interventions—often performed under extreme circumstances—in the crisis of vision in which we find ourselves.

Not only medical care has failed society. One important example deserves some attention. The inadequacy of the education system explains why, within Covid-19, schooling went into an online surrogate mode. Increasingly watered-down teaching practiced in the commercialized education system became the norm. Of course, it could not compensate for the now missing socializing that students got used to. Not only the education of future doctors has suffered, but also the education of engineers, teachers, social workers, etc. The online version proved adequate to becoming a platform for incendiary rhetoric—the loudspeaker for any frustrated voices—but not for meaningful learning interactions. There was no room for addressing the need for a preventive course of action. In the pandemic context, a whole year in the life of the new generation was sacrificed for the sake of maintaining the ever more expensive institutional machine. Every measure was a reaction to the breakdown. Students and teachers counted only as numbers in a database, not as participants in a process vital for the future. Paradoxically, even as the new reality set in, universities kept building—because their economic model is based on more bodies to fill auditoriums, seminar rooms, and various facilities for social activities.

The Covid-19 pandemic continues with more people infected and with a higher number of deaths (the majority in the age group defined as elderly, more men than women), and increasingly longer-term afflictions. Just as persons who suffered from Lyme disease experience the effects for the rest of their lives, the so-called post-Covid (or long-Covid) condition is ruining many lives (Fig. 4).

Moreover, as the vaccination push continues, a new set of concerns—side effects, virus escape, breakthrough, immunity duration, etc.—captured headlines and further increased the feeling of insecurity.

Scientists studying the "transmission dynamics of SARS-CoV-2 through the post-pandemic period" (Kissler et al. 2020) face what seems to be an intractable problem: Will society be different at the elusive end of the pandemic? At this juncture, to post numbers only invites endless readjustments. The second wave of the pandemic brought sad new records with it. The third was not better. The fourth was not yet over—setting new records in some countries—before the fifth (Omicron) started. Variants of increased infectivity made things even worse. One headline captured the situation: "As Covid-19 Ends, Covid-21 Begins" (Hamblin 2021). Following this line of thought, we can go for Covid-22 and 23. The issue of health security, on the minds of many scientists (Ashton 2021), does not yet inform new policies. Suffice it to repeat that hundreds of millions were affected—some infected, others only inconvenienced (in various degrees); and many, too many, lives were lost, or irreversibly changed. A great number of physicians and healthcare professionals, as well as first responders, have died. In previous pandemics, some relatively recent, as well as during all kinds of previous breakdowns (some catastrophic, like the Spanish flu of 1918), lives were also lost, destinies affected, well-being (as relative as it is), sacrificed. Despite the almost limitless data processing capacity available in our days, it will take years before even the quantitative aspects of the pandemic will be properly captured. The meaning, which is more important, will become apparent after a longer time interval—if indeed there will be a competent effort to address it. One requirement stands out: Investigation of the terrible experience has to be detached from the immediacy of emotions, and, more important, from any opportunism (political or commercial). Otherwise, the effort cannot be justified.

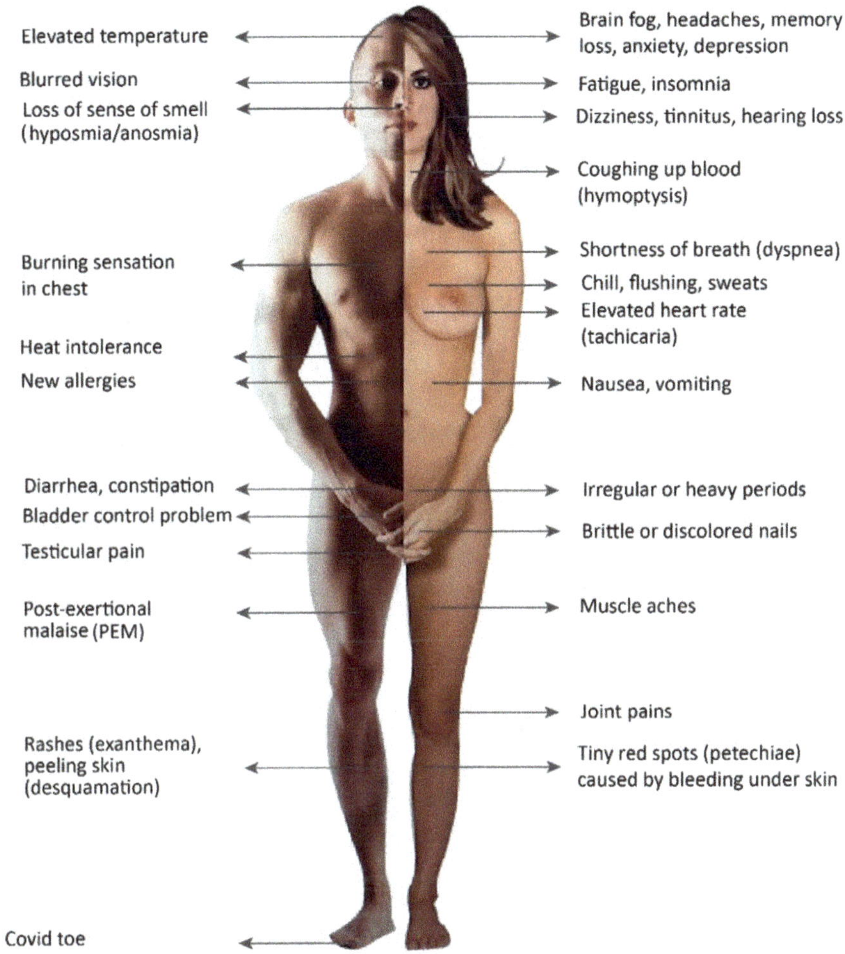

Fig. 4 Long-Covid (or post-Covid) symptoms

The Complementary Perspective

For this to happen, science will have to change. The premise is clear: Definitory of life, anticipation is the complementary perspective, not to the detriment (or in negation) of the determinism of action-reaction, but in conjunction with it. It does not negate either physics or chemistry; it is not supervenient on either. Rather, it adds to our understanding of life what neither physics nor chemistry can contribute. Merely contrasting reaction and anticipation is not productive in the disruption of science. Ascertaining their complementary nature corresponds to a more comprehensive view of the world. Complementarity at its core means that two principles exclusive of each other ought to be considered together in order to understand what neither of them can explain. *If the Cartesian view of a causality anchored in the past (from which evidence is extracted) is complemented by the view of anticipation, then the possible future, into which life unfolds, is effectively acknowledged.* Thus, prevention could become an implicit dimension of human activity, instead of remaining a rhetorical expression resuscitated each time a breakdown is expressed. Doubt, i.e., questioning, a principle that also bears Descartes' signature, is part of such a complementary perspective. Just for the sake of exemplifying complementarity as the foundation of the new science suggested, let us take note of the revolutionary view of light advanced in the context of quantum mechanics: simultaneously wave and particle (Bohr 1958). Only by adapting an inclusive model of science, one that unites the knowledge of the inanimate and that of the living, will society reverse the Upside/Downside ratio of desired outcome decoupled from consequences that undermine the viability of life on Earth.

Causality of living processes is more than, and different from, the non-living matter causality of action-reaction. Above all, it is also purpose—the maintenance and further flourishing of life. The possible future—life-threatening dangers avoided on account of anticipatory action—engenders anticipation. It is also learning, the outcome of which is both anticipatory action, expressed as behavior and the values attached to it. Morality in the first place. Life is the unity of the genome (the inherited) and the exposome ("nature and nurture" in a recent formulation, Wild 2005).

A Conceptual Framework

The gist of the Sustainability Thesis (ST) is that reactions ultimately pertain only to changes in the physics and chemistry of matter, but leave out what is characteristic of life, i.e., its anticipatory condition. It cannot be repeated often enough: Reactions are more expensive, by many orders of magnitude, but also less effective than anticipation-informed pro-active actions. They are part of the reductionist-deterministic legacy of unsustainability, i.e., saving the present at the cost of the future. In the long run, humanity will not be able to afford the cost and the many side-effects of reactions without risking its own viability. The reversed Upside/Downside ratio of benefits vs. cost of damage is indicative of the danger. In order to provide a shared understanding, let us define a conceptual framework and suggest a simple associated formalism (kept to a minimum). This should serve as the premise for further inferences, from experimental or empirical evidence, as well as from logical and methodological considerations.

Definitions:

System: an epistemological construct. Describes a whole (living or non-living) identified as the outcome of interactions among its various entities.

Open system: describes an entity with a dynamic affected by identifiable endogenous (inside the system) and exogenous (with the environment) interactions.

Closed system: interactions in a closed system are circumscribed to the interacting entities.

State of a system: the condition of the network of interactions at any instance in time.

State space: the set of all possible conditions that a system can be in.

Dynamic system: a description of how the state space changes in time.

Current state: the discrete condition of the system at the instance of its observation.

Past state: the condition of the system at any time interval prior to its observation.

Future state: the condition of the system at any time interval subsequent to its observation.

Possible future: one or several, simultaneous or subsequent, conditions of the system.

Predictive system: from past data and models of change to a future path. Predictive systems are closed to possible future causality.

Deterministic system: a system whose current state depends on a previous state:

$$x(t) = f(x(t-1))$$

Corollary: Prediction means to infer from data describing past states to present and to future states.

Non-deterministic system: entailment of unpredictable possible outcomes.

Randomness: a system without any pre-defined or pre-definable state.

Reactive system: a deterministic system whose current state is the outcome of a previous state identified as cause. Reactive systems are deterministic.

Anticipatory system: The current state of an anticipatory system depends upon a previous state, the current state itself, and possible future states:

$$(x(t) = f(x(t-1),\ x(t),\ x(t+1))$$

Corollary: Anticipatory processes underlie possible actions in the open-ended space of contingent futures.

Supervenience: describes the relationship between characteristics of different levels of reality, i.e., what supervenes upon what.

Corollary: the phenotype supervenes on genotype.

These definitions (Fig. 5) are provided with the purpose of establishing a premise for an informed dialog based on shared understanding. *Disrupt science*, like any form of disruption, will not come about by edict. The prerequisite for the disruption is the shared understanding of the concepts involved in redefining science. Awareness of the need for change is not the same thing as the forces at work that actually change the state of a system.

Thesis: Anticipatory action is guided deterministically by previous conditions (of the world and of self) and contingently by possible future conditions (of the world and of self). It is by necessity.

(a) *holistic*—it engages the entire being—and
(b) *contextual*—it takes place in the ecology of the world to which the self belongs and in the recurrent ecology of the self as part of the world ecology. To preserve life is to engage ll that it takes to give it a future. A science unable to contribute to sustainability, but which actually undermines it, has to be disrupted.

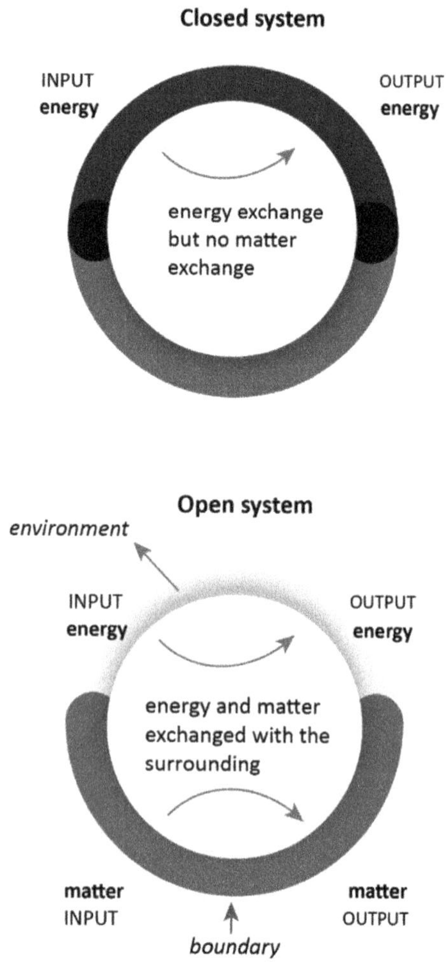

Fig. 5 Closed and open systems

From Principles to Reality

The crystal-structured spikes (the "crown") of the SARS-CoV-2 virus (one of the several known SARS viruses)—the spike (S) protein—interacts with the delicate olfactory epithelium, one, among others, of the entryways into the body (see Fig. 1). It is an interaction between non-living matter and the living receptor cells in the upper respiratory tract (one among others). The living interprets the encapsulated protein as a legitimate part of the organism. The virus genome provides the instructions for being copied. The living cell host, in a state of continuous renewal, has the ability to replicate the pathogen. The virus is a closed system with a limited state space (it encodes 29 proteins). The host is an open system with a continuously changing state space. Therefore, there can never be a unique treatment, a particular drug, or some other singular intervention for Covid-19 (and its successive manifestations).

Current biochemistry shows that the molecular functions of viral proteins and their possible interaction with the host proteome are still ill-defined (Stukalov et al. 2021). In the case of SARS-CoV-2, the living cell hosts produce copies that become the huge number of virions that eventually affect the lungs or the heart, the thyroid, the gastro-intestinal system, and even the brain and toes. Virions from an infected person are the pathogens for the infection of others. There is a difference between the virus and the virions. Because of the infection, injury of the endothelial cells (cells that line blood or lymphatic vessels) can promote coagulopathy (impairment of the bloods ability to coagulate, Goshua et al. 2020). The resulting state of the body might bring life to an end, especially the lives of people in a weakened state by previous medical conditions generically described as comorbidities. The correct term would be "pre-morbidities." This looks and feels like a *choke flu*: as the body continues to replicate the virus, it eventually gets asphyxiated (the lungs give up) or the heart comes to a stop. Research (Pérez-Bermejo et al. 2020; Sidik 2022), copiously confirmed since, revealed how heart cells (cardiomyocites) are affected. As a result, muscle fibers (sarcomeres) that control the heartbeat are rendered inoperative. Since, unlike other cells in the human body, heart cells do not regenerate, even mild cases of Covid-19 are a preliminary to a variety of heart conditions that become symptomatic over time. Those who survive are usually advised not to exercise.

The legitimate question of prevention cannot be approached without understanding that only a holistic view—the human in its environment—can guide anticipatory action. The replication function, which characterizes the living, cannot be simply switched off. Evidently, controlling sources of infection is a beginning. But an efficient course of prevention involves the virus (and its source), as well as the host: precisely, the human condition in a broad sense. The conceptual framework allows us to understand that while vaccination is an anticipatory path, it is also essentially a non-deterministic action: It can fail. All the discussions on side effects or on virus evasion (infection after vaccination) are, in fact, about the non-deterministic nature of anticipatory interventions. Some individuals are born immunodeficient—usually along a chain that connects them to ancestors affected by interventions that undermined their immunity. Others—in the millions (there are more than 300 types of primary immunodeficiency disorders)—became immunodeficient due to medical treatments (e.g., chemotherapy radiation therapy), all deterministic in nature (physical or chemical treatments). For them, prevention has to take a path other than vaccination.

Understanding the complexity of Covid-19 (and successive pathologies associated with the SARS-CoV-2 virus) is useful in addressing the crisis of vision within which the pandemic evolves. The assertion that Covid-19 takes place in the broader context of CoVID (crisis of vision) is based on observation of a confluence of many events (here only partially presented, Fig. 6).

More on the encompassing meaning of CoVID will eventually become apparent within the broader perspective of this book. For the sake of relating the conceptual framework submitted to what defines the crisis of vision, let us relate to some of the most obvious aspects.

The multiplication of experts (and self-declared experts) can be explained in terms similar to those describing the virus's dynamics. The "replication" process in society is structurally similar to that of the cell-generating virions (Fig. 7). Members of the scientific community, but also politicians, the media, the untiring peddlers of anything and everything, took in whatever data that became available or whatever they themselves made up. Those who are measuring the pandemic— where it started;

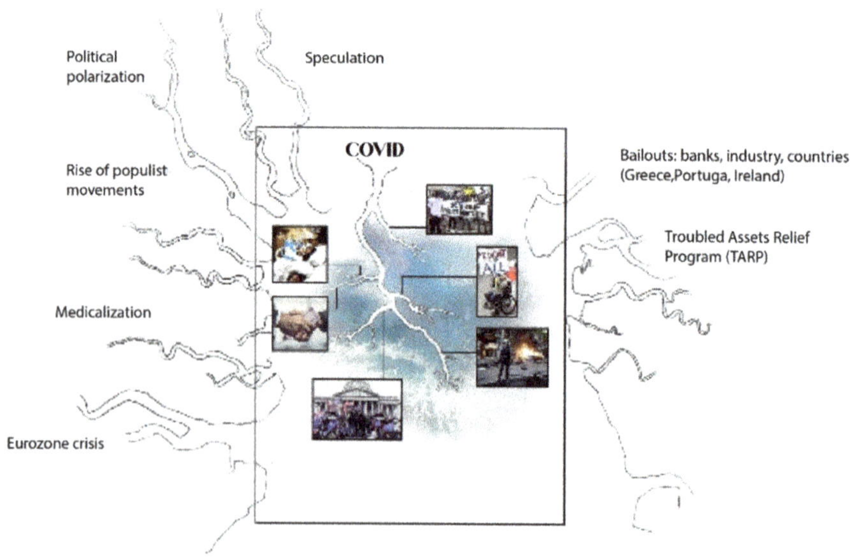

Fig. 6 Time windows of CoVID—the current snapshot is part of a larger picture. A larger time "window reveals that prior to the great recession (2008–2009), policies meant to address various inequities were exploited through predatory lending and lack of oversight of markets. The larger time window reflects the consequences of natural disasters, some of which are unpreventable (earthquakes, volcanic eruptions). Preparedness, reflected in urban development plans and construction codes, could have mitigated costs

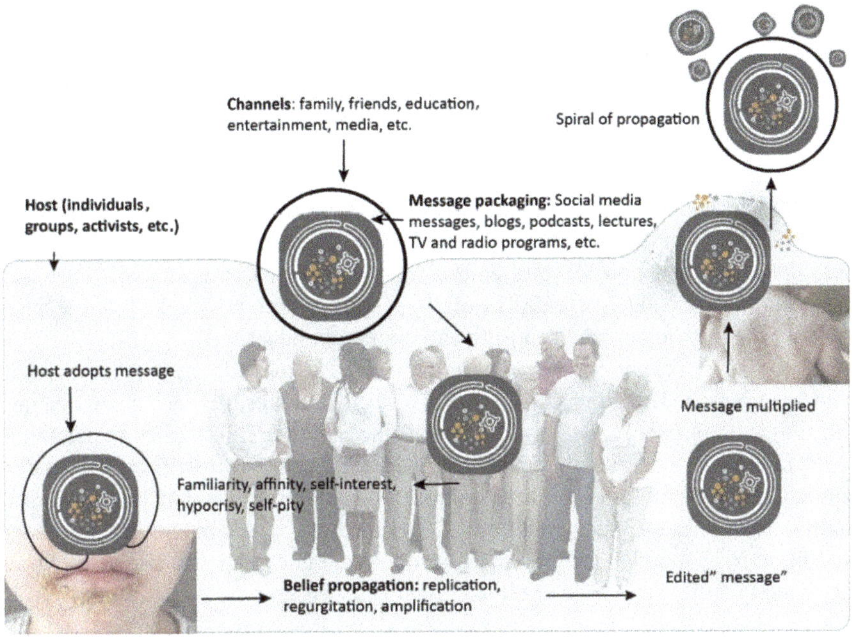

Fig. 7 Message replication in society

the connection to other corona viruses; how it is manifested; the genetic make-up, etc.—were involved in a serious effort, undermined by the implicit reductionist-deterministic premise.

Millions of hours of research and large-scale data processing produced a deluge of publications: *Scientometrics* (February 2021) identified close to 300,000 pre-print and peer-reviewed articles (see also Mazer 2021). Further studies under the same heading show that the number has almost doubled since then. The media added to the deluge. Political actions—some stirred up by social media—as well as marketing—reaching into the absurd—saturated communication channels. The unbridled replication process generated a pandemic of the pandemic, fired up by everyone and anyone able to seize a loudspeaker or a cell phone. Machine-learning-based authoring contributes fake studies with fake numbers and images.

The pandemic became an opportunity. Some experts recognized the challenge to inform and tried, to the best of their abilities, to live up to expectations. Others (the majority, it seems) followed business instincts or aligned themselves with interest groups with no regard of moral and ethical constraints. Truth was sacrificed for convenience. In the global economy of interdependencies that override ethical considerations, the pandemic was like profit-making fire. Lots of fuel—economic and political interests, racism, ignorance, malice, etc.—capable of fanning the flames ever higher was at hand. Deeply embedded in the government machine, experts and self-proclaimed experts influenced not only the initial response, but also the long-term commitment to the reactive mode. Virologists and epidemiologists (rarely seeing eye-to-eye) installed themselves in the driver's seat of a rocket, void of any control function, but well exposed to the television cameras. Players in the healthcare economy (pharmaceutical companies, providers of equipment, testing devices, and services, etc.), public health workers, and politicians ("experts" on everything) echoed their views. They contributed, in most parts of the world, to the blind barrage of reactions that ensued once the pandemic was officially declared. After the financing of the new "business"—a start-up called "Covid-19"—was voted into law—in the USA as well as in Europe, a new wave of entrepreneurship grew into a tsunami. The sickness, echoing the pandemic, affects everyone: society becomes choked in messages covering the continuum between truth and malignant falsehood.

The broader societal crisis predating the pandemic made things worse. Political cannibalism in the USA, but also all over the world, transformed each challenge into demagogical arguments in total disregard of the victims. This applies to pandemic as it applies to floods, earthquakes, hurricanes, disastrous fires, financial market failures, and terrorist attacks. The incompetent handling of the pandemic by public health officials, as well as by members of the medical establishment, prompted the qualifier "social murder" (Abbasi 2021) A quote might explain this choice of words and bring this first part of the argument to an end:

> When politicians and experts say they are willing to allow tens of thousands of premature deaths for the sake of population immunity or in the hope of propping up the economy, is that not premeditated and reckless indifference to human life? (p. 1)

References

Abbasi K (2021) Covid-19: Social murder, they wrote—elected unaccountable, and unrepentant, *British Medical Journal* 372:314. February 4. https://www.bmj.com/content/372/bmj.n314 [retrieved April 29, 2021].

Arias E, Tejada-Vera B, Ahmad F (2021) Provisional life expectancy estimates for January through June 2020. Vital Statistics Rapid Release, no. 10. Hyattsville, MD: National Center for Health Statistics. February. DOI: https://doi.org/10.15620/cdc:100392 [retrieved March 26, 2021].

Ashton JR (2006) Virchow misquoted, part-quoted, and the real McCoy, *Journal of epidemiology and Community Health*, 60:8, 671 (August). Ashton refers to a publication by Friedlander E, Rudolf Virchow on Pathology Education, http://www.pathguy.com/virchow.html [retrieved March 2, 2021].

Ashton J (2021) COVID-21: Just when you thought it was safe to go back in the water, *Journal of the Royal Society of Medicine*, 114:3, pp. 146–147.

Bach PB (2021) CDC estimated a one-year decline in life expectancy in 2020. Not so — try five days, *STAT* First Opinion February 25. https://www.statnews.com/2021/02/25/cdc-one-year-decline-life-expectancy-really-five-days/ [retrieved March 1, 2021].

Bar-On YM, Flamholz A, Phillips R, Milo R (2020) SARS-CoV-2 (COVID-19) by the numbers. *eLife, Epidemiology and Global Health, Microbiology and Infective Diseases.* March 31, 2020. https://elifesciences.org/articles/57309 [retrieved April 5, 2020].

Bohr N (1958) *Atomic Physics and Human Knowledge.* New York: John Wiley and Sons

Cannon WB (1926). Physiological regulation of normal states: some tentative postulates concerning biological homeostatics. In: Pettit A (ed.) *À Charles Riches amis, ses collègues, ses élèves* (in French). Paris: Les Éditions Médicales. p. 91.

Cole D (2021) Bad Science, Bad Faith, *Taki's Magazine* December 21 https://www.takimag.com/article/bad-science-bad-faith/ [retrieved December 26, 2021].

Damasio AR (1994) *Descartes' Error: Evolution, Reason, and the Human Brain.* New York: Gorosset/Putman.

Dawood A A (2021) Transmission of SARS CoV-2 virus through the ocular mucosa worth taking precautions, *Vacunas* (English edition), 22:1, January-April, 56–57 https://www.sciencedirect.com/science/article/pii/S2445146021000078?via%3Dihub [retrieved May 3, 2021].

de Kleer J, Brown JS (1984) A qualitative physics based on confluences, *Artificial Intelligence*, 24:1–3, 7–23.

Deace S, Erzen T (2021) *Faucian Bargain: The Most Powerful and Dangerous Bureaucrat in American History*. Nashville/New York: Post Hill Press:

Descartes R (1637) *Discourse on the Method of Rightly Conducting One's Reason and of Seeking Truth in the Sciences* (Discours de la Méthode Pour bien conduire sa raison, et chercher la vérité dans les sciences). Leiden: Jan Maire.

Frank T (2020) Removing 1 Million Homes from Flood Zones Could Save $1 Trillion, E&E News, April 27. https://www.scientificamerican.com/article/removing-1-million-homes-from-flood-zones-could-save-1-trillion/ [retrieved December 29, 2021].

Futerman A (2020) The Cell Biology of SARS-CoV-2. On the characteristic biological features. Special Reports: SARS-CoV-2 and COVID-19. April 3, 2020. https://inference-review.com/articles/special-reports/ [retrieved April 5, 2020].

Goas O, Gaddam S (2022) Journey of the Mind: How Thinking Emerged From *Chaos*. New York: W.W. Norton & Company

Goshua G *et al.* (2020) Endotheliopathy in COVID-19-associated coagulopathy: evidence from a single-centre, cross-sectional study, *The Lancet Hematology*, June 30 https://www.thelancet.com/journals/lanhae/article/PIIS2352-3026(20)30216-7/fulltext [retrieved August 10, 2020].

Hamblin J (2021) COVID-19 Is Different Now, *The Atlantic*, March 26. https://www.theatlantic.com/health/archive/2021/03/covid-19-variants-covid-21/618427/ [retrieved March 27, 2021].

Hanlon P *et al.* (2021) COVID-19 – exploring the implications of long-term condition type and extent of multimorbidity on years of life lost: a modelling study. Wellcome Open Research https://wellcomeopenresearch.org/articles/5-75 [retrieved March 26, 2021].

Hixenbaugh M (2020) Scientists were close to a coronavirus vaccine years ago. Then the money dried up, *ABC News*. March 5. https://www.nbcnews.com/health/health-care/scientists-were-close-coronavirus-vaccine-years-ago-then-money-dried-n1150091 [retrieved December 29, 2021].

Huang, N., Pérez, P., Kato, T. *et al.* (2021) SARS-CoV-2 infection of the oral cavity and saliva. *Nature Medicine*, March 25. https://www.nature.com/articles/s41591-021-01296-8 [retrieved March 25, 2021].

Kauffman SA (1993) *The Origins of Order: Self-Organization and Selection in Evolution.* New York: Oxford University Press.

Kissler SM *et al.* (2020) Projecting the transmission dynamics of SARS-CoV-2 through the post-pandemic period, *Science* 368:6493, 860–868. May 22.

Koyama T, Platt D, Parida L (2020) Variant analysis of SARS-CoV-2 genomes, *Bulletin of the World Health Organization*, 98, 495−504. https://doi.org/10.2471/BLT.20.253591 [retrieved March 13, 2021].

Longo G, Montévil M (2012) The inert vs. the living state of matter: extended criticality, time geometry, anti-entropy – an overview, *Frontiers in Physiology*, 3:39, February. https://doi.org/10.3389/fphys.2012.00039

MacCandless Farmer B (2020) (March) Dr. Anthony Fauci talks with Dr Jon LaPook about COVID-19, CBS News. https://www.cbsnews.com/news/preventing-coronavirus-facemask-60-minutes-2020-03-08/ [retrieved April 10, 2021].

Marx V (2022) Method of the Year: protein structure prediction, *Nature Methods* 19, pp. 5-10. https://doi.org/https://doi.org/10.1038/s41592-021-01359-1 [retrieved August 24, 2022]

Mazer B (2021) Scientific Publishing Is a Joke, *The Atlantic*. May 6. https://www.theatlantic.com/science/archive/2021/05/xkcd-science-paper-meme-nails-academic-publishing/618810/ [retrieved May 7, 2021].

Moreno JE (2020) Government health agency official: Coronavirus 'isn't something the America public need to worry about', The Hill, January 26. https://thehill.com/homenews/sunday-talk-shows/479939-government-health-agency-official-corona-virus-isnt-something-the [retrieved March 26, 2021].

Nadin M (2013) *Are You Stupid?* Heidelberg: Synchron

Nadin M (2019) Medicine, Anticipation, Meaning, *Cybernetics and Human Knowing*. A Journal of Second Order Cybernetics, Autopoiesis & Cybersemiotics, 26:4, pp. 9-31

Nadin M (2004) *Project Seneludens*. https://seneludens.utdallas.edu/ [Retrieved April 12, 2021].

Ortega y Gasset J (1930) *La rebelión de las masas*. Barcelona: Espasa Lobros S. L. U.

Pérez-Bermejo JA, *et al.* (2020) SARS-CoV-2 infection of human iPSC-derived cardiac cells predicts novel cytopathic features in hearts of COVID-19 patients, bioRx. doi: https://doi.org/10.1101/2020.08.25.265561 [retrieved August 28, 2020].

Peyroux O (2020) Adolescents en migration ou les multiples facettes de la mondialisation, *Administration et Éducation*, Ruptures 2: École et migrations, 2020/2:166, 25–28. https://www.cairn.info/revue-administration-et-education-2020-2-page-25.htm [retrieved December 30, 2021].

Popper KR (1945) *The Open Society and Its Enemies*. London: Routledge

Ritchie H and Roser M (2021) (updated from 2014) Natural Disasters. https://ourworldindata.org/coronavirus-testing [retrieved December 17, 2021].

Service RF (2020) 'The game has changed.' AI triumphs at protein folding, *Science* 370:6521, pp. 1141–1145. https://www.science.org/doi/full/10.1126/science.370.6521.1144 [retrived August 24, 2022].

Sidik SM (2022) Heart disease after Covid: what the data say, *Nature* 608:7 921, August 2, pp. 26–28. https://www.nature.com/articles/d41586-022-02074-3 [retrieved August 24, 2022].

Sterling P, Eyer J (1988) Allostasis: A new paradigm to explain arousal pathology. In: Fisher S, Reason JT (Eds.) *Handbook of life stress, cognition, and health*. Chichester, NY: Wiley.

Stukalov A *et al.* (2021). Multilevel proteomics reveals host perturbations by SARS-CoV-2 and SARS-CoV, *Nature*, April 12. https://www.nature.com/articles/s41586-021-03493-4 [Retrieved April 15, 2021).

Wadman M, Couzin-Frankel J, Kaiser J, Matacic C (2020) How does coronavirus kill? Clinicians trace a ferocious rampage through the body, from brain to toes, *Science*, April 17. https://www.sciencemag.org/news/2020/04/how-does-coronavirus-kill-clinicians-trace-ferocious-rampage-through-body-brain-toes. [retrieved April 20, 2020].

WHO (World Health Organization) (2020) Coronavirus disease (COVID-19): Virus Evolution. December 30. https://www.who.int/news-room/q-a-detail/sars-cov-2-evolution [retrieved March 25, 2021].

Wild CP (2005) Complementing the genome with an "exposome": the outstanding challenge of environmental exposure measurement in molecular epidemiology, *Cancer Epidemiology Biomarkers & Prevention* 14:8, 1847–1850

Numbers and Meaning

Comparing is at the origin of what is termed “to measure.” Before numbers (probably ca. 4000 years ago), language hosted descriptions that stood for what was measured. Comparisons evolved from being dominantly qualitative to becoming quantitative. This process was facilitated by the abstractions expressed through numbers. The language of mathematics has numbers at its core. By extension, the language of sciences uses numbers as their alphabet. All human activities leave a rich record of numbers associated with them, as well as records of purpose, i.e., why they were carried out, what their meaning was. The future, as the open realm of possibilities beyond the present, conjures, in addition to the quantitative representation, the need to understand what numbers stand for, i.e., what their meaning is. Anticipatory action is meaning-driven. To consider the context in which numbers acquire meaning is the premise for defining the knowledge necessary for understanding change.

Science anchored in reductionist determinism promotes understanding reality through measuring it. The measured becomes data, i.e., numbers. In the language of mathematics, these facilitate inferences from the past—causes represented by numbers—to the present, i.e., effects, as yet other numbers. Within this particular rationality, data are raw material for predictions void of understanding. The limited causality path thus made available is legitimate as long as the subject of change has no active role in the process. Stones erode into sand; tectonic plates collide; water evaporates; ice melts; light travels in a straight line (within a given medium)—all without any trace of intentionality. However, that is not the case on the path from stem cell to a new organism, to aging, to eventual death. Or in any process in which the living entity makes choices: mating, food intake, migration cycle onset, and even the migratory path.

In all these examples, understanding is a matter of degree (from misunderstanding to self-defeating certitude) and of context. It is a life-long process, with many pitfalls. Suffice it to recall that pandemic outbreaks of the past were once upon a time associated with sin and godly revenge—an understanding not really helpful in reacting to disease, and even less in preventing it. The meaning of otherwise unexplainable

© The Author(s), under exclusive license to Springer Nature Switzerland AG 2023
M. Nadin, *Disrupt Science*, https://doi.org/10.1007/978-3-031-43957-5_3

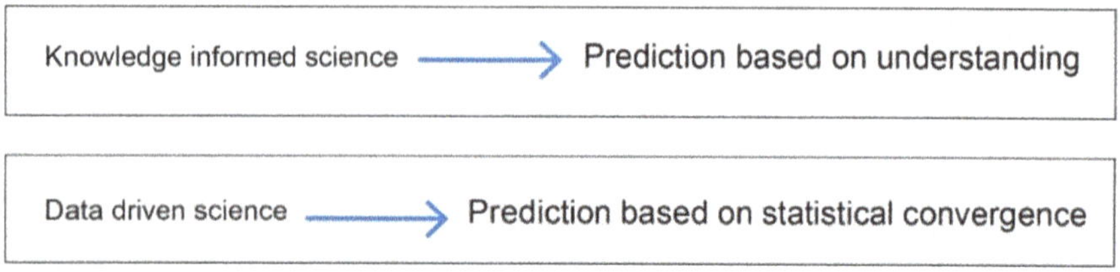

Fig. 1 In 2008, Chris Anderson announced "The End of Theory" (WIRED, 6/23/2008), while David C. Krakauer (Santa Fe Institute) decreed that "…the limits of thought" were reached (2020, Aeon)

death was hijacked into the irrational. Intentionality, attributed to human constructs, such as idols, deities, evil spirits, gods and goddesses meant to explain what seemed unexplainable, obscured the explanation of what seemed unexplainable. In our days, the cult of data—the new deity—processed without any understanding of what they stand for, became the new theology, replacing what the Cartesian Revolution so convincingly debunked. Based on increasingly more sophisticated measurements, science furnishes descriptions of the dynamics of the physics and chemistry of matter. Unfortunately, negating that there is a difference between what is alive and what is not, little progress has been made in understanding the living. Ascertainments about the living (regardless of the specific type) are ascertainments about its future. Life processes are intentional in nature. Therefore, to understand them is to understand their meaning, more than their descriptions as exclusively quantitative phenomena.

The crisis of vision (CoVID) illustrates blindness to meaning, which the technologies of data acquisition do not compensate for. Actually, the broader CoVID is the outcome of giving up knowledge in favor of the expediency of data (Fig. 1). It translates as living on borrowed time. Society references this expediency to the globalization of the Upside/Downside Ratio of benefits vs. harmful consequences of new endeavors, attained about 40 years ago and signaled by the HIV/AID pandemic. It marks the time when reaction literally did away with even considering prevention, not only in medical care, but in all forms of human activity (economic, financial, social, political, environment, ecology, etc.). Prevention does not pay for those who perceive it as demanding too much effort, and taking too long. It is convenient to live off the future. COVID-19 exemplifies the process.

Turning a deeply rooted crisis—the 2019 pandemic—into an opportunity—the post-COVID-19 business boom—explains the obsession with numbers in disregard of meaning. Of course, inflation, unemployment (followed by shortages of available labor sources), bottlenecks in deliveries of retail items (especially food) and services, increased migration (world-wide), beg for immediate remedies, and result in reactions of all kinds. The new globality of risks shared *volens-nolens* with the rest of the world eliminates the possibility of a long-term perspective. Under the pressure of a reality that no one predicted, no room is left for thinking about prevention.

Indeed, the meaningful mapping of human interactions, in a moment of extreme situations such as the COVID-19, could provide significant information for anticipatory actions. Actually, this is the unavoidable path towards effective prevention. Consider only the fact that at the peak of the calamity, speculation in housing, automobiles, and cryptocurrency increased at the rate equal to that at which governments poured money into reactive measures. The American Rescue Plan fueled a shortage in the employable workforce. Gasoline sprayed over all kinds of fires fed higher flames. Speculators benefitted at the expense of those who one day, in some future time, will have to pay for these reactions. Expensive—and only partially adequate—reactions to breakdowns based on data-driven predictions, but not on knowledge, are not only unsustainable, but they also undermine the future. The alternative course of action should be the integration of reaction and anticipation. This could make science both effective and meaningful at the same time.

If someone not captive to reductionism were to take the mathematics used in the representation of physical phenomena as a model for describing the organism, it would have to realize that the number of variables necessary would exceed the number of entities that make up the universe (Kauffman 2019). This illustrates why the predictive power of deterministic science, successful in describing the non-living universe, proves insufficient for describing change in the living. The virus-organism interaction that eventually led to the pandemics is an example. In order to understand it to the extent to which something can be done about it, not only describe it, science will need the appropriate knowledge. While there are much data—no doubt about this—there is little (if any) effective science. The analytical performance of reductionist-deterministic science is beyond doubt. Nevertheless, in the absence of the anticipatory perspective, the understanding of the meaning of change (e.g., new forms of the virus) is missing. The fact that genetics, for example, provides descriptions (in the form of sequencing data) of the make-up of the virus is admirable—but ultimately not consequential for treating the sick. With the pandemic, humankind realized that the exploration of the universe (rockets, satellites, measurement technology, etc.) is trivial compared to the new biological entity labelled COVID-19.

The rocket made it to Mars and generates data about its particular dynamics. The space telescope will generate even more. In the same time interval, people died because current reactive science could not save them. It probably never crossed the minds of those who put their blind trust in the science capable of landing *Perseverance* on Mars, or launching the Webb telescope into deep space, that this same science might, in some ways, be the cause of their SARS-CoV-2 induced anguish. It is a spectrum condition of the human body resulting from its accidental interaction with a virus that existed for millions of years in the ecosphere. It is important to take note of the fact that the corona virus was studied since the SARS (acute respiratory syndrome) epidemic (ca. 2003), but actually known since 1965. According to Dr. Jeffrey S. Kahn (2006), "I don't believe there was a big effort to make vaccines against these because they were thought to be more of a nuisance than anything else." He was not alone in this assessment (Williams 2020). Infectivity, extremely high, goes to the credit (if the expression can be used in the context of a life-threatening condition) of the human being—sick or only carrying the virus—not to that of the agency-free virus.

What also escaped the attention of those desperately seeking to help the victims was the paradoxical fact that the new possibilities opened by science and technology proved to be, at the same time, sources of new risks. In other words, vulnerability is actually self-inflicted. The benefits of activities based on inadequate science are cancelled by the damage caused to us human beings. It is not the gods or idols, or sins that brought it about, but humans themselves, captive to a new theology originating in their own thinking and ways of acting. At this point, a bit of history could be useful in explaining the focus on the WHY? question of the pandemic.

Understanding the *Why?*

John Snow, celebrated as the founder of epidemiology, traced the source of the cholera epidemic (London 1852–1860) not to miasma ("bad air"), but to contaminated well water. He published his findings as *On the Mode of Communication of Cholera* (1855)—which the medical establishment rejected out of hand. John Snow was still a student of medicine when the first cases of cholera in England (1831) broke out. It is not clear what triggered the outbreak at that time. But there is a long history of plagues of all kinds that Snow was aware of. They go back to the Plague of Athens (430 BCE) and the belief that the gods abandoned the Athenians (who in turn faced the curse through extreme indulgence). During the Antonine plague (165–180 CE), almost one-third of the population of the vast Roman Empire perished. Apollo, the "avatar of evil" was of no help. Neither was Christianity. The Plague of Cyprian (249–262 CE) was named after a bishop, not so much for his healing abilities as for "preparing" the victims for life after death.

To reference each and every disaster within different cultures and different religions is to accumulate answers to the *Why*? question that everyone posed. The answers deliver an epistemological portrait of humankind in its evolution over time, revealing what the people in Athens knew, what the Romans started to discover, what the Moslems or the worshippers of Buddha practiced. The Jews, given to handwashing and bathing (based on Biblical injunctions turned into rules defining a way of living), were many times spared lethal diseases—only to be accused of having provoked them. The irrationality of the argument did not make it disappear. In our days of epidemiological research, well beyond what John Snow carried out, there are "The Chinese did it," "Americans spread the virus," and, again, the "Jews [this time Israel] caused it"—answers to the *Why?* question impossible to ignore. They echo assertions that testify to a primitive mentality: find guilt in the darkest assignment of it to those you hate or despise.

But back to England, 23 years after the first cholera outbreak (1854), SoHo, a suburb of London, and a terrible situation: people were dying of cholera. The Bishop's pronouncements, along the line of the centuries-old answer that God punishes sinners, called for acts of redemption. Dr. John Snow found out that "within 250 years of the spot where Cambridge Street joins Broad Street there were upwards of 500 fatal attacks of cholera in 10 days." Measurements led to numbers: data that describe a situation.

There was a hypothesis—the virus was waterborne—and there was anecdotal evidence: the coffee shop served glasses of water from the Broad Street pump along with meals. Nine customers got cholera. A woman who no longer lived close to the pump liked the taste of its water so much that she had some delivered to her new address. She died the next day. However, in the neighborhood, the prison, with 535 inmates, had no cholera. It had its own water source. And finally, there was the grid—measurement of the space in question—on which data were referenced to particular persons. This made the meaning of data evident: there was a contaminated source of cholera infections. Meaning always informs action:

> I had an interview with the Board of Guardians of St. James's parish, on the evening of Thursday, 7th September, and represented the above circumstance to them. In consequence of what I said, the handle of the pump was removed on the following day.

Data, no matter how "Big," do not answer the *Why?* question. They must be referenced to what lies behind them, to the processes from which they are extracted. Practitioners of medicine might not even be aware that one of their precursors, Ignaz Semmelweiss, pleaded (1847) for hygiene—handwashing between patient visits in hospitals—only to be mocked by his colleagues. Doctors were not willing to accept that they themselves contributed to the spreading of disease by dispensing with handwashing when going from patient to patient. But surprisingly, in the context of COVID-19, "Wash your hands!" sounded like a new commandment for those not yet really convinced to do it as a matter of routine. The order was repeated as though no one has learned anything from the lessons of past pandemics, or even yearly bouts of contagious illnesses. Handwashing (part of hygiene), social distancing, mask wearing, diet, and physical activity—anticipatory actions—are elements of behavior (Zuk and Spencer 2020).

The *Why?* question is a composite: Why me? Why not my neighbor? Why us? Why do some die and others do not? And many more. Religious, political, economic, social, cultural takes result in a variety of answers. For instance, those given by the Church, by the Board of Guardians, by merchants in the affected area.

Separating lepers from their community goes back to Biblical times, and probably earlier. This was, of course, in the first place a reaction to the manifestation of the disease. Lepers were shunned as outcasts in order to prevent contamination of the population. For all practical purposes, separation is a primitive anticipatory action, and therefore the affected community tried to prevent contact between the sick and the healthy. In our time, and under current circumstances, China practices the same—to the extreme. Other times, other types of experts, and a variety of contagious diseases that lie in the past. The fact that there was no political correctness to demonize the act of isolation as discrimination is also significant.

Facing a similar crisis (that caused the death of millions of people), humankind discovered what today is called "quarantine," which stands for *quarante giorni*—a limited isolation period way less drastic than the death sentence of being forced to live among fellow lepers (as was done in Hawaii, beginning 1865, on the island of Kalaupapa). The 40-day isolation of the ill was chosen because it took so many days to be rid of the pestilential smells of bodies and rotting goods. Or because of some religion-based reason: the 40 days and nights of Noah's flood, the 40 years that the Hebrews wandered in the desert; according to the New Testament, Jesus spent 40 days in the wilderness; and then there are the 40 days of Lent. Regardless, the pro-active measure was obviously one of containment. No one even knew how to properly formulate the *Why?* question of the disease, not to say how to answer it.

During the lockdowns tried at the climax of COVID-19, and 600 years after quarantine was introduced, people asked if society would relax after 40 days, or however long lockdowns and social distancing remained in effect. Or will everyone turn into a mask-wearing member of the herd—despite the warning that masks (including the N95) are only partially effective? Will life return to normal? "Sorry, but grandmas die and babies are born. So goes the cycle of life," was repeated again and again, in various chats, and in various languages around the world. The descriptive "We don't need to slow the spread. We need to have health organizations throw spaghetti noodles at the wall for a cure and treatment until one sticks," (Gawande 2021) made the headlines. "Don't tell me what to do" encapsulates the misunderstanding of freedom claimed by those who are less free than anyone else. (The phrase quoted comes from someone in the county in North Dakota hardest hit by COVID-19). The "Why should the pandemic affect my freedom?" has been echoed in the anti-vaccine posture of people questioning science: "You brought the damn thing on us, so why should we trust you?".

Einstein advised: *Stay away from negative people; they have a problem (or more) for every solution.* For example: Social distancing and lockdowns will result in a higher rate of births, divorces, alcoholism, weight gain, mental health problems, and social disunity. Well, yes and no. Hundreds of years ago, Newton, caught in a lockdown, came up with his theory of action-reaction (Newton's third law: For every action, there is an equal and opposite reaction), which changed the world, as he enjoyed the unintended benefit of social distancing. Trinity College (Cambridge) sent students home during the Great Plague of London (1655). The focus on how matter (atoms and molecules, cells and genes) changes originates in Descartes' *Method*, and in Newton's physics, which added legitimacy to determinism. Indeed, the machinery of the universe, with its precise clock, does not conjure behavior (or free choice), but rather the explanation of its functioning according to forces described in what are called the laws of physics. Within this context, scientific prediction is a matter of knowing the law and applying it. Data from measurements can confirm knowledge or generate new questions. No pandemic affects the laws of physics. The *How?* of physics, expressed in laws experimentally confirmed, leaves the *Why?* question out. Without understanding the *Why?* of the living, the *How?* remains an incomplete description. The *Why?* of life is closed to experimental evidence because it cannot be reduced to particular aspects. The *Why?* of physics is quite different from the *Why?* of the living.

Pandemics are not physical phenomena, like eclipses and earthquakes. An infectious disease spreading over a vast territory and affecting a large population is biological in nature. This is what those who work under the banner of the Cartesian Revolution have not yet understood or refuse to accept. They ascertain that physics (and, by extension, chemistry) is *the* science of all that there is, living or not. While the laws of physics—as they apply to the non-living— are pertinent for biology, the life-specific causality is different in nature. The outcomes of living processes are not forces of attraction or positions of stars, or of moving vehicles, of stones rolling down a hill. They are not even electrons, atoms, or molecules from which cells are made. They are goal-driven behaviors, intentionality (sometimes fulfilled, sometimes not) across all scales of living matter. Seeking and finding food (a source of energy), mating (reproduction), and escaping predators (survival) preserve life. Preservation of life—which will be mentioned repeatedly in this book—transcends the individual organism (bacterium, insect, animal, plant). Let's take termites (*neocapritermes taracua*) as an example. Aging specimens develop "suicide vests"—pouches containing copper-rich proteins. Intruders endangering the colony are killed as the elderly termites (in what can be called "suicidal altruism") blow themselves up. Nothing of such an expression of awareness can be detected in physio-chemical processes—neither at nano, micro, or macro-level, nor at the scale of stellar processes. The long-distance (in the thousands of miles) migrations of some birds and fish species, for example, is of a nature different from the predefined trajectory of a rocket used in space exploration. There is a determined part—both a living body and a corpse are subject to gravity—and a non-determined part—there is failure (injury, death), and there is a "changed my mind" in the dynamics of life.

After John Snow got the pump handle removed, that particular neighborhood no longer experienced death by cholera. Some people, including members of the Board of Health, declared this to be mere "suggestion." The preventive action—stop access to infected water—was not yet knowledge. It was informed by a perspective that religion and the primitive views of the healthcare community opposed for quite a while. Only when Robert Koch isolated (1883) the bacterium *Vibrio cholerae* did Snow's view of the contamination become part of the knowledge that undergirds the science of epidemiology.

Data and Knowledge

A science based on knowledge acquired under the assumptions of experimental and empirical evidence answers questions relevant to how things change in time (Fig. 2).

Its predictive power reflects the degree of understanding the process of change. The concrete form through which evidence becomes pertinent to the acquisition of knowledge depends on how it was acquired. Measurement—an action—means introducing a "measuring stick" against which what is to be assessed is compared. Humans used their fingers and arms (a cubit is the length from the elbow of a medium-height man to the tip of the finger) and their feet (the Roman foot had twelve

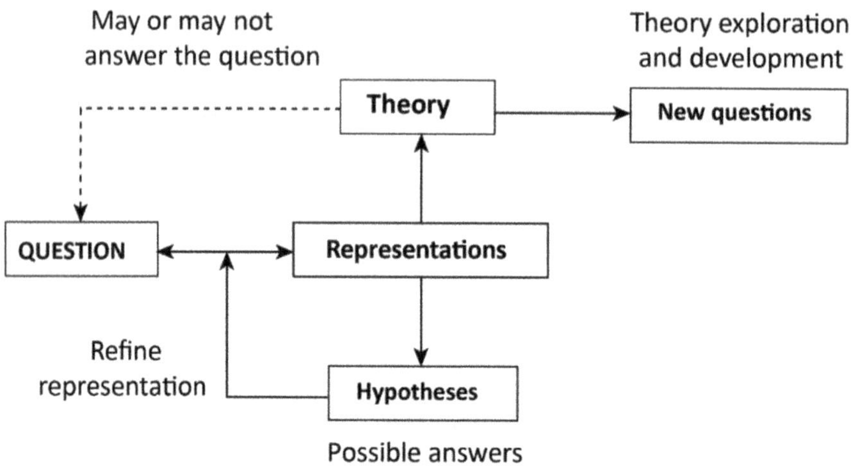

Fig. 2 Scientific law as expression of knowledge. It all starts with questioning, and conjures the formation of hypotheses, measurements, and testing procedures

nuciae) as measuring sticks before they adapted what today are the standard units of length, weight, surface, volume, etc. Measurement has data, approximate or precise, as outcome. Quantitative in nature (how heavy, how large, how deep, etc.), measuring has numbers as its outcome. These are human-conceived cognitive constructs, a "name," a "label" for a shared understanding with pragmatic consequences. The grain of wheat and of barley were units of mass, but also currency. Evidence can as well be qualitative: how pleasant a color, how impressive a melody, how smooth a surface, etc. Of course, the physics of color, or of sounds, or of tactility made it possible to represent them also by numbers. Images, sounds, tactility, smell, etc., everything the measuring person, or a sensor—the artificial "nerve" embodied in some matter—can distinguish, can as well inform science.

The intuitive level at which numbers are assimilated in culture is significant for understanding how measurement changed over time. Pebbles (*calculae*) and the knots (on a whip) used to represent numbers. So were the fingers on our hands and similar direct forms of describing quantities. The abacus is a necklace with beads arranged to "hang" on different rows. The difference among them lies in the way of thinking that is the basis of their use. Indeed, the jump from the concreteness of a particular number of pebbles to the abstract nature of the beads on several rows of wires pertains to the levels of cognition. The automated abacus, with rules for moving the beads on each row in order to perform an operation (addition, multiplication, subtraction, etc.), is the next step in reaching yet another level of abstraction.

In our time, the change from the concreteness of measuring—the grain, the foot, the stone—to its abstract condition is encountered in various attempts to "measure" the pandemic. How many individuals got infected, how many died, how many suffer from various long-term symptoms, were interpreted as numbers. Other numbers describe the virus (physical measurements, for instance), or the antibodies; yet others, of a different nature, are no less concrete: how many beds available for the afflicted, how many ICU units for emergencies, how many masks are necessary, how many vaccines, which dosage, at what frequency etc.

The crisis of vision is to a great extent the expression of the attempt to reduce reality to numbers while abandoning the understanding of what they stand for. The new equity business—monetizing the repair of unequal opportunity in a heterogenous society—uses numbers (usually legislated) as criteria for success. Predicting from numbers not associated with meaning is to make the future a repetition of the past, instead of an opening of new paths for human creativity. For performing such predictions, members of society in technologically advanced countries, as well as in some rather primitive cultures, became targets of an incessant tracking process—quantification of life—that covers from the trivial to the most intimate aspects of life. This reductionism is inadequate. Example: the number of crimes committed, the race, age, and gender of those who committed the crimes and of those of the victims, are not decisive in prevention. Missing is the meaning: why people commit crimes. In a different vein, how high is the cost of living, how many people cannot afford medical care, how many people get arrested and how many are killed, who they are, etc., are quantifiers that entirely miss meaning. Therefore, such data ultimately cannot inform effective remedies.

Measurements can generate data on account of which patterns can be effectively distinguished. The numbers derived from measurements are not only a map of the current situation (how things are at a certain moment in time), but also a reference for defining how fast the change, how deep, and how difficult to handle. Much higher numbers of deaths mean more coffins and the need for increased incineration capacity. This kind of assessment, qualified as prediction, is dependent upon immediate measurements and upon the historic record. As important as it is—and it is—this assessment does not reflect understanding of what happens and why. Improved analytic performance through the automatic processing of numbers (i.e., computation) can lead to more precise descriptions of reality. The thermometer indicating fever (the number expressed in a variety of possible scales) does not understand what might cause it. It does not even understand—neither literally nor metaphorically—that the measurement itself might be faulty when the "sensor" is not in proper working order. Temperature might be a necessary premise for understanding a state (of a machine, of a person, or of the climate), but it is not a sufficient condition for a meaningful diagnostic. It does not substitute for knowledge. In the pandemic, measuring temperature was practiced as a matter of routine. If you need yet another example of Upside/Downside Ratio flipping, this is as good as it gets: the benefit of checking everyone's temperature (never mind the production of thermometers, an activity that leaves a large carbon footprint) vs. the cost (include here the cost of measuring and recording) is minimal. Yes, we can! But should we? Lots of useless data, no knowledge.

These preliminary observations follow the line of the argument concerning the need for a science that makes the shared understanding of what is observed the basis for action in addressing change. Understanding is not in reaction to events but rather the outcome of exploring the possible future into which reality evolves.

This pertains to events detrimental to individuals or communities, but also to those beneficial. Vaccination, as a pro-active measure (per its original meaning), is an example of anticipatory action. So is access to affordable and quality medical care, which changes entirely the premises of a pandemic. It can, in fact, prevent it. Other preventive measures were spelled out in the *COVID-19: Make it the Last Pandemic* by The Independent Panel for Pandemic Preparedness & Response, presented at the World Health Assembly in May 2021.

The pandemic caught the world of science in a critical moment. The relation between knowledge and data resulting from measurements (expressed as numbers) affected the way it was perceived and handled. Medicine adopted measurement for describing the state of an organism long ago. Consequently, since more can now be measured, the variety of medical conditions and associated treatments continues to multiply. This prompted Aldous Huxley's rather ironic take: "Medical science has made such tremendous progress that there is hardly a healthy human left" (cf. https://www.quotetab.com/quotes/by-aldous-huxley/2). Ivan Illich (1975) was even more blunt: "The medical establishment has become a major threat to health." The development of a plethora of measurement methods associated with medical interventions turned the patient into a source of data. The pandemic arose after other major crises (AIDS, SARS, MERS, etc.) occurred, and prompted even more measurements. Some of these new means and methods were conceived for specific purposes: sequencing, testing for infection, clinical evaluation of new vaccines. Most were "borrowed" from physics and chemistry, in particular from attempts to explain life in terms of the causality of the non-living matter, or in within the machine model of life. Unfortunately, the machine model that promised an easier path to dealing with illness—fixing a machine is a relatively straightforward operation—was adopted *in its literal meaning*. Life was reduced to functioning, measured pretty much like the functioning of tools or of other automated devices. The medicine conditioned by this understanding treats the human being more and more like a mechanical contraption (Fig. 3). The broader context for this view is offered by the "new" machine that expresses determinism through the algorithmic processing of data.

Everything that can be described through a recipe (set of instructions) can be viewed as the output of the machine called *computer*. The "mother and father" of all machines contains not only all typewriters (the word processing programs in their variety), all calculators, all control devices for cars, elevators, production lines, electric generators, the power distribution grid, pipelines, etc. etc., but also the "machines" used by doctors: from the blood pressure measuring device to robotics (e.g., joint replacements, spinal surgery, organ removal). In this view, reality itself (pandemics included) is the outcome of computation (or, for some taken by the quantum computer, the outcome of quantum computation).

The crisis of vision within which COVID-19 started is embodied in the total surrender to the *machine of all machines*, which is by necessity deterministic. Circular thinking (also known as hypostatizing): The machine, one in a sequence of machines invented by human beings (recall the windmill, the watermill, the electric engine), generates the reality to which those who conceived them and the machines themselves

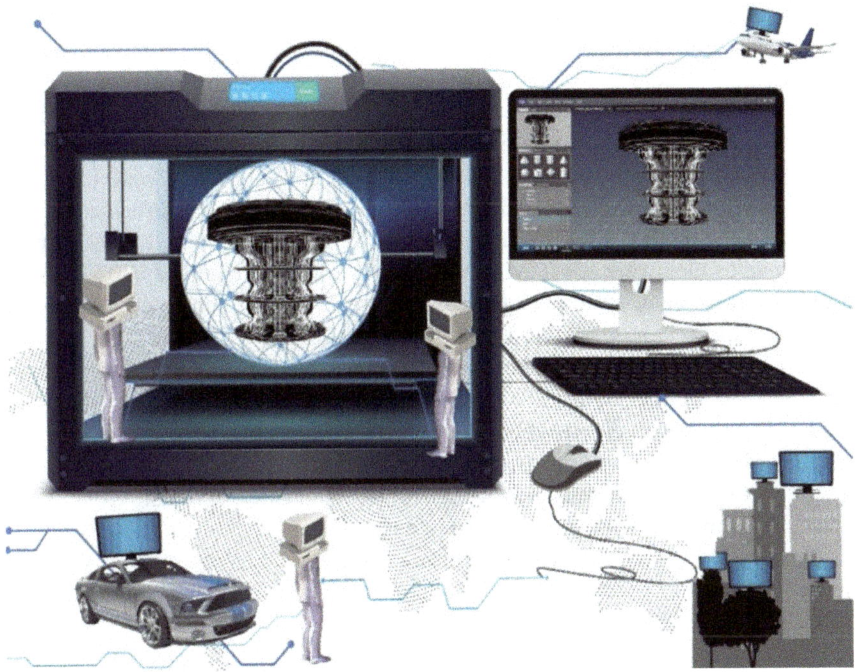

Fig. 3 The "machine" (with a suggested quantum configuration) generates reality (in this illustration as the output of a 4D printer) to which those who conceived them and the machines themselves belong

belong. This is very much like: Humans made God, and then attributed to the divine their own making and the making of reality (free will included or discarded).

Numbers, as representations of quantitative aspects of change, reflect the scale of measurement. The weight of a virus does not register on the kitchen scale supposed to assist in preparing a meal. Collisions, earthquakes, solar and lunar eclipses (part of the dynamics of the universe), or the collapse of a bridge can be observed. They can be completely measured. Temperatures—of flowing lava, or of a human patient—as different as they are in representing the respective phenomena that have nothing in common, are yet another example of what can be measured. Numbers as representations of quantitative distinctions do not originate with a reference to what they represent. In the absence of such a reference, they have no meaning in respect to the phenomena measured. Numbers are part of the *alphabet* of the language of mathematics scientists use to describe change. This description translates the concreteness of phenomena (a specific volcano's eruption, or the flu affecting a person) into abstractions.

The generic notion of data applies to the ever-broader variety of measurement methods and types. Some measurements are direct, some are indirect. In recent years, measurement itself became subject to simulation: Why physically measure when data can be generated from a model? The spiral of escaping the real in favor of settling in the virtual becomes steeper as more computation power become available. Data can fully describe the physics or the chemistry of how matter changes. But: *How do you measure a pandemic?* Or, for that matter, the concrete expression of the "wokeness" of a movement, or the cancelling impetus whose stated goal is to correct the sins of the past (racism, sexism, colonialism, elitism, privilege, etc.)? Some entrepreneurs conceived and produced measuring devices for the pandemic that were legislated under the influence of medical business. Of course, government officials "in the know"—preparing the "gifts" for the industry—took stock (not only figuratively) in the companies they were subsidizing with taxpayer money (Ballhaus 2020). Other entrepreneurs profited from the business of equity—which generated other laws. You want funding for searching quantum computation? Nuclear fusion? Starting a new school? Just show the numbers that reflect "inclusion," the demagoguery that shadowed the pandemic. The long term of shared opportunity was effectively sacrificed for the short-term reactive genuflection. The benefit compared to the cost of reaction (which can only be short-lived) is yet another example of a reversed Upside/Downside Ratio characteristic of unsustainable deterministic reactions. "Woke" itself is reactive, without any anticipatory component.

The disease's spread was quantified through the R_0 (R-zero): the number of people infected by one virion carrier. Furthermore, the number of deaths, assumingly from COVID-19, were tabulated (sometimes in error, sometimes intentionally). Attempts were made to capture the total number of victims. It should be clear, in the context defined so far, why understanding the spread of the disease is quite dissimilar to what is needed to land a device on Mars, or to conceive and deploy a sophisticated deep-space telescope. This realization of their fundamentally different natures contradicts the deeply ingrained reductionist-deterministic prejudices of those who called for more data as the answer to the pandemic. Or, for that matter, more and more data as the answer to all the problems that humanity faces, and even as an explanation of life and its origins. The fact that sustainability is not a matter of insufficient data was proven by the natural disasters that paralleled the pandemic: The Texas "deep freeze" (2021), the successive heat waves in the American Northwest, mudslides in California, the catastrophic floods in Germany and Belgium, followed by extreme heat in the summer of 2022, fires in Australia, volcanic eruptions, earthquakes. They all prompted massive reactions—at the expense of the future, and without any preventive measures. Although everyone talks about climate change and its various manifestation, nobody is prepared for them. The post-catastrophe blame-casting demagoguery flared up, but did not lead to action. The breakdowns, prompted by natural disasters, were all, like the pandemic itself, preventable. A simple decision, such as rebuilding away from flood-prone areas, for example, or prohibiting home construction on unsuitable plots of land, would save lives and money.

A pandemic is different in nature from the physical reality that can be measured. Pandemic or not, life is in continuous change; physical reality, although changing as well, is relatively stable. It takes longer to even notice how a stone erodes into sand than to go from sleeping to getting up from bed, with all its associated changes in physiological parameters. Within the broader crisis of vision, the COVID-19 experience is yet another argument for a different focus: not on numbers, as useful as they can be, but on meaning. It is necessary to identify in the ocean of data the significant data. In checking if hospitals have enough beds available, numbers are important. More important is the focus on meaning: "We have the beds, but do we have the knowledge needed to give the proper help?" The number of vaccines necessary to inoculate everyone in the world is not the same as the numbers describing individual behavior, expressed as the desire to be vaccinated or to reject vaccination. Billions of vaccine doses had to be discarded when they passed their "use by" date. Most of the persons affected by COVID-19 were over the age of 65, especially those in nursing homes and other care facilities for the elderly. The evidence in the USA was in plain view: for the first acknowledged case in Snohomish County, Washington (January 19, 2020; Holshue et al. 2020), it became clear that medical evaluations based solely on numbers without understanding the significance of pathological details was misleading. In Italy, where the first wave was extremely severe, this realization became the basis for addressing the concrete aspects of medical care under pandemic conditions. Later, when measuring genetic characteristics entered the picture, sequencing took the upper hand. The obsession with data continued, focused on virus variants, to the detriment of searching for meaning. The original virus and the BA.5 or the BQS or XBB mutations affect people in different ways. The question of the *Why?* of mutations rarely got attention. This would have meant focusing on how the human condition under the pressure of COVID-19 was affected.

Evidently, the hour of statistics and number processing—sometimes extremely successful—has come, but with no understanding of what the flood of numbers meant. Data acquired from measuring change in the living—healthy or affected by COVID-19—without understanding *what* is measured and *why*, are only incidentally relevant. Temperature screening became a quite revealing example. It took a long time and billions of meaningless measurements before the medical establishment itself acknowledged the fallacy underlying it (Karlis 2020; Wright and Mackowiak 2020). But even in the context of widespread vaccination, checking the temperature of everyone visiting a healthcare facility continues.

Significant Versus "Big"

Since the beginnings of epidemiology—i.e., since Dr. Snow's work in containing cholera in Soho—it became clear that, in evaluating the living, "Small Data," which means little but *significant* data, are more relevant. From an epistemological perspective, significance conjures understanding. There is a "before measuring" time, when observing change is not affected by the means used to measure it. Indeed, before the London pandemic, the water pump was used without reservation. And there is the "after measuring" condition. Under the guidance of knowledge—who was affected by the cholera—the measuring process can be focused on the significant: they all got water from the pump. Let's take oxygen level—associated with COVID-19. In the absence of symptoms (trouble breathing, confusion, bluish lips, among others), the measurement of the blood oxygen level is at most a "Let's make sure!" option.

However, given the physics deployed—a beam of red light through the fingertip—a physician would know that colored nail polish, or tremor, or dark skin color might render the measurement meaningless. Still, each visit to a doctor's office entails the pulse oxiometer measurement. The huge amount of data collected in disregard of knowledge (never mind common sense) might eventually end up in the Big Data-driven generation of models, which rarely inform a meaningful treatment. A 100% oxygen level measured in a patient having a heart attack is not unusual. But how meaningful is this measurement? Or is it only an alibi, i.e., the patient is being cared for.

The need to keep measurement of living processes to a minimum is the correlate of an important scientific axiom: to measure is to disturb (Heisenberg's "Uncertainty Principle" 1927). This applies to neurons, to cells, and to organs, as well as to organisms. Ignoring the meaning of the data leads to conflicting explanations (so-called "models"), and to misinformed predictions. Missing is the awareness of an unavoidable arbitrariness: These are the results of *our* measurements and *our* testing. Or the result of some established protocols. More precisely, if the particular testing method was questionable, the resulting data will not help anyone—except the publishers of scientific (and less than scientific) articles. The fact that enormous amounts of data were at play in the research documented through a tsunami of publications did not make them more significant or effective in addressing the pandemic. The "science" of medicine focused on data acquisition through all means is spectacular in terms of technological deployment. But it has failed those who had to help patients through difficult times, as well as the patients themselves.

When Tests Fail

The subject of testing and of medical science will be revisited as we assess the return on the investment in carrying out measurements. But for now, a reference to the early stage of COVID-19 and Japan: It tested 0.185% of its population. (For a detailed report for the entire world, see Total COVID-19 Tests, begun in February 2020 and kept current.) (https://ourworldindata.org/grapher/full-list-total-tests-for-covid-19). But Japan fared better, during the first wave, than all countries obsessed with quantifying phenomena not even partially understood. It should be added that Japan, with a homogenous population (and a growing number of elderly) has a culture of pro-active behavior informed by its susceptibility to earthquakes and tsunamis. This is an example of knowledge based on empirical evidence, in contrast to data-driven predictions that generalize from statistical patterns. Instead of measuring, Vietnam imposed a way of living—extreme control of behavior—that prevented spreading. China (including Hong Kong) has not ceased testing *en masse*, and has been relentless in forcefully isolating people suspected of infection.

Testing, promoted as justified and critically essential, is as good as the testing method used. All kinds of tests were thrown at the confused population, while those who ordered them warned that many were known to go wrong. The US Centers for Disease Control and Prevention (CDC) made it clear: Serologic tests are not accurate enough for making policy decisions since they are "wrong half the time." The initial CDC corona virus test kit was not only faulty but exemplified the lack of professional ethics that the pandemic occasioned. It was deployed by a government agency that knew that it was defective. A more detailed analysis of high-quality clinical studies brings up the high risk of bias, heterogeneity, and, most important, limited generalizability. Conclusion: "Current evidence does not support the continuing use of existing point-of-care tests" (Bastos et al. 2020). Still, the push towards tests continued. Eventually, the focus became "cheap, rapid virus tests." The US government sent test kits to all those who requested them, despite the skepticism of some epidemiologists (Guglielmi 2021).

This is not a matter of theory, but of life and death. No test informs the physician on the appropriate healing process. Knowledge is required, and data, no matter how Big (from billions of tests), do not substitute for it. At most, it indicates the need for action. Testing and tracking are about spreading the disease, not about treatment. Just to highlight what COVID-19 taught us: Paradoxically, many people have died not because they were infected, but because they ended up being treated. Based on data-informed protocols from previous experiences, some patients were hooked up to machines intended to help them breathe, but which ended up asphyxiating them. "Wrong-pipe" intubation (in the esophagus instead of the trachea) and ventilator-induced lung injury explain the assertion that not COVID-19, but the wrong protocols, based on insufficient knowledge though supported by data, killed up to 80% of people who went through the "treatment." In view of the evidence alone, painfully experienced since the beginning of the pandemic, it is justified to point out the consequences. Measurement without a true understanding of the limitations of various measuring means and methods is of minimal usefulness. What counts is what data mean, and how the meaningful leads to a certain medical procedure. Wrong data can kill.

Governing by Numbers

The larger CoVID crisis is to a great extent the outcome of a similar frame of mind: "governing by numbers" (similar to treating disease by numbers). One author went as far as to postulate: "Democracy, in its modern mass liberal forms, requires numerate and calculating citizens, numericized civic discourse and a numericized programmatics of government" (Rose 1991). This idea became the new reality of political and social life. It transformed the media into a number-driven enterprise. Neither racism, nor gender discrimination, and even less the frivolous rhetoric of privilege and intersectionality—all subject to an upheaval symptomatic of a society in crisis—can be meaningfully quantified. Inclusion, in the form of automatically reflecting number distributions, is mere cosmetics. Data, as opposed to the individuals from which they were extracted, can be manipulated. The virus mutates in its interaction with the organism; data are subject to the same, but at a scale at which they cease to have any relation to the reality they are supposed to represent. The growing vulnerability of societies (democratic, and those claiming to be) is the direct result of this process. The right to privacy and the right to be protected by the law instead of being treated as guilty for even your own thoughts are effectively cancelled under the pressure of achieving the "right numbers."

Numbers, mostly illegitimate—in particular those generated by machines programmed to consolidate the powers in place—create a fake sense of objectivity. When humans are supposed to align with ideologies, instead of formulating them, or challenging those who undermine their sovereignty, the result is a dangerous state of human affairs. Rhetoric cancels thinking; the illusion of freedom is translated into defying the law instead of consolidating it for the benefit of all. No surprise then that tribalism was resurrected in its ugliest manifestations, all preceding the pandemic, but making the COVID-19 experience even more unbearable. Day in-day out, numbers are aggregated in order to justify preconceived notions, as damaging as what society was supposed to fix or get rid of. Regurgitation of misrepresentations as a way of proving a point is as healing as vomiting after binging on cake. Instead of arguments based on understanding whatever has to be fixed, the "Give me the numbers" attitude took over the discourse. It continues to drive the machine supposed to fix everything that's broken. In the post-truth age, these numbers are engineered. This is probably the most egregious aspect of CoVID. But second to this is the abandoning of one's own attempt to understand the meaning of change.

The Covid breakdown is all about meaning in the global world. The rhetorical question, "Should we deglobalize?" (formulated in France and Germany) proves the point. Everyone seeks new sources of profit and resources in the global economy, not vulnerabilities. The investment in the science and technology of a massive pipeline intended to provide gas to Western Europe (begun in the late 1990s) is now returning losses exceeding the once-assumed benefits. The flipped Upside/Downside Ratio is exactly what those in power and those speaking for them ignore when they adopt a science devoid of anticipation. In giving up their sovereignty for the sake of prosperity, individuals sacrifice their right to understand the meaning of change—i.e., machines taking over—and exercising their right to influence the direction of change. There is no vaccine for the crisis of vision. Education, the institutionalized cure-all, is compromised to such a degree that it can only make matters worse. Hijacked by reductionist-deterministic science and technology, the social "immune system" itself proved to be extremely vulnerable once global interdependencies were established. Short-term reactions by governments, in the form of all kinds of costly economic packages meant to fix this or that tear in the fabric of society, were no different from the primitive intubation rushed into the throat of COVID victims. But of longer-term consequences. Sustainability, which can only be global, was once again sacrificed.

Looking at Reality Through the "Eyeglasses" of Models

Scientists use statistical data in order to make models for the purpose of assessing the large numbers of people who will die. From all those infected—a number inferred from testing—some will not make it. Statistics says: Based on the past (how many infected people died over a certain timespan), we infer to the present (how many could die). Under the assumption of perfect medical care—which would be the consequence of perfect science and ideal social conditions—all the infected, if unaffected by some other condition, would be healed. In the absence of such an ideal case, the prediction ultimately informs only about the inadequacy of currently available treatment—and inequities. But even this is not necessarily convincing. The most prestigious institutions (e.g., Imperial College of London, Institute for Health Metrics and Evaluation, Harvard University, Hong Kong University) hosted various approaches to modeling. The death rate for Italy, Spain, Germany, and the USA were established with acceptable accuracy. But not so for Bangladesh, for example, and not good enough for a great number of African countries. That fewer people died in such places is the good news—if the reporting is accurate. For example, Zambia recorded fewer than 400 deaths in 2020, while the prediction was for 20 to 30 thousand; Nigeria, 13,000 deaths, in contrast to the 200,000 to 418,000 predicted; Ghana, 300 deaths vs. 7,500 predicted. Malta, almost completely vaccinated, was not expected to register the hundreds of new infections caused by a virus variant. In the next waves, accuracy was even lower, because the assumptions expressed in the models proved false.

The fact that the predictions were, and still are, way off the mark deserves attention because it points to the type of science on which they are based. There was no, or too little, knowledge involved, and therefore no focus on meaning. The data-based prediction has no underlying understanding of the many differences among countries that might be part of the global economy but are not part of the global society. Their cultures are different, and so are their medical care systems (if they have any). Why Omicron began in South Africa and how it spread remain rather unclear. So is the high incidence of the BA.5 in Japan. But no one predicted it. If the prediction pertained to the metaphors of "dying stars" or to "expired" satellites, or to any other physics-based process (e.g., faucet dripping, computer breakdown), there would be no reason to question it. To repeat: measuring the physics or chemistry of non-living entities (e.g., lunar and solar eclipse, space satellites) is different from measuring living processes. Physical law, pretty well defined, guides the understanding of the process. Example: The measurements (of coordinates) provide the data necessary to infer from a past position (of a car, a rocket) to a future position—change as movement. This is how the Mars landing was guided, and how the deep-space telescope was launched. But a pandemic, and for that matter anyone's illness, are subject not only to the laws of physics and chemistry, but also to the dynamics of biology. The change in the condition of a living entity can even affect the outcome of the underlying physical process (Ellis 2005). Neither molecules nor electrons bring a smile to one's face. But the sound of a bird, or of music, or the image of a flower or of loved ones, do. Healing is not the mechanical outcome of a successful chemical process, but of complex processes in which the condition of matter can be influenced by the condition of the mind. Remember that a falling person has a degree of control over the event. If trained properly, the individual can even affect the trajectory of the fall (Fig. 4). The cat falling (a behavior documented many times) is yet another example of how gravity and the goal-oriented (i.e., how to prevent hurt) motoric expression "work together," not one to the exclusion of the other (Fig. 5).

This is reaction and anticipation dynamically integrated. Again: it is behavior, guided by intentionality, not just how matter supervenes on its components (i.e., the elements of which they are made). And it is, most of all, learning. There is a great difference between those (patients, physicians, scientists) who are caught by surprise by an unprecedented condition—COVID-19 is an example—and those who faced it once they learned more about it.

Modeling, as a specific form of knowledge expression, provides mappings from the current reality, captured as data from measurements, to the future under the assumption of probability: the numbers describing not what is known, but rather the level of ignorance. A 50% chance of rain—one way to describe the probability—means only that the knowledge available is half of what it would take to predict rain

Fig. 4 If trained properly, the person can control the fall. It is *learned* behavior, not just a question of how matter supervenes on its components

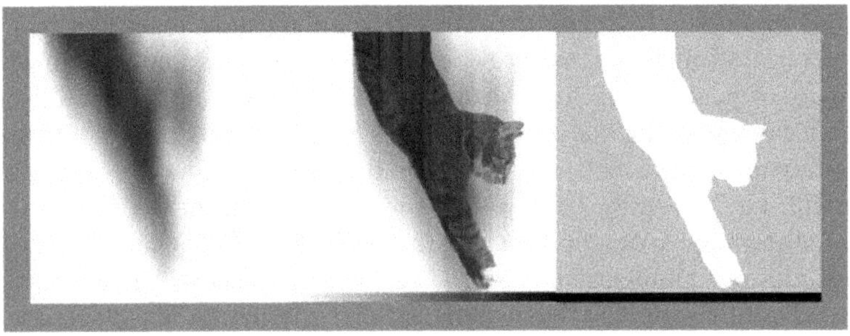

Fig. 5 A cat, even if dropped face up, will always land on its feet

in a foolproof way. Mathematics is the language in which physical processes can be precisely described. Wigner (1960) discussed the *unreasonable effectiveness of mathematics in physics*, without specifically explaining it. In respect to probabilities, the effectiveness is above the threshold of the unreasonable. Almost all technologically grounded human activities are indebted to probability assessments. The failure of the probable rain to occur is less consequential than all probabilities at work in driving a car, flying in an airplane, the functioning of nuclear power plants (to name a few examples).

However, mathematical descriptions (abstract in nature) are only a subset of the open-ended ways in which humans describe reality. Images, sounds, but also emotions, thoughts, etc. are, under certain circumstances, engaged in such descriptions. Semiotics is the discipline that studies the manner in which representations are generated, and it can guide in the selection of the appropriate means for generating them. The goal of a poem describing one out of infinitely many possible experiences of reality is quite different from the goal of generating data necessary for controlling a rocket. It subsumes mathematical representations, or any other kind (computational, aesthetic, genetic, etc.). Mathematics, by definition related to measurements of quantity, adequately represents how the non-living changes over time. In the language of mathematics, probability theory is defined exactly in order to allow access to how events of varying importance partake in an outcome. Probability generalizes from statistical data associated with physical phenomena, or for almost anything involving choices (e.g., elections, betting, shopping). It was adopted in the study of the living as yet another outcome of a deterministic view of life. Medical practice, for the longest time interested in the symptoms of deteriorating health, gave up on the rather impressive semiotics of disease, diagnostics, in particular. It also shifted its focus from what symptoms might mean, to measurements, i.e., to quantities supposed to be associated with the state of the organism.

In our days medicine is grounded more and more in the rationality of mathematics. Why this understanding became dominant is justified in many ways. Mathematical descriptions are useful in determining interventions: the intensity of radiation therapy (a physics-based intervention) for cancer; help in identifying the precise part of the brain the surgeon should operate on; guiding knee replacement surgery to be performed by a robot. In respect to the health of a population, mathematical models quantify various influences and provide variations for interventions. However, in the process of generalizing from quantitative descriptions to a course of remedy, they often fail. Given the nature of the living, measurements of life processes are by necessity incomplete and, moreover, devoid of meaning. Since life processes are meaningful, i.e., driven by goals, assessment of an organism's condition only on account of quantitative descriptions remains incomplete. And it can be misleading.

Pandemics are biological processes in which an infective pathogen affects a great number of people. Dr. Snow, back in Soho, discovered that cholera was not spreading through human contact and that miasma (the smell) was not leading to death. His discovery was informed not by numbers alone, but guided by evaluating their meaning, i.e., what conditions led to the numbers. The physics of the spreading was not known. "Waterborne" was a hypothesis until the handle of the pump that delivered contaminated water was removed. But that was 19th-century England. SARS-CoV-2 already had some precursors (MERS, SARS) that left behind many kinds of tails of numbers. Still, because reaction blinds prevention, the medical establishment did not know for sure how the virus was spreading. The physics of the spreading of the SARS-CoV-2 was analyzed in meticulous detail—often with contradictory results that led to requirements (e.g., six-foot distancing) that made no sense. Generalizing from particle physics to the spreading of infectious diseases resulted in elegant curves (infected persons not practicing social distancing compared with infected persons who did; Vrugt et al. 2020). No knowledge was gained in order to serve "as a basis for political decisions"—a goal specifically specified—due to the false premise: People who observe social distancing can be modeled as particles that repel one another because of the same electric charge. Corona virus-related medical experience of the past was sufficient for avoiding the absurd exercise in reductionism. It could have guided in the effort of identifying the infected, the susceptible, and those who might be immune (in one way or another). This did not happen because once the focus is on numbers, the connection to the reality from which data are extracted is lost. Living cells are a locus of meaningful processes that statistical physics, inferring from the past, cannot describe.

COVID-19 was the opportunity to test modeling—a powerful mathematical procedure as long was what is modeled is also understood. In its automated version—performed by computers, data replaced understanding. Big Data technology—a deterministic machine for processing huge amounts of data—provided the means for automating the crunching of more data than ever before, and faster—of course using a lot of energy. That helped determine pathogen spread but did not help to identify the source—a subject still open to speculation. Other measurements helped quantify the efficacy of specific interventions, the infection rate as it applies to different groups, fatality rate, etc. These results and more in the same vein became the parameters of sophisticated models mechanical in nature. They were supposed to help authorities in engaging the population (tracking) and to help the medical system in addressing needs. The expectations were high—and rarely fulfilled. In contrast, modeling the placement of a space station on Mars—by no means trivial—was successful, and deservedly celebrated, even in the context of the pandemic. Likewise for the 10-billion-dollar space telescope. The fact is that for COVID-19 the results—actions based on modeling—are mixed, at best (Metcalf et al. 2020), despite the scale at which reaction was practiced. The explanation is straightforward: the simplistic view of deterministic causality. The nature of life is different from physical phenomena, whether at the microscopic level or at the grandiose scale of space exploration.

In the USA alone, one model (the March 13, 2020 CDC assessment) projected between 200,000 and 1.7 million people killed (a global view). A subsequent model (Gu 2020) projected (more accurately) a death toll of 225,000 by November 1, 2022, and one close to 500,000 once vaccination started. Data, models, projections—all claiming credibility on account of how successful the statistics associated with measurements (data) of physio-chemical phenomena are—became the focus and fueled the political discourse. In fact, data-driven science underwent, probably against the intention of those who practice it as the panacea for everything, a major test: how statistical data can approximate life. It became the crystal ball disguised as science. The *Resolve to Save Lives* model (in conjunction with the Council of Foreign Affairs) saw half the USA population infected, and between 163,000 to 1.6 million dead from COVID-19. Some models were focused on the demand for hospital beds (one prognosis, 30 times higher than the supply); on ICU (intensive care unit) demand; others on medical equipment (from masks to ventilators), and on recovery facilities and post-trauma care. None predicted the long-COVID—eventually characterized as a healthcare disaster because nobody knew what to do with the COVID survivors who remained affected (in a variety of ways). The Columbia University model established (or claimed to have established) which countries would have their healthcare facilities overwhelmed (as mentioned, they were wrong); the Institute for Health Metrics and Evaluation (HME) model, displayed as a "mark of approval" at some of the White House briefings, projected deaths at 60,415 by August 4 (in July, the number was over 150,000). The same facility claimed that the number of COVID-caused deaths would be way higher than the headcount (Noh and Danuser 2021). None of the models was validated prior to the pandemic, although the scientific community was aware that there was a possibility for one. And none has proved to be accurate during the pandemic or in respect to the number of victims—never mind, in the successive waves; one model focused on virus variants and how vaccination can help. (For those interested in details, there is a convincing COVID-19 Forecast Hub at https://github.com/reichlab/Covid19-forecast-hub, based on the CDC-funded UMass-Amherst Influenza Forecasting Center of Excellence at the Reich Lab.) It turns out that Omicron, with its many variants, proved them wrong. But the daily announcements continued to pollute the discussion of what should be done since vaccines offer only limited protection. The waning after three months is a fact yet to be properly understood, even by those who "sell" boosters as protection against infection.

The models from which such prognoses are derived are as good as what they are made of. The slogan "Garbage in, garbage out" (GIGO) applies to computers. Models are, after all, computer representations of statistical representations based on the data at hand. It reads more and more as "Garbage in, Gospel out"—the

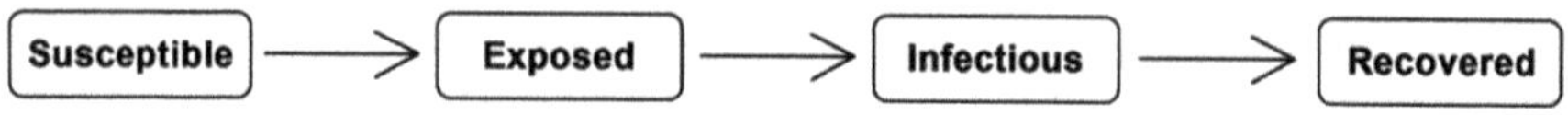

Fig. 6 The underlying SEIR model simulator behind the You Yang Gu model (https://covid19-projections.com/). It uses an abstracted state machine that tracks the probability transitions between the 4 states of SEIR: **S**usceptible-**E**xposed-**I**nfectious-**R**ecovered/deceased. Gu added an exposed (E) period accounting for the reported incubation period of COVID-19 during which individuals are not yet infectious. This was a personal initiative of a passionate young data analysis professional for whom Covid-19 became a test of his abilities (plenty of available data)

new dogma of inferring from data which might or might not be significant, not to say defective. This description conjures the long-lasting concern exacerbated by ever higher computer performance. It goes back to Charles Babbage's question: "If you put into the machine wrong figures, will the right answers come out?" COVID-19 has entailed an inflation of models vying for the public's attention to the detriment of seeking knowledge. Models—SEIR/SIR, named after the variables considered (Susceptible-Exposed-Infections-Recovered)—agent-based or curve-fitting extrapolation, are used to estimate incubation time or transmissibility (Fig. 6).

The much-discussed R_0 ("R-naught"), introduced to the public, remained a mathematical abstraction behind which lay the reality of individuals infecting one another because they did not know any better. Recently another revelation further undermined further the rationality of defining the R_0: "The takeaway from studies is that most people with COVID don't get other people sick, but a few people get a lot of people sick," (Sawyer et al. 2021); moreover: "2% of people carry 90% of COVID-19 virus"—the newest headline of confusing science (Yang et al. 2021).

Forecasts—regarding number of infected persons, number of deaths, number of hospitalized patients, and the number of necessary interventions means (masks, respirators, oxygen tanks, etc.)—are also based on models. Statistical in nature, such representations fit a curve to data and extrapolate. In times of crisis, governments are helped by models of potential outcomes of interventions (e.g., social distancing, tracking, isolation), as well as projections (the *if:then* scenarios). It is quite telling that the medical community grew skeptical of models (Freedman 2020; Jewell et al. 2020), not the least for their minimal impact on healing patients (mainly through projections of necessary equipment and pharmaceuticals).

Let us imagine the model is perfect. Its most important benefit turns out to be in monetizing the pandemic, not in guiding those who treat patients. Indeed, behind each patient, as much as behind each victim, looms a number: the amount of money to be made, in this particular case, from the government (the taxpayers' "deep pockets") in treating those affected. The market economy, within which medical care (especially in the USA) is a high profit-making enterprise, is a machine that embodies the reductionist-deterministic perspective. Intubation, drug production, vaccinations, masking, hospital costs, morgues, cremating bodies, etc. are part of the calculation that models have facilitated. Even social distancing was quantified. When suddenly, in reaction to the pandemic, a subset of medical care—i.e., treating COVID-19 and covering associated expenses—becomes a budget item (again, taxpayers' money), the LifeTime Value of a patient—the cost of taking care of a person—increases well over the $600,000 level reported by actuarial data. Even what is not COVID-19 becomes COVID because the reimbursement value is higher. Someone else's money pays for it.

The Reactive Path is Ignorant of Values

For the record: As generalizations, all models are agnostic of the biology of the process. Being an expression of the reactive perspective, they are amoral. There is no room for ethics in the cause-and-effect sequence. As applied mathematics, i.e., statistics, they offer inferences from assumptions based on the data acquired. In the absence of a reference to reality, they never return any meaning in respect to the pandemic itself. If the data "feeding" the model is faulty—there are no perfect data—the model will not be better than what inspired it. It is quite telling that some measurement methods, such as the serology test, returned data only 50% valid.

To react means to adopt the limiting rationality of cause-and-effect sequences, even in the absence of proper identification of the problem. What for physics is a clear-cut distinction that can be modeled, for the living, where moral considerations (among others) are relevant, this is less distinct. The pandemic reproduced the painful reality of inequality, racism, and gender bias because the reactive path is ignorant of values. The continuum of living processes and the discrete nature of measurements—sampling at no matter how often—rarely come close to each other. Not knowing *what* to measure and *when* is reflected in the questionable test results and in their amoral condition. In the absence of addressing issues of anticipation—i.e., how the possible future affects the present—numbers acquired through tests focused on reaction are as good as the limited understanding of what scientists and physicians actually want to find out.

Within an action-reaction view, scientists indoctrinated in the theology of determinism look for numbers that meet their preconceived notions. These are the "eyeglasses" of models. One hypothesis: It is good to have everyone infected—the controversial herding immunity model (borrowed from animal care); after that, we are immune. Would that mean that the process of gaining immunity by having COVID-19 is a kind of natural triage? New York, with an extremely high lethality rate, would be a testimony to this hypothesis. The Swedish State Epidemiologist was of the opinion that pursuing herd immunity is "futile and immoral" (Daum and Simmank 2020, in an interview with Dr. Anders Tegnell). Of course, immunity acquired through vaccination is a different goal altogether, if indeed it can be attained. The virus variants made this assumption quite questionable. Israel, a country that became the test facility for Pfizer, eventually decided to give up fighting Omicron in favor of letting the "dice fall where they may" (the Swedish model). That means herd immunity at the cost of lives, not only the lives of people opposed to vaccination. Actually, those who want to be vaccinated, those who are hesitant, and those who refuse vaccination are part of the larger picture of an extremely segmented society. Add here: those who can afford it (in places where it is not free), those who can get it (the majority of the world's people), and those who have to accept it, such as in China, known for refusing, for a long time, to accept the mRNA-based vaccine from Pfizer and Moderna.

This image of contradictions and prejudices can stand for the "portrait" of CoVID. It is a world proclaimed as global, but actually fractured, and essentially extremely egotistic. Poverty, the inadequate education system, fanaticism (of all kinds), ideologized science and medicine, media activism (left and right), profit-driven medical care, and intolerance testify to a state of humankind for which the reductionist-deterministic premise can be held accountable. But not in a context in which accountability was cancelled. Having abandoned responsibility towards the future for the delusional reward of prosperity that excludes the many, society is torn apart.

The reductionist-deterministic view cancelled anticipatory actions and thus the moral dimension of social life. From this broader perspective, the contradictory attitude towards vaccination—an anticipatory path that medicine has not abandoned despite the pressure of the action-reaction perspective—can be explained. For instance, it turns out that the specific immunity acquired either by being affected or being vaccinated is not as easy to characterize as the medical community wishes. In many cases, it can be short-lived (Zohar and Galit 2020). Whether a vaccinated subject can still transmit virions, or can get sick, is a major concern not appropriately reflected in the research.

Speaking of numbers and their meaning, another angle was tried: keep the number of infections to a minimum. This is how you minimize the number of deaths. Singapore and Israel, as well as Taiwan and Hong Kong, kept track of every individual. Still, they could not avoid the disease's spreading. In Israel, an easy fill-in form, with short questions meant to capture identifiers (e.g., fever, cough), was widely circulated under the assumption of gaining more data through self-reporting (Rossman et al. 2020). For individuals able and willing to report on their condition, the daily health picture of the population was useful, although at least the fever symptom proved to be non-consequential. If someone loses the sense of smell or taste, chances of being affected by the virus are higher than with coughing or fever. The brutal second wave of the pandemic undermined what had been celebrated as an innovative approach in acquiring more numbers by eliciting self-reporting.

Indeed, statistically projected futures are as good as the long-term record upon which they are based. But there is no such long-term record for COVID-19. When the bright idea of the fill-in-form (heavily milked for publication purposes) made it into social media's self-publicity mechanisms, Israel had fewer than 100 cases; in the second wave, it was headed toward 300,000 (lethality doubled within the following two months) until the second lockdown was imposed. And after that, vaccination loomed large on the horizon, and Israel vaccinated itself out of the pandemic—until the virus variants entered the picture. The doubling of numbers of Omicron-related infections every two to three days could not be reduced to the mechanical performance of tabulating. Numbers do not inspire new ways of thinking. South Korea's experience with MERS (Middle East Respiratory Syndrome) prompted the building of an emergency response system (Ahn 2020), which in itself is an expression of prevention. The authorities measured the temperature of everyone leaving from or returning to an apartment building. In retrospect, this seems ridiculous, but this practice increased awareness of the danger: an example of establishing correlation, but not a causal explanation.

In the COVID-19 wave of variants, even South Korea failed. Actually, even the pandemic's second wave challenged initial assumptions of the extent to which a country of free citizens can be controlled. Americans, more than their government, gave a clear NO to the possibility; Germany dictated a new law, granting the Federal government a limited ability to control its citizens. Infected parents were fined for insufficient protection of their children; France was not shy in imposing curfews. Using the same model, China, again and again, locked down immense areas; individuals had no choice but to accept the rule. Only later did the fact that temperature is not a conclusive identifier of COVID-19 infection became known later. In respect to the numbers recorded—the results of testing—they proved to be useful in containing the pandemic's spread in China because laws were in place that allowed for actions unacceptable in other parts of the world. The moral component: sacrifice for a greater good. China was criticized for its actions. But millions of lives were saved. The USA and, for that matter, many Western countries, had several plans for possible pandemics, but nothing in place for properly assessing the usefulness of the measures to be taken.

A pandemic does not announce itself. Let's take one characteristic: Since most COVID-19 contaminations are not symptomatic—and there is no way to know in advance if non-symptomatic subjects spread the virus—what should be measured? Molecular tests (R-PCR), or antigen tests, exemplify the reductionist focus: detect viral genetic material or, alternatively, specific proteins from a virus particle. The sensitivity of such tests depends on the assay (sample collected) itself, the site from which it was collected, the time, equipment condition, etc. Paradoxically, while "digging" for even small particles of the pathogen in order to determine whether an infection occurred (or may occur), low oxygen levels (leading to death by choking) were not noticed, not even by patients themselves, until it was too late.

Did the Chinese, the South Koreans, and the Vietnamese**,** so successful in containing spreading during the first phase of COVID-19, check the sense of smell? It turns out that the virus penetrates the olfactory epithelium. Some individuals lose the senses of smell and taste. The numbers of those who lost their sense of smell are more difficult to determine than the numbers of bodies—COVID-19 related or not—in the morgue. How do you measure the lost sense of smell or taste? As predominant neurological symptoms (Cooper et al. 2020), smell and taste are by now subject to inquiry (Brann et al. 2020), although predominantly from the limiting deterministic perspective that focuses on chemistry (Kahn 2006). but ignores the anticipatory component. The meaning of anosmia (smell disorder) and ageusia (loss of taste) becomes easier to understand when we realize that olfaction and gustation are interconnected: the organism is aware of meaning—onion tears are different from those associated with pain. Moreover, even the sense of touch is involved. All this points to the need to examine cognitive aspects. COVID-19 taught the world many lessons—soon to be forgotten. Unfortunately, the reductionists manage to get the upper hand even within the context of crisis: They formed the global Consortium of Chemosensory Research, as though the abovementioned symptoms were indicative of deficient chemistry (thus calling for reactive treatment), not of skewed anticipatory characteristics.

Ignoring the Meaning of Data

The countable numbers of those who succumbed to COVID-19 are only partially helpful in understanding the pandemic's dynamics. Together with the count, identifiers—men, women, age group, medical history, race, living conditions, mobility, marital status, sexual orientation, education, etc.—should be taken into consideration. Especially, as it eventually became evident, comorbidities. The meaning of each of such identifiers is relevant for understanding the pandemic and its consequences. A person undergoing transgendering, which entails a lot of medication, generates virions different from those of a gardener who spends time in the fresh air and avoids any medications. In Milwaukee, early data showed that African-Americans became contaminated and died at a higher rate than other segments of the population. Likewise for New Orleans, Chicago, Detroit, New York, and New Jersey. With this observation in mind, the numbers, associated with a context (i.e., a reference) start to mean something. COVID-19 is not a race pandemic and not a gender marker (even if more men than women have died). It does not "see" skin color or country of origin, or race, but reveals plenty of less visible aspects. For instance: poverty as the major characteristic of those who got infected and died. Anticipatory considerations would have helped understanding the many implications of the fact that the life expectancy of African-Americans is 14 years lower than the USA average. In what one scholar (Camara Phyllis Jones, a physician herself and an epidemiologist) calls frontline jobs—hospital orderlies, meat packers, home health aides, warehouse workers, and similar—Blacks, and sometime Hispanics, as well as Asians are overexposed. Blacks and Hispanics, Asians, too, are also subject to quite a number of poor health conditions, rarely, if ever, met with the appropriate health care, not to say preventive measures. Those who, through poor choices, end up in jail or in homeless shelters are even more targets of disease. "Black people have 60% more diabetes and 40% more hypertension" (Wallis 2020). It is exactly in these details that meaning can and should be conjured by referencing data to context. Ethical aspects—the significance of poverty for the future of the poor—become evident once society takes note of the type of discrimination, not codified by law, that characterizes social life in many countries (Fig. 7).

Such figures reveal the reality of what is meant by co-morbidity—a word that the crisis brought up: risk of chronic conditions (asthma, diabetes, heart problems). Poverty, inadequate medical care, lower educational focus, etc. are all parameters of behavior that an anticipatory view would highlight. Anticipatory awareness, more than Big Data, is the premise for designing and applying pro-active measures. The high rate of affected Hispanics, mostly immigrants in low-paying jobs with poor job

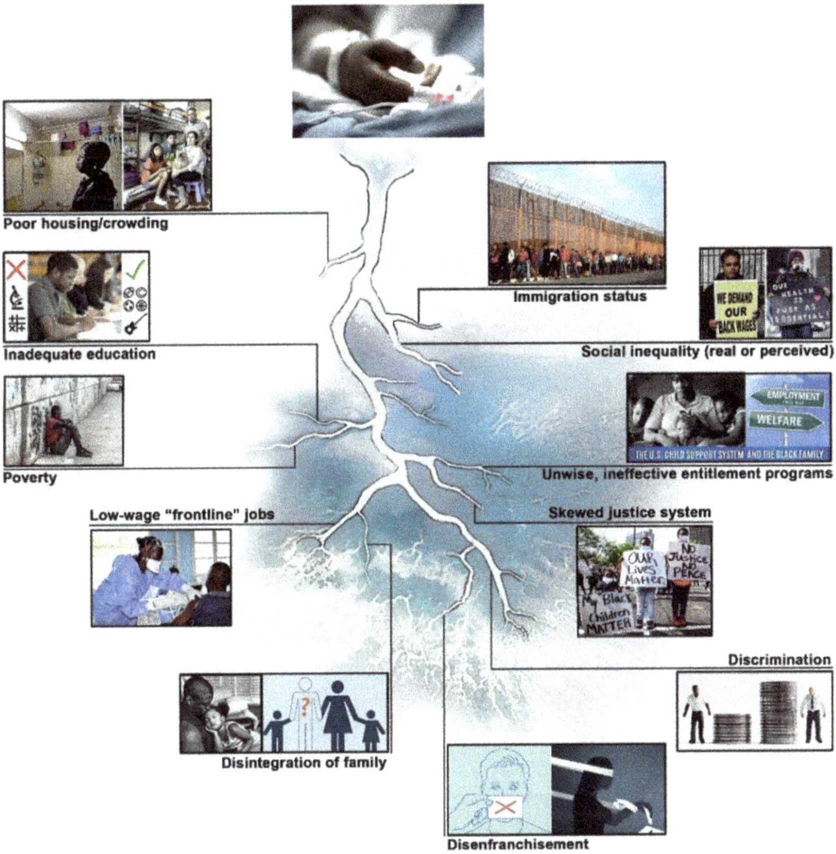

Fig. 7 The high rate of Blacks and Hispanics, as well as Asians (Chinese, in particular), affected by the pandemic is the consequence of a confluence of factors: discrimination, poverty, disenfranchisement, inadequate education, skewed justice system, disintegration of family, etc. Poverty is the major factor

security, is also relevant in this sense (Blumenthal et al. 2020). Anti-Asian sentiment, provoked by associating the virus with China, explains not only aggression, but also forms of discrimination that affected the health of many. In Europe (France, Germany, England, etc.), refugees, treated with either suspicion or in a manner that qualifies as disguised racism, were especially affected. So were guest workers around the world (in the Emirates, East Europe, Singapore, Israel, etc.).

The Value of Life

If instead of focusing on quantitative aspects only, researchers were to consider the meaning of data, they could discover quite a number of other important distinctions. The so-called gig economy—of very competitive work and sacrifice of time—benefits start-ups, but also results in vulnerabilities. Individual success is quite often paralleled by the demise of family life and disregard of offspring. Awareness of meaning—the extreme pressure accepted in the expectation of making it big—is the prerequisite of anticipatory considerations leading to actions that can help prepare for future challenges. Another example: residents of nursing homes and retirement communities were affected more than the rest of the population. The question arose: How does society address this loud cry of danger regarding those who, given their age, became captive to the industrial model of aging? No good answer has been given. In the absence of moral considerations, which determinism eliminates, to age is nothing more than a problem of resource allocation. This issue becomes one of numbers in disregard of individuality.

It seemed, at times, that the pandemic became a method for cleansing society of that part of the past that costs too much to be fixed. For prison inmates, similarly subject to higher numbers of COVID-19, some governments have opted for home confinement, or counted on the good will of those willing to accept them in their living quarters. For naval vessels, the military developed rapid isolation procedures in order to protect crew members from contamination. But all these were measures in reaction. Anticipatory actions, based not on the physics and chemistry of numbers but on their meaning, could have guided society to an understanding of entrepreneurship, of aging, of disease, of the criminal justice system, of education that is not deterministic. The value of life itself is ignored when all that counts is the Life Cost Expectation associated with how much it will cost to make it from cradle to grave. Instead of centralized "factories" for transitioning from old age to death, society could practice distributed models and encourage families to try them. The wish for the return to the nuclear family of the past is an illusion, especially in a time when family became too expensive to maintain. It is conceivable that alternatives corresponding to the faster cycles of life and activity can be worked out. Of course, monetizing aging within the economy of scale—deterministic optimization regardless of consequences for the people affected—corresponds to the broader framework. Submitting every aspect of life to the industrial machine model, corresponding to a science that acknowledges only the deterministic perspective, de facto eliminates moral considerations. The same holds true for prisons. CoVID made it plenty clear that modern society produces and reproduces its own vulnerabilities. Instead of seeking new ways to deal with critical aspects, society reacts instead of acting in an anticipatory manner. Soon the elderly—the post-World War II generation (those born between 1946 and 1964)—will be the majority of the population. But society is not prepared for what this number of aging persons entails. On this account, disruptive alternatives deserve to be considered. Disruption of the science on which the current practice is based is probably the unavoidable starting point.

Just for the sake of illustrating the thought of alternatives (without suggesting that these are to be found only in the past): The Amish family—quite different from the rather broken modern family life—believes that caring for parents and grandparents is a moral responsibility. Instead of sending the elderly to some servicing facility, the Amish developed a culture of providing for their elder family members. The *dawdihaus* (a smaller house often attached to the main house) provides not only shelter, but also a sense of continuity. Three generations interacting—a learning experience for everyone. When kindergarten tuition can be as high as 25% of one's income, the Amish model looks pretty promising. Japan, with the world's highest population of elderly, stayed away from geriatric care—*obasuteyama*, which means "granny-dump mountain." The culturally shared family responsibility is anticipatory: everyone is subject to aging. Of course, it does not mean that all the elderly in Japan end up happy. Models from the past are not answers that can be generalized, especially in a day and age when family, as a common ancestor entity, lost its traditional meaning and economic function. But neither are the last-stop-before-the-grave facilities, "factories" for aging out of society's sight.

Albeit: it is not about atoms, molecules, and genes to be treated with drugs, not about numbers but about *behavior to be changed*. Preaching to those who grew up hating the older generation or some minority group, or Whites—more recently declared to be responsible for all society's ills—is useless. Responsibility cannot be delegated to facilities. It will not be automatically assumed by a generation educated to live by expectations (far exceeding their own contributions to fulfilling them). The high numbers of the COVID-19 infected or dead, young or old, are to a large extent the result of unequal living and learning opportunities (Gray et al. 2020). The machine called *capitalist economy* took over; humanity—values that should be important to each and every one—is becoming increasingly irrelevant in the profit-making process.

Although society reacts—such as in the CoVID crisis—to such flagrant inequities, it does not understand, i.e., anticipate, their long-term consequences. The "post-COVID-19 economy," with everything in high gear, is the proof: living on resources borrowed from a future that might not fulfill the expectation of more and more at any price. COVID-19 turned out to be a magnifying glass of social and political shortcomings that cannot be ignored. Medical treatment in a private hospital in New York City was civilizations away from how a city hospital, or community facilities in fast-dissolving cities and towns even in the USA, treat patients. Before the pandemic "specialty-focused behemoths" emphasized making money over providing care. The hospitals "created more capacity for highly specialized care," which means more advantageous reimbursement policies (Fihn 2020). With the pandemic, instead of profit, losses mounted: over one million jobs gone, and around $500 billion lower income for the medical-industrial complex. But the real loser was society: All those in need of medical care were abruptly asked to take care of their own illness and pain. Sometimes they were left to die—collateral damage in the war against the virus. COVID-19 was not the equalizer, as demagogues of all stripes proclaimed, but a symptom of social comorbidity—poverty in the first place. Divisive impulses, fueled by political demagoguery, further undermined—and continue to undermine—social unity. In the lockdown, meant to prevent spreading of the pathogen, excuses were found to justify marching for causes that sometimes proved to be as questionable as exposing oneself to COVID-19. This is why it remains unclear how many of those who died found their end due to irresponsible behavior than to their weakened condition. It is still open how many died because of the pandemic (the Centers for Disease Control and Prevention identifies only six percent of those dead as being the victims of COVID-19) and how many because there was no medical care available. "One way to gauge the toll of coronavirus pandemic is to count the number of deaths that exceed what is seen in typical years" (Viglione 2020; Fig. 8).

Those who mention prevention, when all means are focused on reaction and nothing else, end up being, if not ignored, ridiculed. It is in this sense that a holistic understanding of medical care, including social and economic factors becomes urgent. Reaction is always opportunistic: repair what everyone can see; the rest can wait. Medical care that transcends the deterministic practice (the mechanics of fixing) is consubstantial with the holistic expression of anticipatory processes. It explains the confluence of the pandemic, economic breakdown, and social unrest (Egede and Walker 2020). This is why COVID-19 cannot be treated separately from the broader CoVID within which it emerged. Society, already morally and politically weakened, unable and unwilling to address accumulated inequities, was caught unprepared, embracing reaction to the detriment of anticipation.

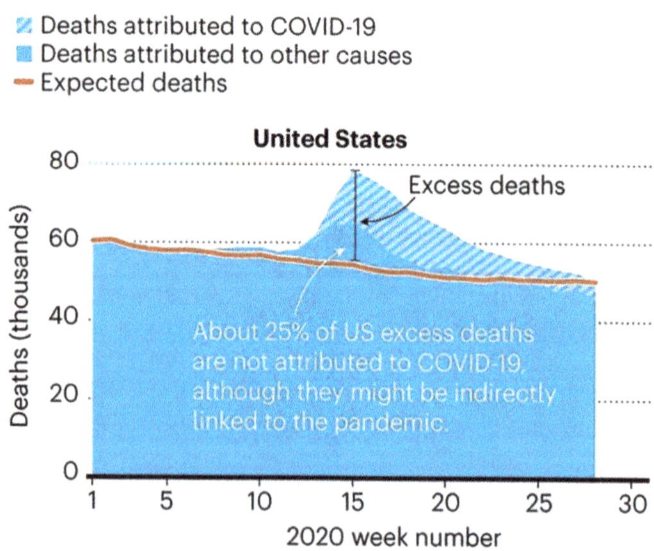

Fig. 8 Gauging the toll of COVID-19 (cf. Centers for Disease Control and Prevention) by counting the number of deaths in excess of typical averages. This is yet another example of numbers for their own sake and in disregard of their meaning. For instance: how many died because their condition, (e.g., cancer, heart problems, kidney failure could not be addressed because COVID-19 closed access to the appropriate medical care. Moreover, when the government reimburses COVID-related death at higher rates, fake numbers cannot be avoided

Who Owns Our Genetics—The CoVID Narrative

Early on, empirical evidence informed governments and physicians about the vector of contamination: people traveling, or what the geneticists found out about globe-trotters (Fischetti and Krzywinski 2020). The disease's spreading is a legitimate target for researchers seeking methods to contain it. For example, knowledge about spreading can help in designing effective containment, e.g., preventing travel to and from areas evincing more than accidental infectivity. A map of how spreading takes place helps in preparing anticipatory actions: getting ready for what might ensue. The conclusion—eventually reached after huge amounts of data from gene sequencing were processed—that people who travel have contributed to the relatively rapid pace of the pandemic is actually common sense. But it was presented as proof of what science is able to deliver: the miracle of sequencing as a data-producing machine!

It took a long—very long—time (Wells et al. 2021) before sequencing afforded some significant findings. After all, it might be that the inception of COVID-19 was imprecisely determined: Wuhan, the now infamous open air "wet market," New Years' Eve 2019. Someone came up with a headline that can be reused for so many aspects of the Crisis of Vision we are going through: "How rumors about COVID-19 origins led to a narrative arms race" (DFR Lab, Atlantic Council 2019). Did the Chinese, or the Americans, or the Russians, or whoever one wishes to pick on have a hand in it? Do they have a hand in the rumors about vaccination? Another headline: *A group of prominent biologists say there needs to be a "safe space" for asking whether the coronavirus came out of a lab* (Jacobsen 2021). None other than Jeffrey Sachs, the Chair of *Lancet*'s COVID-19 Commission, claimed that the virus came from a biotechnology laboratory (Harrison and Sachs 2022). The subject does not go away—COVID fades, speculation is still high.

Misinformation led to an *infodemic* (Scales et al. 2021) that continued unabated—never mind the fake papers generated through LLM-based (large language model) text generation. Follow the money—the catchphrase popularized in 1976 docudrama film *All the President's Men*—became an effective method for tracing the origins of misinformation that defines CoVID. The question, "Did the Chinese, or the Americans, or the Russians have a hand in this?" expanded into several suspicions: Did some corporations or political groupings have a hand, in the financing the Black Lives Matter marches? The LGBTQIA events? The attack on the German Reichstag (2020)? On the Capitol (2022)? The *estallido* in Chile (2019–2022)? Elections (in the USA and many other places)? Are the various labs that work on genetic weapons (creating devastatingly effective new viruses) have a hand in the narrative arms race? Is Big Pharma calling the shots in requesting mandatory vaccination? These are rhetorical questions, of course. But justified, since expansive sequencing was supposed to identify "Patient One," just as once upon a time the HIV/AIDS first case was eventually discovered. Until now, the beginnings of COVID-19 have not been unequivocally determined.

Lipsitch and Inglesby (2014) highlighted the risks involved in research under the "Change of Function" label (generously funded by the US government, but not only). As recently as October 14, 2022, a pre-print (in bioarXiv, Chen et al. 2022) described publicly funded research that resulted in a chimeric strain. It was based on the original virus, but the spike was different from the Omicron mutation). That the resulting virus was not more dangerous does not change the reality that deterministic science plays with fire while trying to calm the public: We have extinguishers on hand and ready! The did not work with COVID-19; they do not work in the broader CoVID context. The fact that some scientists (of course, from the "sequencing elite" dependent on government money, and mostly from the military) claim that nature itself is notorious for "Change of Function" (CoF) arouses suspicions (Derek Lowe cites Florian Krammer, a virologist at New York's Icahn School of Medicine-Mount Sinai 2022). Let us not speculate on the possible source of the devastating virus, or of the many events that shook the world since at least 2007–2008, if not 20 years earlier. But rejecting speculation should not lead us to ignore that sequencing, while important in detecting virus mutations, is rather deficient. Identifying why SARS-CoV-2 eventually landed on a cellular receptor (ACE2) that replicated it as highly infective virions remains an open task. Recently, University of Hong Kong researchers (Chu et al. 2022) showed that the reproduction of corona viruses (SARS, MERS, COVID-19) is facilitated by the antiviral defenses of patients. The enzyme caspase-6 seems to catalyze viral replication. Inhibitors of caspase-6 proved to be effective against all seven topes of corona viruses. As with many other attempts, it remains to be seen whether this would help in containing corona virus-related infections. Some experts in bat studies (yes, a specialized field) dispute even the connection between the virus and bats (Margit 2021). Such facts ought to be considered in the broader context of the crisis of vision within which individual sovereignty diminished while control through the state and economic and financial institutions have gained strength. With each crisis, governments expand their powers and the citizen loses in terms of self-determination.

Population surveillance is made possible the *machine of all machines* whose "raw material" are numbers, processed independent of their meaning. The typewriters of the past processed letters of the alphabet. The mechanical contraptions did not understand one word. Sometimes, even the typist (like some of the scribes of the remote past who copied holy writings) did not care about the content. In its embodiment as printing press, the "type" writer outputted papers and books. Who owned the press owned the truth. In our days, the owners of the TV cameras, radio stations, and the so-called social media are originators of the "truth." (What an oxymoron: "owner of social media.") The narrative of CoVID extends to who owns the "digital press" that outputs not only fake science reports, but also genetic sequences—i.e., the numbers corresponding to the chemistry of deoxyribonucleic acid (DNA) and of ribonucleic acid (RNA). The reverse of "writing" is scanning (i.e., inputting of a text in a word processor); in its generalized form, it is the foundation for of genetic surveillance. An individual's unique DNA is practically taken over by those who own the sequencing technology. Chemical reduction benefits from computation: huge amounts of data describing the "matter" of what is alive can be processed, stored, and compared. Surveillance by numbers is the unavoidable consequence of surrendering to the machine model of life. It is ubiquitous—from the DNA of the cell to the organism, to the individual in society, to the entire society. The ownership of these new "presses" comes with the power to control, to censor, to engineer life.

As different from genetic sequencing (analytic activity par excellence) as CRISPR-Cas9 is, it belongs to the same reductionist-deterministic perspective. It is technology-focused modifying of gene expression. CRISPR was often described as inspired by the processes through which micro-organisms attack viruses. For all practical purposes, it is a form of engineering life. Not surprisingly, it is already deployed: from the rather simple production of beneficial probiotic cultures (for making yogurt, for example) to the "making of twins" presumably resistant to the HIV virus affecting the parents. And even some kind of cancer vaccination (Miri et al. 2020).

In the absence of anticipatory awareness, the ethical dimension of human activity is practically absent. Within the science of action-reaction, if something can be done, it will be done, regardless of immediate and/or long-term consequences. Numbers disconnected from meaning can only support inferences from the past. Therefore, the reactions based on them are not only shortsighted, but also costly. Nevertheless, reactions, in respect to the pandemic or social unrest, need not be downplayed. They have an important role—at least under the expectations of immediate remedy. But the answer to the *Why?* question, on which lives depend, addresses questions that transcend reaction—and are not quantifiable through numbers alone. Corona viruses from the subgenus *Sarbecovirus* (known for a long time) are here to stay. The real question is: How can another pandemic be prevented? Or, to return to the larger crisis of vision: lying, cheating, inequity, racism, fanaticism, ignorance, censorship, and surveillance will be with us as long as society fails to address them from an anticipatory perspective.

The amount of data acquired in tracking changes in the RNA (the nucleotides making up protein shells) is enormous. In sum, they proved useful in alerting to the dangers of a more infective agent: the virus variants. But not in containing them. Opportunities, such as studying the complex processes behind the production of variants, as huge as the risks such variants represent, deserve as much attention as the sequencing effort, and the attempt to capitalize upon it. Under normal circumstances, i.e., before the pandemic, travelers were advised to vaccinate before visiting areas known for certain diseases. And whenever a traveler got infected, he/she would have to do whatever it took to prevent the spread of the pathogens. To make disease less transmissible cannot be legislated. Through some form of containment, which is an anticipatory goal, the process can be slowed down or stopped. But neither sequencing nor CRISPR-Cas9 is a path to prevention. No one ever doubted that people's mobility explains the worldwide spreading of the virus. Genetic sequencing only confirmed common sense when it deployed the "heavy guns" of modern technology. But in doing so, it did not generate knowledge. It became a goal in itself: good for scientists' publication record, for advertising the technology used, for more funding of Big Science, but not necessarily consequential for those who are threatened or affected by a virus subject to mutations. The reason for the massive effort is, in hindsight, quite evident: expensive sequencing technology had to be used, and more of it had to be produced and sold via the public financing of science. This is how biology became part of *Big Science*—at whose scale biological processes are studied using data processing methods—and which parallels the Big Science of the space program and of particle physics (CERN, for example). They are as impressive as they are unsustainable.

The carbon footprint of space exploration by far exceeds what the entire population on earth contributes. Only the greenhouse gas emissions linked to ground- and space-based telescopes is equivalent to about 20 million metric tons of carbon dioxide. This extreme example of flipping the Upside/Downside Ratio (immediate or the long-term benefits versus the amount of various costs, including undermining life on Earth) is characteristic of CoVID. Determinism and reductionism are part of a self-feeding loop: the more specialized, the more resource-consuming. Only the big players can compete for funding and for new talent. Pfizer, Merck, and Moderna are now part of the BIG league, literally crushing the competition. The new "waves" in the post-vaccination era were met with more and more sequencing, but no valid science regarding the process through which variants come into existence. This is yet another proof that data-driven science cannot substitute for meaning-informed knowledge.

The Price of Individualism

People depending upon each other results in a sense of community that no amount of preaching, activism, or state decree can reach. But along the path of progress associated with the reductionist-deterministic take on the nature of change, conditions were created for independent survival and success: individual success (of one, the "rugged individual" of American mythology) gets the better of community (the interconnected many) viability. The age of ever more individualism unavoidably results in the CoVID. Among other instances through which it manifests itself, individualism is tragically expressed in the COVID-19 pandemic and its associated consequences. This being the case, it is legitimate to address extreme individualism as yet another instance in which the benefits (e.g., initiative, self-motivation, willingness to push the limits) are cancelled by the damage: the common good sacrificed for personal gain or satisfaction, usually at the cost of society welfare.

The organism is embodied in biological matter. The reductionist bottom-up causal chain, starting at the genetic or molecular level, within which COVID-19 was approached, offers no clue as to why Blacks and Hispanics were disproportionately affected. Or why those suffering from some disease (e.g., high blood pressure, asthma, hepatitis), or from being overweight, were more susceptible to being killed by the virus. The factual answer—i.e., poverty across races, gender, or any other qualifier—affords a window to understanding what happened: why the change (virus exposure) affected people so differently. Along the sequence atom-molecule-substance, as represented by numbers, science found methods for making things and fixing things. This is how bridges and cars are made, and, when they fail, are fixed or replaced. But from atom to molecule to organelle to cell to tissue to organ to organism and eventually, following the same logic, to community and to society, there is no knowledge to be gained regarding why during pandemics the disadvantaged got hit harder. And even less—but no less important—what should be done to protect everyone to the best of society's ability.

Life expressed as behaviors, not just as matter subject to measurement, is actually the issue. In the reality of living in the USA or in Western Europe, behaviors are conditioned, among other factors, by unequal living and learning opportunities. Tracking the virus through formal symptoms—the syntax—such as temperature (which proved to be irrelevant), but not through specific criteria—loss of smell and taste, for example—affords no knowledge. In China, where the QR code for each individual is scanned, the argument is simple: "*fei chang shiqi*," i.e., extraordinary time. In practical terms: everyone in an area of infection was and still is questioned, in patient conversations, regarding their habits and whether they experienced changes in their behavior. There is no privacy to account for. The spreading was contained, deaths avoided. Community interests prevailed over individual rights. In Wuhan, apartments inhabited by infection-suspected individuals were taped shut (Hessler 2020). Food was delivered to them—this is the community effort—but they were not allowed outside their residence. Hospitals were built in a hurry and medical assistance was provided. Individual freedom became less important than the well-being of the community. For their own sake, people accepted intrusion into the private sphere and tracking. Whether we like it or not, the new normality is one of being monitored. To the advantage or disadvantage of the observed?—that is the question!

Although the numbers are changing, a point can be made referring to what is known. Behind the record numbers of Americans infected and dead—the USA has 4% of the world population, but over 30% of the population affected by COVID-19—there is a behavioral reality (Shackelford 2020) that needs to be discussed openly. Some experts (e.g., Lipsitch 2020, Noh, Danuser *op.cit.*) estimate that the actual number of people infected in the USA is way higher than reported. That reciprocally responsible behavior cannot be dictated, and even less attained by some miraculous consent for which America the divided is not prepared, should be obvious. Let us assume that Americans would decide to invest in anticipatory actions—education, access to medical care, fair distribution of wealth, etc. This is not the same as saying that in a society conditioned to expect instantaneous reward, long-term processes would all of a sudden become acceptable after COVID-19. And even less, that everyone would automatically become responsible. As long as everyone is in a hurry to get their individual share of immediate satisfaction, the goal remains utopic. America's future depends on it, but so far America has not come up with any meaningful course of action. The European Union, with a centralism the USA does not tolerate, chose a long-term plan that integrates ecological and climate considerations. This plan sets concrete goals to be achieved. A next pandemic, if not avoidable, might find Europe better prepared. The rest of the world—a huge chunk of humankind—is in a more or less wait-and-see condition. For a time, Sub-Saharan Africa seems doomed. Everyone—China, the USA, Europe, Japan, Middle East countries—wants its resources, but nobody cares for those who live there, where survival is at risk. Often mistreated by their own leaders, the native population is cruelly exploited in order to provide the rest of the world with cheap raw materials for its digital processing machines and for other high-tech equipment (including war technology). Sacrificing sustainability for the immediacy of rewards (i.e., prosperity), the path leads from the lost sense of community, in its traditional condition, to disregard for the larger environment in which humankind lives.

The workshop on *Biodiversity and Pandemics* (organized by the *Intergovernmental Science Policy Platform on Biodiversity and Ecosystem Services* (OPBES), meeting virtually (July 27–31, 2021) specifically approached land use, biodiversity loss, climate change, and trade in wildlife from the perspective of pandemic preventions. "The existential threat to the health and welfare of people across our planet" can be met only by prevention. The experts agreed that "human ecological disruption and unsustainable consumption drive pandemic risk." Sustainability itself is at stake. This fact has escaped the attention of those who, in their own research, actually continue on the path that led to more rapidly succeeding epidemics and pandemics. They justify the reactive use of therapeutics that often aggravate the situation, using disingenuous arguments. Of course, the word "profit" is never mentioned.

The anticipatory perspective was ignored by the well-meaning scientists who warned about disasters yet to come: From 631,000 to 827,000 "currently undiscovered viruses" could have the ability to infect humans." Humankind continues to "disrupt natural interactions among wildlife and their microbes." While all these pronouncements are correct, those who are fighting the pandemic face questions that science has failed to address. It took a long time until some hypotheses were formulated regarding several questions: why more men than women became infected; why so many in nursing homes, in jail, on cruise ships. But more than these: How come non-symptomatic subjects, including children, can transmit the virus? How come vaccinated persons can infect others? Remember: If you stated this, on Twitter, for example, your account would be cancelled. And more yet: The "long COVID" that remains with survivors and causes them to suffer a number of varied painful conditions is still not properly addressed. Some question this; others warn about the extent to which society as a whole will be affected. Yet others get rich on all kinds of fake therapies.

The percentage of deaths in respect to the entire population or to those infected, is indicative of how successful the measures taken are or how willing the population is to share in concerns and responsibilities. China, Vietnam, and New Zealand were for a long time emblematic of an alternative that few other countries would even consider. The high mortality rate in the USA, considered against the background of government actions (at federal, state, and local levels) is indicative of the profile of Americans. Individualism is extreme—grounded in the still young experience of living on the frontier and stimulated by the capitalist economy of extreme competition. Rejecting government, but at the same time becoming increasingly dependent on it, the country—not the only one in the world—pursues what is deemed as individual freedom, even if this means a low (or no) sense of solidarity. Someone went as far as to state, "The American pursuit of individualism is indistinguishable from a mental illness" (Guye Guion, on Twitter). Others claimed that "Exceptionalism is Killing Americans" (Konyndyk 2020); or the opposite: "The unraveling of America reduced to tatters the illusion of American exceptionalism" (Davis 2020). In 2020, John Walker adapted Aaron James' 2012 book, *Assholes: A Theory*, for a documentary: "How America Became a Country Full of Assholes." The qualifier points to the deterioration of the moral fabric. Deep political divides, distrust of authority and of science, and incompetence explain the truly irresponsible response to the pandemic. The ten-day motorcycle rally in South Dakota (the infamous Sturgis super-spreading event linked to 260,000 infections) is one example among many (Majra et al. 2021). Similar events have continued to take place even after it became clear how dangerous large gatherings can be.

Although the USA as a nation started as an anticipatory project, with the future of hope and liberty driving her beginnings, it soon became the expression of the social Darwinism (called "competition") view of capitalist society (Nadin 2013). Government or community actions are a matter of the social contract in which people enter. But in the America of our days, freedom of the individual takes priority over community concerns and shared well-being. Americans reject government. There is no such thing as a state or country called America. A collection of states—economic entities in the first place—does not add up to a nation. It constitutes a dysfunctional entity in which even medical concerns (such as life and death) take second place to a rhetoric of individual rights disconnected from the reality of a rather disempowered member of society. To speak of the failure of a government conceived from the get-go to be powerless is to question the reason for the economically united states, but politically, as dis-united as ever.

The pandering pandemic—involving a fractioned people following blindly whoever waves a flag or issues a memorable slogan—is congruent with the COVID-19 pandemic; and together they define the crisis of vision, spreading, like any new American fashion, over the rest of the world. Super-spreader events—e.g., a church in Daegu, South Korea, the Diamond Princess cruise, Oktoberfest 2022—brought up the K-factor that describes clustering behavior. The Edmonton (Canada) prayer meeting, mass religious events in Israel, the Tönnies meat processing plant in Gütersloh (Germany), political rallies in Spain, the dormitory for asylum seekers in Halluin (France), among other instances: these testify to a toxic individualism that extends from the USA to the rest of the world. India, Brazil, and England mirrored the same social reality.

Imitation of American postures, from chewing gum to cowboy fashion, to rap is nothing new. Movies and television, and lately social media, broadcast an image that would be comical, if not rather deplorable. Along this tendency, the painful experiences of racist incidents and of intolerance were displayed without an understanding of the meaning of words, actions, gestures. The Black American football player kneeling as the national anthem was played had an authenticity rooted in a pain similar to, if not the same as, that of those who suffered or lost lives due to their skin color. It was an expression of protest and solidarity. The kneeling of German and French soccer players was opportunistic, at best. British actors, as well as performers in the Netherlands and Canada, got into the business of a solidarity that was as empty as it was probably well intended. Their reality, which they are as much aware of as they are of racism in the countries where they are paid to entertain spectators, not to enlighten them, is one of privilege. One soccer player in Liverpool was asked why he, earning millions of dollars and enjoying a life of privilege, is in solidarity with the Black victim from Minnesota: "It looks good on TV!" Black Lives Matter signs at the entrance of mansions and expensive residences—where no Black (except in servicing functions) ever set foot—have as much to do with solidarity as preventing COVID-19 with some bogus cure (freely touted in a free market economy).

The Game of Lying Numbers

For those interested in the deterministic aspects of disease (i.e., for those who carry a "hammer" in hand to hit the "nail," i.e., bacteria, inflammation, virus, accident, etc.), numbers are nails waiting to be hit. Numbers, i.e., data from measurements from which they are derived, actually lie. They replace the continuum of phenomena through discrete labels. The organism's condition at the time of measurement is considered as permanent as the numbers spit out by the measuring device. Numbers literally eliminate time—more precisely, the future—from the image. They reflect the metric adopted. Examples: temperature (expressed in some units referenced to physical phenomena such as boiling water); blood pressure (millimeters of mercury, a chemical element used in the past for weighing the "weight" that the blood pumped by the heart exercises on the arteries); weight (corresponding to the interaction between the body and gravity). The metric has a reference in physical phenomena—but does not justify the generalization from one phenomenon to others. Behind each assessment of quantity through measurement there is the physics embodied in the respective devices. In other words, numbers correspond to the implicit perspective: in measurements we see the world through the lenses of predefined views. There is a built-in loop corresponding to a circular view of causality (cf. von Foerster and his *Anthology of Circularity Principles*, 1995).

The deterministic-reductionist reaction to COVID-19 meant, for instance, the successful use of highly sophisticated genetic analysis procedures. The fact that each tool embodies views on which it was conceived and built is rarely made explicit. Thus, the illusion of the objectivity of such tools is never questioned—and even less the degree of appropriateness of the means used. It is almost absurd that genetic sequencing, used to determine the building blocks (bases) that make the DNA molecule, was applied in order to figure out how the virus spread.

Scientists can use numbers to tell the truth only if they go through the effort of discovering their meaning: What is measured? What is ignored? How is it measured? And most important: Why measure? All of which means: Only if numbers are referenced to specific phenomena, i.e., put in context, do they become meaningful and become information about such phenomena. Otherwise, the formal characteristics of the number (the syntax of adding and subtracting) become the goal, to the detriment of obtaining an appropriate understanding of the phenomena they stand for. When something is represented as a number, its formative history—the biography of a dead person, for example—is discarded. In other words, numbers do not properly account for process. They are snapshots: At moment X, the person's weight is Y. This is quite helpful in finding out the weight of a product: for instance, in figuring out if the foundation upon which it will be placed will support it, or for figuring out the price of shipping it. But in assessing someone's metabolism—e.g., weight before lunch and after, weight before and after a bowel movement, and many more measurements of all kinds—they are only partially significant. In the living, *process* defines its dynamic. Health or the lack of it is not reducible to numbers. It reflects the characteristics of the process called "life" and which is affected not only by the past (*anamnesis* as a record), but also by the possible future (behavior dependence).

Yet again: for individuals captive to a lockdown or to "shelter-in-place," what do numbers mean? To the people allowed to go outside their homes for some limited time, as was the case with France during the first lockdown, it was unclear why they were literally "arrested" in their homes, and how the number of those infected or those who died made sense in their own lives. In Romania, the military issued daily instructions to the citizens, as one might expect during wartime. In Iceland, the police (actually a troika of agencies) were in charge. If businesses died—as many did all around the world—because the COVID-19 emergency measures made it impossible to keep them open, why would the owners care about the numbers attached to the history of similar disasters that killed millions upon millions? The most recent flu epidemic (2019–2020) infected a very high number of people around the globe (between 39 and 56 million) and resulted in deaths, estimated at 62,000; but the world economy barely took note of them. The HIV/AIDS pandemic continues, with the number of victims reaching a higher order of magnitude than COVID-19. Moreover, the H3N2 ("Hong Kong flu"), as deadly as COVID-19 (between one and two million succumbing), did not cause the havoc that the SARS-CoV-2 is given credit (or maligned) for. Individuals faced with a barrage of numbers for justifying mask-wearing and lockdowns learned that heart disease and cancer continue to kill more people than epidemics do. Those conditions are not infective, but, like all chronic conditions, entail a high price. However, COVID-19 soon surpassed these and was declared a medical crisis. New numbers made headlines: joblessness, personal and business bankruptcies, people marching for various causes, people dying because they could not be treated, schools and universities closed, theaters and music halls reduced to webcasting production sites. The pandemic by now became a part of the larger social and political crisis (Fig. 9).

The incompetence of those in power, and the arrogance of those reporting on political breakdowns, were met by the growing expression of lack of confidence in those who usurped the people's sovereignty. A minimal effort was made to help, in a calm and competent manner, members of society to understand what all this means. Sensationalism translated as numbers—each cause generated its own numbers—sells better than knowledge. Politicians make hay from anything that causes the people to feel insecure, not to say powerless. Unfortunately, practitioners of medicine, too often engaged in the political arena, became activists, cheering ("Oh Vermont, you are doing so well!") or booing (especially Florida and Texas) in a manner recalling football games. In the absence of a professional education that should involve anticipatory considerations, physicians continued to operate almost exclusively in reactive

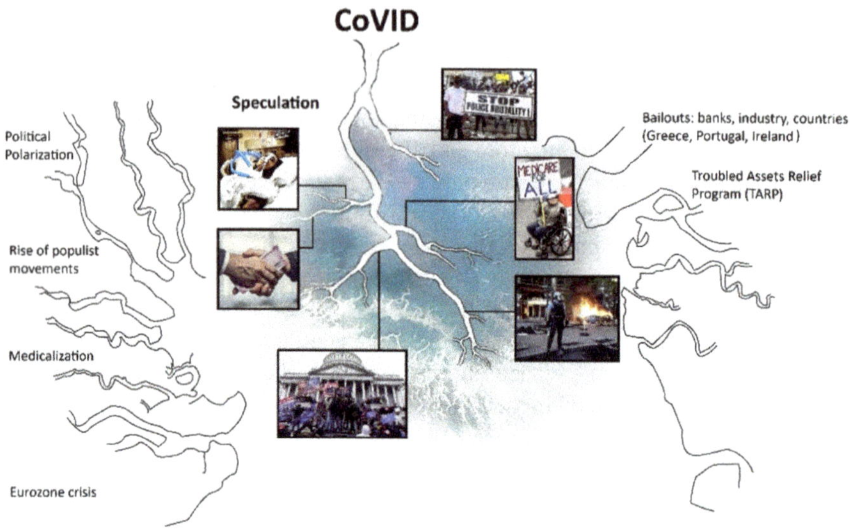

Fig. 9 Pandemics as part of the larger crisis

mode—proving or disproving one or the other faction—despite the fact that the outcome is rather disappointing. If before COVID-19 medical treatment was the third cause of death in the USA, during the pandemic it became the second (if not the first—when we consider the statistical modeling suggesting that "COVID-19 has claimed between 7 and 13 million lives. That's up to four times higher than official estimates" (*The Economist*, May 14, 2021, "Covid-19: how many people have really died?").

The understanding of life according to the principles set forth by Descartes in 1637 leads to the conditioning of those who accepted them in that spirit. Individuals—regardless of their identity (race, gender, religion, political inclination, age, education, etc.)—acquiesced to being treated like machines. Built on principles that translate action-reaction into quantitative descriptions, machines can be monitored by using numbers as representations of their performance. Pandemic or not, people in need of medical care look for "better mechanics," i.e., the specialists, using better tools, to get to the bottom of the matter—to atomic, molecular, cell, and gene levels—under the assumption that the fix is hidden there. Indeed, in the spirit in which Descartes described the body, the patient is reduced to inert matter—and is treated as some *thing* that needs fixing. The fact that lifeless matter does not know what pain is, and does not die, is simply ignored when life is reduced to the physical. In the pandemic, it became once again obvious that the anticipatory dimension of life has been even more often overwritten by the reactive view, thus making each and every person more vulnerable—and more dependent. In the absence of the moral dimension—anticipatory by its nature—society abandoned meaning. The bridge from individual vulnerability to global vulnerability was again crossed.

The reward that Western governments offer to their citizens for giving up freedom—"lockdown" is a weak synonym for this—is the huge relief program—more numbers—thrown at the COVID-19 pandemic within the cause-and-effect model. More—a lot more—in terms of bribing the citizens through various programs will follow in the years to come. The meaning of throwing money at the crisis was entirely lost. In fact, the money associated with COVID-19 was used for unrelated purposes—in the USA and abroad. The second wave of the pandemic came not only with a higher number of infections, but also with bigger bribes. And so did the third wave. All of which means taking money from the future in order to keep the illusion of normalcy in the present. In saving the financial markets, various relief funding (in the USA, European Union, Japan, etc.) move huge amounts of money around. A lot of money was being printed, mortgaging the future, with the excuse that the people accept these actions. The relief funds sweetened life in the digitally connected cell of loneliness and also helped to avoid or dampen social unrest. But more important, they have led to higher profits for those who made a fortune in building the vulnerable economy that the virus brought to a standstill. So far, the pandemic has led to one of the greatest wealth transfers in history (Cramer 2020). "No crisis should go to waste" means different things to different people. In our days, the one percent own over 50 trillion dollars in assets; the bottom half of Americans have more debt than assets. COVID-19 widened this chasm even more. The larger crisis, CoVID, brought this untenable situation into the headlines. Will this settle the matter? Or will this eventually lead to uncontrollable conflicts? The question is inevitable.

A Test for the Obsession with Testing

Headlines (of articles that physicians wrote), such as "Your Risk of Getting Sick from COVID-19 May Lie in Your Genes" (Brown and Cortez 2020), or variations—risk lies in your biome, in venous thrombosis, in gastritis, in overweight, in hair loss, and in so many more—are indicative of the many angles of identified vulnerabilities. All converted into numbers because this was the hour of numbers. TEST! TEST! TEST! became the new refrain of the old song of a primitive view, according to which measuring is supposed to save humanity. And when testing returned worrisome numbers, the slogan adopted was "Flatten the curve."

A hypothetical: everyone has been tested. COVID-19 will still be with us. Testing is not a cure, and not testing is not the aggravating factor it has been declared to be. Viruses, and for that matter the ACE2 "port of entry," do not read newspapers. Testing is like surveying the landscape: in the absence of a cure, or of a vaccine, or of antibodies, which can be a treatment (of shorter effectiveness than a vaccine), tests inform government decisions, and even individual choices. But not all decisions are meaningful. It turns out that the most deployed test (the PCR) initially returned false positives almost 90% of the time due to the way in which its sensitivity was set. If tests are deficient, to test becomes an expensive futile activity, providing fodder for political bickering but not useful data (Michael Mina, interviewed in *Harvard Magazine*, see Shaw 2020). There is some history to recall: In 2007—14 years back—after an intense effort to test for pertussis (whooping cough) in New Hampshire and Vermont, deploying the polymerase chain reaction (PCR) molecular test—now considered the most reliable test for COVID-19—the following conclusion was reached: "The big message is that every lab is vulnerable to having false positives.... No single test result is absolute and that is even more important with a test result based on P.C.R" (Kolata 2007).

In general, confusing messages from the medical community created some of the panic associated with the pandemic. After declaring masks ineffective—a position that was official in Holland, for example—the same high-profile experts started preaching mandatory mask wearing. Basically, masks should be worn by people who are infected, in order to prevent spreading the virions. To assume an effective protective function is to ignore physics—exactly by those who promote physics as the science of everything (Verma et al. 2020). It is easy to state that behavior regulation—social distancing, wearing masks, lockdown, quarantine, etc.—and the allocation of resources for addressing the needs (ventilators, for instance) of those affected should be based on objective evaluations. But when the regulators involved have a stake in the regulation—be it economic or political—the premise of objectivity is compromised. Medicine has to free itself from participating in the poisonous political bickering of this time or of any time. It has to live up to its obligation to care for the patient's health, not for the outcome of elections or the profit of the pharma industry and medical facilities. The COVID-19 experience proved that the medical care system needs not more and more numbers, but *a focus on meaning*, and on the actions to mitigate risk. And it needs a clearly spelled out conflict of interest clause in order to save medicine from corruption. Ethics, possible only within an anticipatory view, is not optional in the practice of medicine. Unfortunately, society as a whole gave up on ethics in favor of more efficiency ("profit" would be a more honest word for this).

"Will 35 million Covid-19 tests per week, or even per day, in the US solve the problem?" New York University economist Paul Romer (2020), a Nobel laureate in economics, posed this question. It was echoed and became raw meat in the political arena. In an area of high infectivity, pool testing—take samples from a group, and if no infection is detected, you increased the numbers of persons tested and saved on the cost—was correctly deemed as delusion. The fact that testing cannot flatten the curve that is actually the outcome of testing—notice the circularity of the situation—was lost in the high-pitched arguments. You don't lose weight "testing" on the scale repeatedly. You have to change your behavior. Your blood pressure is affected by having it measured, but the condition leading to hypertension is not. Changing behavior helps.

Data from the past were used as premise for models, but this is by no means a method for containing a new pandemic. Bright minds, or at least well-intended professionals, argued for testing as a method for limiting the number of victims. New data had to be generated in order to fuel the predictive engine. In reality, flattening the curve—which means reducing the rate of increase in infections and thus death—could be achieved only through actions. Indeed, measurement in coordination with data-informed actions is meaningful. In the absence of actions, it is a useless technical exercise.

Americans reject government control mechanisms. The Vietnamese and the Chinese had no choice but to accept them. Great Britain hated the lockdown, but it helped. The Europeans tried to legitimize each decision: some, such as compulsory mask-wearing, were rejected by the population. Numerous demonstrations—in Germany they were never-ending (until the news that Russia shut off its gas supply, leading to severe energy shortages in Germany, became more important than COVID)—as well as non-compliance are on record. New Zealand imposed a strict lockdown, embraced by over 80% of its insular population. The island voluntarily separated itself from the world. Testing only documented how successfully "the doors were locked." Facing the second wave of the pandemic, the much-praised prime minister of the island country exclaimed "Not again!" China was even more radical in imposing a lockdown that no country calling itself a free democracy could pursue. Shanghai, with its size as a country, was literally "arrested." Measuring does not change—or should not change—what is measured. Japan produced the proof for this (Sayeed and Hossain 2020). Of course, isolation and distancing, stricter hygienic behavior, self-control, and accepting tracing and tracking (through all means) converge in a controlled behavior that limits the propagation of the infection and reduces the fatality rate.

Behavior is the result of comprehension: life is a good teacher, but a lesson learned late might be costly. The premise of a successful education of the public is credibility. Neither the medical establishment nor the governments (with some exceptions), and unfortunately neither the education system, proved to be credible.

This observation brings back the critical issue of the meaning of numbers: What does it mean that Singapore, South Korea, Taiwan and Vietnam kept the number of victims low (at least until the virus variants became the focus)? How come the Baltic republics fared as well as they did in the first phase? What does it mean that China, with its 1.5 billion inhabitants reported over 2.3 million cases and only slightly less than 15,000 dead? Authoritarian measures (e.g., isolate, keep track of everyone, have everyone wear a mask, limit mobility, cordon off parts of the country), which could not be imposed in the USA or in many other countries, are not compatible with commercial democracy, i.e., capitalism and consumption. Within capitalism (especially the American version), driven by profit-making, the economy is a machine. The accelerated movement of money in the economy translates into higher profits. Otherwise, money loses value. Obviously, in the absence of workers, neither products nor services are available at the level at which consumers have been conditioned to expect them. Citizens kept captive at home cannot consume, even if China—the "factory" for a world seeking cheap labor—will continue to produce regardless of the virus, and the tariffs.

Regulating behavior, sometimes the wrong way, and throwing money at the pandemic, are reactive solutions. The method has been applied to each crisis that humankind has gone through in recent times (e.g., hurricanes, earthquakes, flooding, massive fires, financial breakdowns). The premise for the reactive path seems to be "Bribes wash all sins." In reality, it takes a long-term investment, not only monetary, but probably moral in the first place. To get to the roots of the evil and create circumstances for its recognition as such, and for a course of action from which victims and perpetrators can learn is a difficult task. Fire management is a good teacher. The potential for disaster is there as long as combustible matter accumulates, so apply preventive measures. It's time to understand that, even when successful, each reaction to a terrorist attack, or each reaction to cyberterrorism, each reaction to a financial crisis is not really a victory, but a disaster delayed.

An Anticipatory Perspective

The alternative to expensive and unsustainable reaction is *the anticipatory perspective* as the necessary complement to the reactive path. Prevention means, for instance, eradicating the conditions that lead to committing acts of aggression. Reaction is always as short-term and incomplete as the lack of understanding of what caused a failure. Duration of anticipation-based predictive activity is as long as the time it took for a crisis to emerge and unfold.

Of course, this begs for some explanation, especially since the notion of anticipation, kept out of academic discourse and marginalized in scientific dialog, is often misunderstood. Anticipation, as opposed to guessing and prediction, is expressed through action. It is the actual outcome, the realization of change (the imminent, the possible) and what it might further entail, morality included. Indeed, in hindsight, the new virus and the pandemic, as well as the subsequent breakdown of the economy, never mind the social and political turmoil within which it erupted, were all in the realm of *possibilities*. So was the moral aspect: inequities, discrimination, greed, social and political polarization. Failed democracy is not reducible to one or another symptom. It involves the system in its entirety.

There is a record of various attempts, worldwide, at preparing for a possible pandemic emergency. The World Bank Group president, Jim Yong Kim argued in favor of creating a facility for dealing with economic and financial aspects (2014). The factual observation, "When it comes to health emergencies…our institutional toolbox is empty," describes a reality that includes governments and international agencies. Almost at the edge of the crisis (in October 2019), a high-level simulation exercise for pandemic preparedness took place in New York. The Johns Hopkins Center for Health Security, the World Economic Forum, and the Bill and Melinda Gates Foundation—all credible brands—underwrote the event. Given the premise—grounding in reactive thinking and no discernible anticipation-informed contribution—it is not surprising that the simulation exercise proved inconsequential: one month later, COVID-19 became a global reality. At the same time, the Center for Strategic & International Studies, issued its own report on an anticipatory approach to global pandemics, concluding that "An ounce of prevention is worth a pound of cure" (Carroll 2019). In January 2020, the Davos Conference, taking note of 440 people infected in China and only 17 deaths, decided, like the World Health Organization, that this was not a pandemic. Whenever scientists made recommendations, as in some European countries (France, Germany, England), the governments ignored them. The *probability* of this specific pandemic, of extremely high infectivity and toxicity, was close to zero. The SARS outbreak in China (2002), the MERS (Middle East Respiratory Syndrome) in Saudi Arabia (2012), and other corona viruses (there are many of them) could have alerted the world to the danger. But each strain is different, and the probability of a comeback of a sort was downplayed. Deep in the health crisis, Deborah MacKenzie (2020) drew attention to the title of her book, *The Pandemic that Never Should Have Happened and How to Stop the Next One*. Her view, that experience with pandemics could have served as a premise for avoiding it, is quite telling of the dominant Cartesian perspective—fighting a new war like you fought the last one. That is the implicit limitation of all deterministic processes. Closer to the tail end of the story, Hotez (2021)—known for having developed a vaccine ten years before COVID-19 began—took a broad view of the many factors involved in pandemics and preventing them.

Pandemics occurred many times in the past. And the economy faltered repeatedly—the inherent crises of the capitalist system—for reasons other than pandemics. Natural disasters, of increasing intensity, accentuated the challenges to any form of reaction. In the not far distant past, it was Pandora's box (from Greek mythology), hiding so many dangers; or the stars, the gods, and quite a number of other naïve suppositions (e.g., "bad air," miasmas) that were associated with the interruption of the continuum of progress. In more recent times, with spectacular progress in measuring everything, different culprits were found—poor hygiene, poverty, limited medical competence, mediocre levels of education, and limited access to medical care—but no sign of culpability.

On the timeline of its history, humankind arrived at a stage of higher and higher expectations, based on scientific and technological accomplishments, often bordering on the miraculous. But this self-proclaimed progress entailed risks at the same or at higher scale. Consider in this context only that the global economy of higher-than-ever efficiency and the global pandemic are hand in glove. Einstein (again), among others, went as far as to claim, "The tragedy of modern humankind is that they created conditions of existence for which, from the perspective of its phylogenetic development, it is not adjusted." Unpacking this sentence could afford some promising ideas for a science that makes development possible but creates also means for adjusting to such developments. The courageous space explorers are not simply deployed into deep space, but trained to meet the challenges of new conditions of life and work.

Evidently, defining CoVID as a confluence crisis that integrates the pandemic, the economic breakdown, the financial meltdown, as well as the social and political failures it engenders, makes it more of a label instead of facilitating a better understanding—the premise for action. Therefore, let us place CoVID against the background of ecological abuse, self-destructive in the final analysis, and against the progress, unsustainable, supposed to justify capitalism and its many shortcomings. Illness (the choke flu) and social and political illness, like tributaries, merge. The streets of Paris, London, Berlin, as well as in cities in Asia, Australia, and South America, not to mention New York, Portland (Oregon), Seattle, Chicago, Kenosha,

Rochester, Baltimore, and Philadelphia became the backdrop for intense expressions of frustration bordering anarchy. In Amsterdam, police treated people in ways reminiscent of Nazi brutality. China practiced the same. As a reason for fearing the unknown, the perception of the SARS-CoV-2 forced the future onto a stage of confrontations.

It is not a matter of whether one can survive COVID-19, but rather if humankind can survive its own rapid move towards a future empty of future. To end poverty, hunger, to ensure healthy lives, to achieve racial and gender equality, and to provide quality education are honorable goals within the *2030 Agenda for Sustainable Development*. The same holds true for ending the abuse of water resources and opening access to affordable drinking water. And for a meaningful understanding of work, family life, and learning. All of this hides the stark contrast between ideals and the reality of being conditioned to expect more and more, at lower price. Never questioning the cost of prosperity, the morality of abundance in a world in which the vast majority lives in below poverty standards takes a toll on humanity itself. Those who have (the "haves") feel entitled to have more: a bigger home, a second and third residence, a luxury car, several televisions with huge monitors, and a dog (rather, dogs). For the well-being of pets, all with a large carbon footprint, more is spent yearly than what the world's poor (the "have nots") earn in a lifetime. Billions of dollars are spent for weight loss while a large segment of the world population suffers from malnutrition.

The professionals of determinism in science and technology fail to understand the meaning of some of their accomplishments, in particular the shortcomings and drawbacks, especially abandonment of morality. The return is not only disappointing, but also suggests abandoning values and self-respect. Doing away with humans—through the spectacular AI-based machinery bound to replace them (physicians included)—AI scientists claim to accomplish the task for the sake of improving the human condition. Imagine a society in which guaranteed minimum income (GMI) is paid for by AI and robots deployed in factories and farms. The idea is not new—it goes back to the Persian and Roman Empires where poor families received various kinds of rations. In more modern times, inequities were to be alleviated through some redistribution of taxes. The GMI claim is present in the current upheaval. Opportunistic immediacy dictates the high tone of justifications and reparations but does not place them on the solid ground of a new system of values. The long-term moral consequences are simply ignored.

Questioning the legitimacy of new technology, or the legitimacy of expectations, would place CoVID in the wrong perspective. It is not impossible (though extremely difficult) to engineer a biological entity that lives longer than human beings live today, or that never gets sick. But would it be a human? The "sinister combination of ecosystem alteration, wildlife exploitation, and global connectedness" (Nuñez et al. 2020) was clearly leading to the possibility of the disaster experienced in our days. Smith et al. (2014) and Afelt et al. (2018) were specific in their warnings about the consequences of "biological invasion." But no one took the time to deal with the long-term impact of humanity's own doings—and even less to try to alert society. Coping with the risks generated by science and technology is rarely a subject of interest in any university or research facility. Long-term consequences, in particular the distorted ethical profile, are a taboo subject. Governments ignore them. So do the people who took to the streets to denounce what they consider justified grievances, but which others see as expressions of selfishness.

A holistic understanding, implicit in the anticipatory perspective, makes possible the observation that we cannot really separate individuals from the medium of their existence: they are fused. This applies not only to the microbiome (a recent re-discovery of the gut that has consequences for the practice of medicine). Only such a holistic understanding can make it possible to take pro-active measures in order to mitigate risk. "We must not be in a position of playing catch-up ever again," argued David Ho, who, together with Alan Perelson, changed the way HIV infections are treated (Strogatz 2019). The variety of pathogen-informed efforts in reacting to them—such as antibiotic treatment—or in anticipatory actions—such as safe sex practices, vaccination, when possible—was never seriously studied. Bacteria, prions, and protozoa, behind other epidemics, invite different approaches. Mad cow disease is avoided by preventing contamination of meat in the first place. Malaria (associated with protozoa) and the focus on mosquitos as factors of infectivity is in the process of eradication. The simple use of mosquito netting around beds has prevented most new cases of malaria in areas prone to the disease. But it took fifteen months—or 60 years (Molteni et al. 2021)—to find out how SARS-CoV-2 propagates—via aerosol or droplets—despite the enormous effort in genetic sequencing. Walls and tables, doorknobs and everything else humans touch were generously disinfected, not because it was necessary, but rather out of ignorance. Almost 200 years ago, Dr. Snow was faster in identifying how contaminated water gave cholera the "fuel" to spread in Soho.

From among the ca. 200 identified viruses that might result in medical conditions, probably only ten inspired treatments in the form of antivirals. Scientists are preparing the "mouse trap" for the newest virus in the form of an artificially made ACE2-like receptor. This artificial substance could absorb the virus before it touches the living ACE2 (Chan et al. 2020). Some scientists called this "a phenomenal accomplishment." To give credit to such verbiage would give it legitimacy. Exaggerations of such scale sound like parodies of science. Many, way too many, such parodies made it through the media to the population at large.

What about disenfranchisement? A generation is on its way towards uselessness, enamored of and dependent on technology to the extent of no longer recognizing its own identity—gender, race, culture, ethnicity, etc., not to mention moral profile. The same exaggerations, coming from other prophets of the frivolous, made many young people invest their lives (and taxpayer money) in future professions that actually have no future. To multiply the useless is by no means an act of justice, but rather of shortsightedness. It is obvious that plenty—the name for prosperity, as unevenly as it is distributed—does not transcribe into a sense of future. Poverty, violence, dependence on entitlements—at the very rich side of the spectrum as well as among the very poor— are as insidious as a virus. CoVID, as a crisis of vision, reflects the fact that pandemics do not unify, but rather question the false sense of unity that society's members seem to enjoy, or expect. Young and old clash, suspicious of each other since the malaise does not treat them as equal. The poor, the Blacks and people of color, the Asians (of various ethnicities), and all those who feel they are discriminated against clash with the apparently and actually rich and privileged. Those who are part of the establishment—in politics, science, medicine, etc.—forced their views upon the new generations. They judge them according to how they fit into the picture, not what they bring with them, especially their ability to challenge the doctrine in place.

There was dissatisfaction and there was violence in the world before the virus. In the last ten years, the number of riots almost tripled. Strikes, mainly in Europe, expressing opposition to particular aspects of capitalist practices, grew in number by 821%. Most worrisome is the increase in violence (Brzozowski 2020). In the no longer recent past, the ideologically confused *gilets jaunes* in France took over Paris. The virus brought them back to the streets, as it brought thousands in Munich and in several cities in Italy. Thailand rose up in arms against its monarchy. At least five coup d'états are on record, some ending with many persons dead on the streets. In pursuing its own economic ambitions—and receiving vaccinations without having to wait for the blessing of the bureaucrats in Brussels—England experienced the alienation involved in its separation from the European Union. Belarusians grew tired of the tyrant who declared himself ruler and took their protest to the streets (followed by the irony of Belarusia's support of the war against Ukraine). The Neo-Nazis took to the streets as well (and got permission—in Germany nothing can happen without some permission), and even tried to storm the Reichstag (House of Parliament), a name reminiscent of the days when Germany was a Reich, an imperium. Until prosperity—at the expense of others—was threatened by Russia's hold on gas delivery over Germany's support of Ukraine. Suddenly, Germans faced the reality of a higher cost of living, and demanded subsidies because they feel entitled to more and better than anyone else. For a while, dysfunctional America got used to shootings and violent demonstrations, to lootings and businesses set on fire, to language and physical abuse from different groups taking advantage of the troubled times. Then the Russia-Ukraine war got the military-industrial complex going again—with billions of dollars at stake. Spend public money on military aid with the purpose of keeping the economy going. At this juncture, intolerance has the upper hand. Supremacies are questioned by those wishing to establish their own (Chavez et al. 2020, writing about the NFAC, an armed all-Black group).

As expressions of revolt, the anti-government protests in Hong Kong, Belarus, and Kazakhstan, the protests in Israel, the protests in Iran, the unwillingness to accept inequality (Ecuador, Bolivia, Chile) and corruption (Lebanon, the African countries, Russia, Ukraine) are, of course, quite distinct from each other. Still, they seem to suggest a confluence, made even more evident once the pandemic invited, instead of a common front, a return to national borders. Under the threat of virus, the borders were closed because nobody understood that globality cannot be reversed. In some parts of the world, the ugly hour of nationalism arrived. Without identifying a common target—the enemy, the culprit—all those caught up in revolt demanded change. Of course, no one posted "*Future* matters!" although in a time like ours, it became clear that the future is the real subject. It was for their future that young (and often less young) once marched together (e.g., 1968) or violently collided, defying not so much the virus, but the lack of perspective. And even more the current perspective of the future. When the choice for national political leadership was between a deranged amateur politician—an aberration—and a professional politician portrayed by the opposition as on the verge of senile dementia, we see CoVID unfolding. The *lack of choice*, characteristic of a lack of perspective, is the real culprit—in the USA, and, for that matter, all over the world. The mayhem becomes a festival of tribalism, for which the media, all over the world, provide the instigating drumbeat, as noisy as the numbers used for scaring society.

Cleansing or cancelling the past is different from affirming a right to the future that transcends numbers (mostly fabricated). The right to defining new values is the right to a meaningful future. People know what they don't want—the infection, the disease, racism, intolerance, violence, instability. But not what they want! Except that it better be the magic bullet. Determinism, together with the expensive obsession with measuring everything, still has the upper hand—just as the blind faith of religion used to have before the Cartesian Revolution. The one-sidedness of reaction, in disregard of anticipation, has a high price tag. Instead of the idolatry of numbers, the anticipation perspective ascertains the right to meaningful living.

References

Afelt A, Frutos R, Devaux C (2018) Bats, coronaviruses, and deforestation: toward the emergence of novel infectious diseases? Frontiers in Microbiology 9:702, April 11.

Ahn M (2020) How South Korea flattened the Coronavirus curve with technology. April 24. https://www.nextgov.com/ideas/2020/04/how-south-korea-flattened-coronavirus-curve-technology/164884/ [retrieved July20, 2020].

Ballhous R (2020) As Covid Hit, Washington Officials Traded Stocks With Exquisite Timing, *Wall Street Journal*. https://www.wsj.com/articles/covid-washington-officials-stocks-trading-markets-stimulus-11666192404 [retrieved December 21, 2022].

Bastos ML *et al.* (2020) Diagnostic accuracy of serological tests for covid-19: systematic review and meta-analysis, BMJ 2020,370:m2516, July 1 [retrieved July 3, 2020].

Blumenthal et al. (2020) Implications for the Health Care System, *NEJM* July 22. https://doi.org/10.1056/NEJMsb2021088 [retrieved July 22, 2020].

Brann DH *et al.* (2020) Non-neuronal expression of SARS-CoV-2 entry genes in the olfactory system suggests mechanisms underlying COVID-19-associated anosmia, *Science Advances* 6:31, July 31. https://doi.org/10.1126/sciadv.abc5801. [retrieved January 3, 2022].

Brown KV, Cortez MF (2020) Your Risk of Getting Sick from COVID-19 May Lie in Your Genes, *Bloomberg News*. April 16. https://doi.org/10.1126/sciadv.abc5801 https://www.bloomberg.com/news/articles/2020-04-16/your-risk-of-getting-sick-from-covid-19-may-lie-in-your-genes [retrieved June 3, 2020].

Brzozowski A (2020) Civil unrest has doubled, and COVID-19 could make things worse. *Euractiv*. June 10 2020. https://www.euractiv.com/section/defence-and-security/news/civil-unrest-has-doubled-and-covid-19-could-make-things-worse/

Carroll A (2019) "An ounce of preventions is worth a pound of cure" *Commentary*, Washington DC: Center for Strategic and International Studies. October 24, 2019 https://www.csis.org/analysis/ounce-prevention-worth-pound-cure [retrieved July 1, 2020].

Chan KK *et al.* (2020) Engineering human ACE2 to optimize binding to the spike protein of SARS coronavirus 2, Report, *Science,* August 4. https://science.sciencemag.org/content/early/2020/08/03/science.abc0870 [retrieved August 5, 2020].

Chavez N, Young R, Barajas A (2020) An all-Black group is arming itself and demanding change. They are the NFAC. *CNN*, October 25. https://www.cnn.com/2020/10/25/us/nfac-black-armed-group/index.html [retrieved November 2, 2020].

Chen D-Y *et al.* (2022) Role of spike in the pathogenic and antigenic behavior of SARS-CoV-2 BA.1 Omicron, bioRxiv. October 14. https://doi.org/10.1101/2022.10.13.512134v1.full [retrieved December 15, 2022].

Chu H, Hou Y, Yang D, *et al.* (2022) Coronaviruses exploit a host cysteine-aspartic protease for replication. *Nature*. August 3. https://doi.org/10.1038/s41586-022-05148-4 [retrieved August 10, 2022].

Cooper KW *et al.* (2020) COVID-19 and the chemical senses: supporting players take center stage, *Neuron*. https://doi.org/10.1016/j.neuron.202006.032. [retrieved August 15, 2020].

Cramer J (2020) Jim Cramer: The pandemic led to 'one of the greatest wealth transfers in history,' *CNBC Report*, June 4. https://www.cnbc.com/2020/06/04/cramer-the-pandemic-led-to-a-great-wealth-transfer.html [retrieved June 5 2020].

Daum P, Simmank J (2020) "Herdenimmunität anzustreben ist ethisch nicht vertretbar," (Pursuing herd immunity is ethically futile and immoral), *Zeit Online* October 27. https://www.zeit.de/wissen/2020-10/anders-tegnell-corona-lage-schweden-sterberate-hygienestrategie [retrieved October 29, 2021].

Davis W (2020) The unraveling of America, *Rolling Stone*, August 6. https://www.rollingstone.com/politics/political-commentary/covid-19-end-of-american-era-wade-davis-1038206/ [retrieved August 10, 2020].

Egede LE, Walker RJ (2020) Structural Racism, Social Risk Factors, and Covid-19—A Dangerous Convergence for Black Americans, Perspective, *NEJM*. July 22. https://doi.org/10.1056/NEJMp2023616. [retrieved July 22, 2020].

Ellis GFR (2005) Physics, Complexity, and Causality, *Nature* 435:743.

Fihn SD (2020) COVID-19—Back to the Future, *JAMA Internal Medicine*, 180:9, 1149–1150. doi:https://doi.org/10.1001/jamainternmed.2020.2498 [retrieved September 18, 2020]

Fischetti M, Krzywinski M (2020) Virus Mutations reveal How COVID-19 Really Spread. Globe-trotting humans were the culprits. https://www.scientificamerican.com/article/virus-mutations-reveal-how-covid-19-really-spread1/ [retrieved June 5, 2020] See also: How COVID-19 Spread Like Wildfire, *Scientific American* 322:6, 80 (June 2020).

von Foerster H (1995) Anthology of principles propositions theorems roadsigns definitions postulates aphorisms etc. May 17–21. https://www.uboeschenstein.ch/texte/Foerster1995.pdf. http://www.cybsoc.org/heinz.htm (retrieved December 20, 2022).

Freedman DH (2020) A Prophet of Scientific Rigor and Covid Contrarian Ideas, *WIRED*, May 1. https://www.wired.com/story/prophet-of-scientific-rigor-and-a-covid-contrarian/ [retrieved July 15, 2020].

Gawande A (2021) Don't Tell Me What to Do, *The New Yorker*, February 15 & 22. pp. 36–47.

Gray DM *et al.* (2020) COVID-19 and other pandemic: populations made vulnerable by systemic inequity, *Nature Reviews Gastroenterology and Hepatology.* June 15 https://doi.org/10.1038/s41575-020-0330-8. [retrieved June 16, 2020].

Gu Y (2020) Covid-19 Projections Using Machine Learning, August 13. https://covid19-projections.com/. [retrieved August 14, 2020].

Guglielmi G (2021) Rapid coronavirus tests: a guide for the perplexed, News Feature, *Nature* 590, February 9, 202–205.

Harrison NL, Sachs JD (2022) A call for an independent inquiry into the origin of the SARS-CoV-2 virus, *PNAS Opinion Microbiology*, 119:21. May 19. https://doi.org/10.1073/pnas.2202769119 [retrieved August 22, 2022].

Heisenberg W (1927) Uncertainty Principle. Über den anschaulichen Inhalt der quantentheoretischen Kinematik und Mechanik, *Zeitschrit für Physik*, 43, pp. 172–198.

Hessler P (2020) How China controlled the Coronavirus, *The New Yorker*, August 17, 28–39.
Holshue ML *et al.* (2020) First Case of 2019 Novel Coronavirus in the United States, *New England Journal of Medicine* 382, 929–936, March 5.
Hotez PJ (2021) *Preventing the Next Pandemic: Vaccine Diplomacy in a Time of Anti-Science* Baltimore: Johns Hopkins University Press.
Illich I (1975) Limits to Medicine. Medical Nemesis, The Expropriation of Health. London: Marion Boyars.
Jacobsen R (2021) Top researchers are calling for a real investigation into the origin of covid-19, *MIT Technology Review*, May 13. https://www.technologyreview.com/2021/05/13/1024866/investigation-covid-origin-wuhan-china-lab-biologists-letter/ [retrieved May 1, 2022].
Jewell NP, Lewnard JA, Jewell BL (2020) Caution Warranted: Using the Institute for Health Metrics and Evaluation Model for Predicting the Course of the COVID-19 Pandemic Free (Ideas and Opinions), Annals of Internal Medicine April 14, 2020. https://www.ncbi.nlm.nih.gov/pmc/articles/PMC7197035/ [retrieved July 23, 2020].
Kahn S (2006) The widening scope of coronaviruses, *Current Opinion in Pediatrics*, 18:1, 42–47, February.
Karlis N (2020) Are temperature checks effective, or just "medical theater"? *Salon.* October 15. https://www.salon.com/2020/10/15/are-temperature-checks-effective-or-just-medical-theater/ [retrieved March 13, 2021].
Kauffman SA (2019) *A World Beyond Physics: The Emergence and Evolution of Life.* New York: Oxford University Press
Kim JY (2014) Speech by World Bank Group President Jim Yong Kim at Howard University: "Boosting Shared Prosperity." October 1. https://www.worldbank.org/en/news/speech/2014/10/01/speech-world-bank-group-president-jim-yong-kim-howard-university [retrieved May 2, 2020) See also: Prosperity for All: Ending Extreme Poverty April 10, 2014. https://www.worldbank.org/en/news/feature/2014/04/10/prosperity-for-all-ending-extreme-poverty. [retrieved May 2, 2020].
Kolata G (2007) Faith in Quick Test Leads to Epidemic That Wasn't, *New York Times*, January 22. https://www.nytimes.com/2007/01/22/health/22whoop.html [retrieved August 30, 2020].
Konyndyk J (2020) Exceptionalism Is Killing Americans. An Insular Political Culture Failed the Test of the Pandemic, *Foreign Affairs*, June 8. https://www.foreignaffairs.com/articles/united-states/2020-06-08/exceptionalism-killing-americans [retrieved June 10, 2020].
Lipsitch M, Inglesby TV (2014) Moratorium on Research Intended To Create Novel Potential Pandemic Pathogens, *mBio*, 5(6):e02366–14. American Society for Microbiology, December 12. https://mbio.asm.org/content/5/6/e02366-14/article-info. [retrieved April 28, 2021].
Lipsitch M (2020) Estimating case fatality rates of COVID-19, *Lancet* Infectious Diseases, July, 20:7, 775. March 31. https://pubmed.ncbi.nlm.nih.gov/32243813/ [retrieved May 2, 2021].
MacKenzie D (2020) *The Pandemic that Never Should Have Happened and How to Stop the Next One.* New York: Hachette
Majra D *et al.* (2021) SARS-CoV-2 (COVID-19) superspreader events, *Journal of Infection*, 82:1m pp. 36–40 November 25. https://www.ncbi.nlm.nih.gov/pmc/articles/PMC7685932/ [retrieved May 2, 2021].
Margit M (2021) COVID-19 not connected to bats, Israeli biologists say, *Jerusalem Post*, The Media Line, July 11. https://www.jpost.com/health-science/covid-19-not-connected-to-bats-israeli-biologists-say-673548 [retrieved July 18, 2021].
Metcalf CJE, Morris DH, Park SW (2020) Mathematical models to guide pandemic response, *Science* 369:6502, 368–369, July 24.
Miri SM *et al.* (2020) CRISPR-Cas, a robust gene-editing technology in the era of modern cancer immunotherapy, Cancer Cell International, 20, 456. September 15. https://doi.org/10.1186/s12935-020-01546-8 [retrieved December 20, 2022].
Molteni E *et al.* (2021) Illness duration and symptom profile in symptomatic UK school-aged children tested for SARS-CoV-2, *The Lancet Child & Adolescent Health* 5:10, 708–718, October.
Nadin M (2013) *Are You Stupid?* Heidelberg: Synchron

Noh J, Danuser G (2021) Estimation of the fraction of COVID-19 infected people in U.S. states and countries worldwide., PLOSON, February 8. https://doi.org/10.1371/journal.pone.0246772 [retrieved April 27, 2021].

Nuñez MA, Pauchard A, Ricciardi A (2020) Invasion Science and the Global Spread of SARS-CoV-2, Trends in Ecology & Evolution, May 19. https://www.cell.com/trends/ecology-evolution/fulltext/S0169-5347(20)30134-8?rss=yes [retrieved June 10, 2020].

Romer P (2020) Paul Romer on the COVID-19 Pandemic, The Library of Economics and Liberty, May 15 https://www.econtalk.org/paul-romer-on-the-covid-19-pandemic/ [retrieved May 17, 2020].

Rose N (1991) Governing by numbers: Figuring out democracy, *Accounting, Organizations and Society*, 16:7, 673–692. Amsterdam: Elsevier.

Rossman H *et al.* (2020) A framework for identifying regional outbreak and spread of COVID-19 from one-minute population-wide surveys, *Nature Medicine* (Correspondence) 26, pp. 634–638. April 9, 2020. https://www.nature.com/articles/s41591-020-0857-9 [retrieved April 15, 2020].

Sawyer SL *et al.* (2021) Just 2% of SARS-CoV-2-positive individuals carry 90% of the virus circulating in communities, (preprint) March. medRxiv 2021.03.01.21252250; doi: https://doi.org/10.1101/2021.03.01.21252250 [retrieved July 6, 2021].

Sayeed UB, Hossain (2020) How Japan managed to curb the pandemic early on: Lessons learned from the first eight months of COVID-19, Journal of Global Health, 10:2, 020390, December. https://jogh.org/documents/issue202002/jogh-10-020390.pdf [retrieved May 23, 2021].

Scales D, Gorman J, Jamieson KH (2021) The Covid-19 Infodemic—Applying the Epidemiologic Model to Counter Misinformation. Perspective, *New England Journal of Medicine*, May 12. https://doi.org/10.1056/NEJMp2103798 [retrieved July 21, 2022].

Shackelford E (2020) The Price of American Arrogance, SLATE, May 22. https://slate.com/news-and-politics/2020/05/covid-american-arrogance-south-sudan.html [retrieved May 23, 2020].

Shaw J (2020) Failing the Coronavirus-Testing Test, *Harvard Magazine*, August 3, 2020 https://harvardmagazine.com/2020/08/covid-19-test-for-public-health [retrieved September 1 2020.

Smith KF *et al.* (2014) Global rise in human infectious disease outbreaks, Journal of the Royal Society Interface, 11:20140950. http://dx.doi.org/https://doi.org/10.1098/rsif.2014.0950 [retrieved June 10, 2020]

Strogatz S (2019) How Calculus Was the Unsung Hero in the Fight Against HIV, *Scientific American*, 320:4, 70–73, April 2019.

Verma S, Manhar D, Frankenfeld J (2020) Visualizing droplet dispersal for face shields and masks with exhalation valves, *Physics of Fluids* 32, September 1. https://doi.org/10.1063/5.0022968 [retrieved September 2, 2020].

Viglione G (2020) The True Toll of the Pandemic, *Nature* 585, 22–24, September 3. https://media.nature.com/original/magazine-assets/d41586-020-02497-w/d41586-020-02497-w.pdf [retrieved September 3, 2020].

Vrugt M, Bickman J, Wittowski R (2020) Effects of social distancing and isolation on epidemic spreading via dynamical density functional theory, *Nature Communications* 11, 5576. November 4. https://doi.org/10.1038/s41467-020-19024-0. [retrieved January 3, 2022].

Wallis C (2020) Why Racism, Not Race, Is a Risk Factor for Dying of COVID-19, *Scientific American*, June 12. https://www.scientificamerican.com/article/why-racism-not-race-is-a-risk-factor-for-dying-of-covid-19/. [retrieved June 12, 2020].

Wells HL *et al.* (2021) The evolutionary history of ACE2 usage within the coronavirus subgenus *Sarbecovirus*, *Virus Evolution*, 7:1, January 2021 February 5. https://doi.org/10.1093/ve/veab007 [retrieved February 28, 2021].

Wigner EP (1960) The Unreasonable Effectiveness of Mathematics in the Natural Sciences, *Communications in Pure and Applied Mathematics*, 13:I (February), 1–14. New York: John Wiley & Sons, Inc.

Williams, S (2020) A Brief History of Human Coronaviruses, *The Scientist*, June 2. https://www.the-scientist.com/news-opinion/a-brief-history-of-human-coronaviruses-67600 [Retrieved May 13, 2021].

Wright WF, Mackowiak PA (2020) Why Temperature Screening for Coronavirus Disease 2019 With Noncontact Infrared Thermometers Does Not Work, *Open Forum Infectious Diseases* 8:1, January 2021. https://academic.oup.com/ofid/article/8/1/ofaa603/6032722 [retrieved March 13, 2021].

Yang Q *et al.* (2021) Just 2% of SARS-CoV-2−positive individuals carry 90% of the virus circulating in communities, *PNAS*, May 25. https://www.pnas.org/content/118/21/e2104547118 [retrieved June 12, 2021].

Zohar T, Galit A (2020) Dissecting antibody-mediated protection against SARS-CoV-2, *Nature Reviews Immunology,* 20, 392–394, July 2020.

Zuk M, Spencer HG (2020) Killing the Behavioral Zombie: Genes, Evolution, and Why Behavior Isn't Special, *BioScience*, 70:6, June 2020, 515–520.

Matter and Living Matter

Everything embodied in lifeless matter is subject to entropic processes: it decays. Lifeless matter and how it changes can be completely and consistently described. Living matter is negentropic: things alive tend to maintain, and even increase, their state of order. No living entity and no life change can be fully and consistently described. These characteristics are reflected in the specific manner in which everything changes: in reaction or in anticipation of the future. As a result of the long-term change from a universe devoid of life to one in which evolution occurs, life is by now dominant in the world.

Explicit knowledge about change, as part of science, can explain it after it took place; implicit knowledge, such as experience, can assist in purposeful life, i.e., what in retrospect appears as meaningful change. Reaction suffices for addressing breakdowns of anything embodied in lifeless matter. The dynamics of living matter is undecidable. Change in the living is non-deterministic in nature: it is defined by multicausality, although it can also be a causal. Anticipatory action is the necessary, but not sufficient, condition for the prevention of breakdowns. In contradistinction to prevention, reaction is not sustainable.

Is There Only One Nature?

In the spirit of the Cartesian Revolution, mathematical descriptions of physical phenomena are expected to apply to the all-inclusive Nature (Newton 1686). This was Newton's understanding of reality, a premise which science has promoted since his time. Mathematical descriptions of living phenomena—as rare and abstruse as they are (Chauvet 2004; van Hemmen 2021)—suggest that such processes are quite different from those that pertain to lifeless matter. Physical and chemical interactions, corresponding to the dynamics of lifeless reality, convincingly explain lightning and thunder, rain, earthquakes, and how stone becomes sand. But they fail to indicate

© The Author(s), under exclusive license to Springer Nature Switzerland AG 2023

M. Nadin, *Disrupt Science*, https://doi.org/10.1007/978-3-031-43957-5_4

how a stem cell becomes a unique person, and, in general, even to take note of the uniqueness of individuals, regardless of species or kind. Thermodynamic laws, which describe how order and disorder are connected, help in the understanding of the role of energy in all processes of change. The time scale of change in the world corresponds to the type of interactions and to the energy involved. It takes a long time for a stone to turn into sand, unless extraordinary force—energy at work—is applied to it. It takes a short time to perceive a moving object, or to avoid being hit by one. The living, having its own time scale—actually, a multitude of them—is driven by purpose, pursued in a reality that integrates matter and living matter. As opposed to positions of stars in the universe, the specific unfolding of an organism cannot be predicted. There is no mathematics to describe it, and there is no physics or chemistry, at least until now, to explain it. Moreover, originality must be taken into account: each organism is unique. But there is no mathematics of uniqueness because mathematics is about sameness, generalizing what things have in common, in disregard of what differentiates them.

How does the process of life unfolding take place? The age of the universe might be 13.7 billion years. Furthermore, assuming that the elements we are familiar with—listed in the Table of Elements—were generated by the universe itself (something along the line of the self-organization principle (for details see Kauffman 2014, p. 8), we could infer to a dynamic corresponding to the laws of physics expressed in mathematical terms. If we consider proteins, nitrogenous compounds of long chains of amino acids, which make up the essential part of living entities, we realize that not everything made of matter behaves the same way. Electrons, protons, neutrinos, etc., as well as the making of chemical elements, are relatively well described. The making of what constitutes living matter continues to escape similar descriptions The fact that lifeless matter and living mater change differently requires that we give up Newton's inclusive Nature—reality as one entity—and seek to characterize the physical and the living as different. There is no one inclusive Nature; but there are at least two distinct states: non-living matter and living matter. Accepting this premise will further help in better understanding how they change over time. The richness of their interactions transcends the rather limited scope of physical and chemical interactions.

Our current level of knowledge reveals that proteins are made up of 20 kinds of amino acids strung together in a linear sequence by peptide bonds. Since a "typical biological protein" consists of ca. 300 amino acids, Kauffman arrives at possible protein numbers in the range of 10^{260}. Moreover, in a universe of about 10^{80} particles, the making of various proteins of length 200 would require, even considering the fastest time scale (the Planck scale of 10^{-43} s), an age of the universe in which the assumed life (13.7 billion years) would have to be raised to the 37th power—to make them only once! This explanation, based on the best physics currently at hand, unequivocally proves that matter and living matter are by necessity different. Ultimately, Kauffman (and those accepting his views) ascertains that the physics and chemistry of the universe is of a nature different from biological processes. Long before him, Elsasser, attempting "a theory of organisms" (1987, and in publications preceding his book) made a similar observation, arguing for physics in the quantum mechanics understanding as a foundation for the life sciences.

In Elsasser's work, four principles are to be considered for a scientific foundation of a science of organisms:

1. The living consists of structurally different entities, expressed in the infinite variety of cells, as opposed to the homogeneity of what makes up the atoms of the physical substratum. This is the principle of *ordered heterogeneity*. Individuality, which is not identifiable in physical entities, is the outcome of ordered heterogeneity.
2. The heterogenous entities making up organisms unfold in a very large space of possible futures. The selection of choices is said to be always unique—as each organism is unique. This is the principle of *creative selection*.
3. Each new entity (cell, protein, organism, etc.) resembles earlier patterns without being their copy. (Think of offspring resembling parents, but think also of style, associated with various forms of aesthetic expression or human typology.) This is the principle of *holistic memory*. While individual aspects remain unique, their aggregate is such that it appears to exist as a community of shared traits. The memory is not stored as data, but rather transmitted as meaning.
4. On the biological level, in which almost all processes are autonomous, whatever is shared among entities becomes subject to interpretation. Replication and reproduction are conceived together. The trigger is the meaning, not the data. This is the principle of *operative symbolism*.

They are kept together by the holistic condition of change in the living.

As Bohr observed (1932), the physical substratum of the living and the nonliving is the same. Different is the nature of the organization of the elements making up matter. Short of assuming that somehow the proteins came from outside the universe (a hypothesis named *panspermia*, entertained by distinguished scientists such as Kelvin, Arrhenius, Hoyle, among others), we'd better look for models that can explain not only their presence, but also those aspects of the living that no physics, not even quantum mechanics, can justify—at times at the price of ignoring those aspects. For example, evolution (which Bohr considered): the emergence of design (the successful species) without a designer. (On this topic see Dawkins 1986; Ayala 2007) The understanding mentioned and the authors referenced suggest that the living and the physical are the same, that life is reducible to matter and its laws of motion. They entirely miss the fact that the living co-designs its existence. And that anticipation is the process that underlies the continuous creation of one's self and maintaining one's individuality. Only with this understanding can we finally free ourselves from the epistemological chokehold of reductionism dominating the philosophy of evolution and its extension into explaining the living as nothing but a machine.

A Distinction that Makes a Difference

To assist in understanding the distinction between reaction—in the here and now of everything embodied in matter—and anticipation—preparedness of living matter, from the cell level to whole organisms, for possible futures—let us start with evolution (treated here as an example).

The theory of evolution convincingly explains the dynamics of the living at all its scales and in all its embodiments. For those who are blind to (or agnostic of) anticipatory processes, the process of evolution is determinism in action, in the form of physics and chemistry processes. In this view, it is a bottom-up process: from the matter in which life is embodied to the emergence of life and, furthermore, to how it has continuously diversified over time. If this assumption were to hold, it would mean not only that life can be made from the elements, but that hypotheses regarding how life changes could be experimentally tested. Evolution, however, is not subject to experimental validation because, among other reasons, open-ended processes cannot be replicated. Since they are open-ended, we could never finish measuring them. They are intractable. This is not just a logical assessment; it is also empirically confirmed along the timeline of change in the living and in the ecology within which it unfolds. At best, we can measure intervals. However, they do not stand for the unfolding whole, i.e., for the open-ended process. To reproduce in a laboratory examples of how the living behaves (such as sexuality in its many forms, metabolism, self-repair, predatory behavior) with the purpose of discovering the laws governing behavior means to generalize from samples. To explain the whole in a reductionist manner, i.e., by explaining its parts, is a convenient, but misleading, way to treat complexity. Parts removed from the network of interactions within the whole to which they belong are representative only of themselves.

Electrons, neutrons, protons, and other particles make up the matter into which both living and non-living entities are embodied. To measure their characteristics (size, weight, geometry, velocity, trajectory, etc.) is relevant for the physics and chemistry of change. For the physicist, molecules, atoms, subatomic particles, and the energy involved in their movement are all that there is to account for. However, atoms, molecules, and particles integrated in living matter are only the necessary, but not sufficient, condition of life. Purpose, i.e., functioning in anticipation of, not only in reaction to, changes in the ecology is not the outcome of particle dynamics, but of epiphenomena, which transcend the limitations of matter proper. Therefore, the physics or the chemistry of living matter making up organisms (regardless of the kind—bacteria, plants, insects, vertebrates, etc.), as important as it is, is not sufficient for understanding what life is. Chemists search for the elements. Even though they realized that living matter, which they call "organic," is made only from a subset of all chemical elements (hydrogen, oxygen, nitrogen, carbon), they expect them in the pure form in which they are defined. But in living matter, the elements are present in molecular configurations that are in continuous change. Physio-chemical causality is based on the assumption of sameness (homogeneity). Hence, it is different in nature from the causality characteristic of organisms, which are made of things that are different (heterogeneity).

Determinism, as the underlying premise of physio-chemical causality, confirmed, through measurements in experimental settings, the characteristics of the particles, or of the elements, from which cells, tissues, and organisms are made. But even this make-up is not permanent. What escapes, by necessity, such measurements is the understanding of functions, such as metabolism (acquiring the energy for maintaining life), self-repair, and reproduction. Evidently, these have the nature of purpose, expressed as process. Neither protons, nor electrons, nor particle interactions explain motoric expression, and even less thinking, the continuous remaking of living matter (creativity), and language. Not even the alternate "particles," i.e., biological entities such as cells, genes, and proteins, support the reductionist method. "Love is in the genes" is a headline (or a metaphor) far removed from the complexities of genetic expression. The determinism of heredity is only a part of the larger process of reproduction. There are non-deterministic aspects to it, empirically documented through phylogenetic variations, as well as through ontogenesis. The epigenetic (i.e., phenotype changes no involving DNA) is only a partial description of the unity between the sameness and variation characteristic of reproduction.

Acceptance of the physio-chemical reduction implies the expectation of acquiring knowledge about life exclusively through replicable experiments in physics and chemistry. Not surprisingly, this leads to the prevalent view according to which the living is determinism embodied in the matter of automata. The clock, the hydraulic machine, the electric engine, the computer are actual embodiments of deterministic laws. Theoretical constructs, such as the mathematical entity known as *automata* express the same. Validation of knowledge through experiment is ultimately confirmation of sameness. However, the living is always the expression of uniqueness. The bar-tailed godwit that flies across the Pacific Ocean—from Alaska to New Zealand and back—is evidently different from a well-engineered bullet, a rocket, or a drone traveling the same distance. It is not a machine. Therefore, its performance cannot be described in the mathematics of machines. The path pursued is from the figurative metaphor of the machine—the clock as embodying the functioning of the universe—to the literal interpretation: birds (or anything else alive) are not only "like" machines, they are machines. Since no experiment at the scale of migration is possible, models of all kinds are tried in order to explain the process. Engineers—aware of the laws of physics describing change as moving an object from one position in space to another—probably could design a mechanical migratory bird. They could even program the artifact to return not on the direct route over the water, but following the continental margin of Asia. The birds do, from time to time, change the migration route, for reasons that escape deterministic explanations. "It's only a matter of programming," is the mantra of such attitudes. To build a so-called "intelligent object" capable of making decisions—such as when to begin the journey—based on awareness of predators, of wind direction, of the moon's position is only a matter of time, they would argue. And they might almost be right. But the bar-tailed godwit migratory patterns of behavior are more than the record of the change in coordinates, and the associated cost (energy, data use, adaptive processing, etc.). Anticipatory processes, extremely subtle in nature, are at work. Those who study bird migration became aware of them through empirical data: the record of the flight itself (no flight is the same), one year after another, under variable ecological conditions. The birds prepare themselves for the journey to an extent not matched by any predictive model.

How and why from an open-ended set of possible choices corresponding to ecological challenges (environmental aspects being only a subset) only some are pursued cannot be explained from a particle physics viewpoint. Neither can it be inferred from a chemistry viewpoint—usually known as the genetics perspective (the cavalier expression, "It's all in the genes"). There is purpose—preservation of life—and there are unpredictable choices—easy to explain in retrospect, but impossible to fully describe since the phase space (i.e., the variables involved) changes. Some adopt the "harnessing of stochasticity" as an explanation (Noble and Noble 2018); others opt for "self-fabrication" (Hofmeyr 2017, pp. 1–15). No one acknowledges the fact that living matter is matter endowed with anticipatory functions. There is no predictive model (as in Rosen's 1985 definition of anticipation) to guide the birds; there are only rich interactions and a competitive motivation: survival (from the cell level to the individual organism and to the large-scale migratory "communities" of birds, butterflies, fish, etc.) in search of a supportive environment. The genetic code might indeed link the chemistry of the DNA sequence to the protein structure, but there is no individual protein that provides a specific function, as there is no particular cell to which a characteristic of life can be traced. Moreover, there are no identical cells. As already suggested, heterogeneity defines living matter, in contradistinction to the homogeneity of lifeless matter. The making and remaking of the cell itself is affected. There is robustness: living in a world of noise and perturbations of all kinds, impossible to ignore without a certain resistance to change. More important is adaptively, which enhances survival and reproductive performance.

Suppose an automobile could respond to a rough road by thickening its tires, stiffening its springs, and respecifying its gear rations (Sterling and Laughlin 2015). This is quite an engineering challenge—but not in the realm of the impossible. It turns out that adaptive processes are integral parts of the process called "life." They are defined as "Each fresh experience helps prepare for future need. One more quote: "Where skin sustains rough wear, it thickens to callous; where muscle and bone sustain mechanical tension, they strengthen."

The subject of how life can (or cannot) be emulated through artificial means (physics, chemistry, computation, etc.) evinces various epistemological aspects. The unrelated subject of the impossibility of certain mathematical operations suggests that claims such as "It is only a matter of programming" or "Only a matter of performance" are unfounded. To understand why, let us draw attention to the science of the impossible. The problem of doubling a cube using a ruler and compass (Kak and Ulam 1968) is a classic example. The famous square-the-circle problem, and the problem of trisecting an angle with a straight ledge and a compass come to mind. Impossibility is not of the nature of a limitation (such as "It cannot be done now, but it can be done in the future") regarding means and methods, or current abilities. There is no such thing as a better compass or ruler that will eventually make it possible to solve the impossible problem of doubling a cube; or, for that matter, the impossibility of finding a mechanical process for deciding whether a theorem is true or not—the Turing proof (1937 *op. cit.*) in answer to Hilbert's *Entscheidungsproblem* (1928 *op. cit.*). Not even the most sophisticated computation (quantum or another) could.

Nothing should, in principle, preclude the making of an artificial migrating bird (such as the bar-tailed godwit or some other migrating bird) that behaves like a real bird. The impossibility is in respect to awareness. It concerns the anticipatory grounding of the adaptive behavior of the living. A bullet, whose "behavior" can be fully explained through physics, does not "know" anything. The makers know its physics and the chemistry of combustion. Engineered by its designers and makers, it moves according to the laws of physics, but not as a result of understanding what such laws describe. An "intelligent" weapon is, in this respect, as dumb as a bullet, as dumb as any atom or molecule of non-living matter! Its movement is programmed, predefined. A bird—lightweight body, small brain (but all it needs for survival), no notion of physics or of metabolism—in a flock, of many thousands of other birds of a feather—acts in awareness of its target. Its activity generates possibilities, and it makes choices—it adapts continuously, throughout its life. Autonomic functions related to altitude—how high to fly—and oxygen intake, to stored "energy," to sleep patterns, etc., are part of the anticipatory behavior. They are expressed as necessary. Speed, influenced by flock formation, by humidity, by wind, etc., is not the outcome of atoms or molecules, not even of the chemistry of cells or proteins, but of choices. There is no way to ever describe their behavior fully and consistently. This proof of impossibility is by necessity different from mathematical proofs for the simple reason that it pertains not to measured quantities, but to the meaning of living in anticipation. Purpose invites representations that transcend measurement. Harvesting energy while flying (for fish, while swimming) keeps all cells alive, but it is different in nature from the interaction of molecules in a given cell. There are many components at work. It is impossible to single out a specific protein that would explain the choices made in preparing for or performing the migration.

Data from experiments (closed systems of measurement) in emulating capabilities of living entities inform us only that mimicking nature requires a lot of power. In the living, the relation between energy use and outcome is a measure of viability. If a bird dies of exhaustion during or at the end of a difficult flight, there is no justification for performing such a flight; the minimum necessary for survival is not reached. Migration is intended for preserving life: seeking sources of nourishment, an environment propitious to reproduction, and beneficial to offspring; avoiding harsh conditions, predators, and other dangers, finding sources of nourishment. Migration is not for suicide, even though great numbers of migratory organisms (from plankton, to butterflies, to birds, to fish, to larger animals such as bison) end up dying. The fittest survive. An artificial device does not have such an existential motivation or limitation; it can go on and on, as long as energy (let's say as provided through batteries) is available, or until the goal is achieved, at any price. Or even if the goal is not achieved. No meaning is attached to the machine's performance. No *Why?* question is formulated, never mind answered. It is data processing performance for the sake of performance.

Minimum Energy Principle (MEP)

Entities embodied in lifeless matter succeed more and more in emulating activities associated with living matter organisms. This is the domain of the artificial—of AI, in particular, and more recently of Machine Learning, as a particular form of AI. There is even discussion of the Singularity (Vinge 1993; Kurzweil 2005), the presumed state in which the artificial outperforms the living. Large Language Model (LLM) developments integrate the attempt to use language in triggering performance, such as text-to-image generation. Therefore, it is justified to see how the distinction between matter and living matter can help in understanding what this all means in the perspective of sustainability.

In living matter across scales—from cells to organisms, to species—activities for the preservation of life cannot consume more energy than what the metabolism affords. This is the Minimum Energy Principle (MEP). The MEP does not hold for the human being. From the entire realm of the living, only the species *homo sapiens—thinking* being the ultimate identifier—consumes more energy in its self-preservation than what metabolism alone contributes. What became known as culture—i.e., the tamed nature within which human activity takes place—is the outcome of progressively increasing energy use, and thus the continuous remaking of oneself. No other form of life on Earth has this behavioral pattern. The human being redefined itself in respect to physical abilities—augmented by tools and machines—and to thinking—cognitive abilities, associated with a larger brain. The augmented capabilities are energy dependent.

Intelligence-driven evolution of humanity reached the level at which resource consumption became an issue of its survival. Especially since endowing non-living matter with capabilities that can be associated with natural intelligence—the cutting edge of science today—is an energy hungry endeavor. Outsourcing natural functions to artifacts starts with the use of tools. They are in anticipation of their use, i.e., a way to multiply future possibilities. Tools date back to the first identifiable human forms of activity. Their development is in full swing in our days, when intelligence itself—in the form of processes emulating intelligent behavior—is expected from machines. From their hard condition—matter made into artifacts—to their soft condition—programs to activate various machines—they represent knowledge put into action. The immediate result of this pattern—from hardware to software—is the disconnect between means of existence—ecological sources of energy—and progressively reduced natural expression, i.e., declining anticipatory action. With the focus on understanding how change takes place—including their own changes over time—humans effectively substituted their innate anticipatory abilities with artificially constructed models of the future inspired by the past. The minimum energy threshold characteristic of survival was effectively overwritten by the optimistic principle of Everything Is Possible (EIS)—at the expense of the ecological system. The human species thus lives at the expense of the rest of the ecology. And it is the only species *devolving* into overpopulation. No other living being could afford activities in which the outcome is less than the effort. The flipping of the Upside/Downside

Ratio, the negative yield of human activities, is characteristic of a new stage in the life of societies. It documents the assertion that human beings live more and more at the expense of the future.

Although the MEP does not hold for humans, it is justified to define intelligence in the perspective of the MEP, in conjunction with its particular expression as Data Minimum Principle (DMP). Taking in reality through the senses consumes energy. Given the behavior conditioned by the MEP, it follows that the living, through perception, measures reality to the minimum possible. This means that data pertinent to life interactions cannot be less than the minimum it takes for the preservation of life.

Given the obsession with higher performance in imitating life, it is justified to evaluate the performance of artificial means, in order to assess their viability. Based on this understanding, an evaluation principle can be formulated:

> Artificial entities could justifiably claim intelligence if, in executing a task, they would use as much energy or less, and as much data or less, than a living entity performing the same task.

Energy use and the necessary data are a good metric for comparing natural and artificial performance: how much energy and how much data used in a well-defined activity. Some machine learning applications (chess, or *Go* playing, cf. Dickson 2020, Labbe 2021) can use as much electricity as a small town over the duration of the performance.

The energy consumed by the miniscule bar-tailed godwit during migration is acquired through metabolism. The data processed is acquired through "measuring," i.e., sensing the environment. (Data acquisition also involves energy expenditure.) The take-off, ascent, gliding, bonding, soaring, and continuous forward flight through flapping wings are energetically different. Some actions more "expensive" than others. Altitude is yet another factor: less oxygen, for example, addressed by a different motoric for consuming less of it (Alexander 1998, 2002). For the sake of the discussion, it suffices to mention that a flight power of 4.3 watts is actually used (Wikelski et al. 2003). They are by some orders of magnitude less than what would be needed to guide an artificial bird of similar size and weight. Winning or losing, for that matter a game of chess or of *Go* would not require a power plant if performed by a human being. Getting a robot to dance or to collect samples from Mars, or getting a submersible to collect samples from the ocean depths requires lots of data and lots of power. The data processed by humans in playing the games is in the order of kilobytes. This is way smaller than the huge data amounts (order of 10^{120}) guiding the artificial playing machine. The human brain operates on 20–30 watts—less than an LED source. Even the plankton inhabiting the oceans is much more intelligent than what the most sophisticated machinery, based on the deterministic science dominating civilization, can achieve. This pertains to the energy used and the data collected, and is expressed in the adaptive performance. Indeed, in anticipation of adverse conditions, swarms of migrating birds or of fish change, respectively, flight altitude or swimming depth. The plankton navigates the oceans under terrible conditions, finding survival niches for which we do not have names.

What makes the difference is the anticipatory component of the activity. Migratory behavior exhibits adaptive characteristics associated with anticipatory processes. The timeline (migration start) and the trajectory are fine-tuned to possible storms, as though the migrating birds, or migrating fish or animals, are prescient of what might affect their respective journeys (on the predictive performance of Veeries, see McGlashon 2019, one reference from among many). Artificial entities embodied in non-living matter are "fired up" with energy from the outside and with data from measurements of similar activities. To be precise: the living senses the environment. Sensing involves energy use: to measure is an activity engaging the entire living being. When the available energy acquired through metabolism or stored is too low, the living ceases to take in "reality." Even under the most generous assumptions of scientific and technological progress, performance comparable to that of living entities in a continuous state of anticipation is not even on the agenda of current science and technology. Such performance is as impossible as doubling a cube using a compass and ruler or squaring the circle.

The living is in a state of anticipation from the start of life until its end. It is a continuous state, with various forms of expression and variable intensity. It engages the entirety of the organism, at all its levels, and in this sense, it depends on metabolism and on perceptual activity. Avoiding danger, as opposed to reacting to it, is, from the perspective of the data involved and the energy consumed, quite different. Outperforming others in the context of the competitive nature of life and in securing evolutionary advantage takes place also on account of energy use and data processing appropriate to the circumstances. Inventions qualify as examples of activities driven by anticipation. Anticipatory processes effectively extend awareness of cause-and-effect into the richer sense of causality that integrates past, present, and possible future. "Sensing" the future, i.e., virtually living it before it becomes real means awareness of consequences. From an energy and data perspective, this is different from the practice of predicting it on account of measuring reality and inferring from a current state to a future state. It succeeds (or fails) if the energy expense undermines life. That is, if the data goes beyond what a specific living entity can afford acquiring, there is no future state to account for. The minimum energy for the human being is not predicated by the threshold of life, i.e., what is needed to maintain life, but rather by gaining independence from environmental limitations. The human being is able to extract from the environment more energy than what is needed to survive. Humans also to acquire more data than what would be needed to maintain life.

Life, the Living, and Complexity

A comprehensive view of the world should integrate the reactive dynamics of inert matter, described in the laws of physics, together with the dynamics of anticipatory life expression characteristic of living matter at all scales. In this broader view, life itself is the outcome of change. To repeat: the living—particular material instantiations of life (animals, plants, bacteria, etc.)—in contradistinction to the non-living, is defined through purpose rather than only through its material characteristics (mass, composition, structure). The provocative statement of those who study life as an aggregate of relations, and not only as a characteristic of the material, says it all: "Throw away the matter and keep the underlying organization" (Rosen 1991, echoing Bohr 1933). Indeed, regardless of the matter in which the living is embodied in unique forms, organisms are similar in their anticipatory expression, a necessary condition for survival. Sexuality, i.e., the underling interaction of reproduction, is essentially an anticipatory process. Migration (of birds, fish, insects, animals, human beings) suggests that this is a pattern of behavior that involves each particular species in its entirety. It also reflects the richness of interactions with the environment.

The properties of all physical objects (i.e., those not endowed with life) can be associated with the elements they are made of. Interactions with the environment, including those with living entities, are also important. In the dynamics of the non-living, i.e., how things that are not alive change over time, the interactions among the elements making up a physical object are consistent. One does not negate the other. The properties of living entities, regardless of their scale, can be associated with the elements they are made of—a small subset of all elements. Ninety-six percent of all molecules of living matter are made up of carbon, hydrogen, oxygen, nitrogen, phosphorus, and sulfur, and of essential minerals (iron, copper, zinc, and others). However, their being alive is not the exclusive outcome of the properties of these particular elements. Those chemical elements present in living matter are the same as those in non-living physical objects. But the cells are endlessly different, and they are (with a few exceptions) continuously remade so as to remain different. They make up proteins, carbohydrates, lipids, and nucleic acids involved in interactions corresponding to the timeline of life. However, to recall some of the examples given, the navigation abilities are not reducible to the make-up of insects, birds, animals of the ocean, etc. Neither is bioluminescence, essential for living in the darkness of deep oceans, only a matter of specific chemistry. Navigation, bioluminescence, specific physiology, and diverse forms of motoric expression correspond to meaningful actions related to nourishment, sexuality, and self-defense. Devotees of reductionist-determinism will check bees for magnetic elements, or would blind their subjects (birds, insects, fish) to discover the role of vision in their peculiar journeys. Pigeons fitted with contact lenses in order to research the role that sea landmarks play are turned into agents for testing the optics, but not for learning the meaning of their flight. Such "blinded" scientists will measure without realizing that they quantify not navigation abilities, but rather how the disturbance which they put their subjects through affect their behavior. Empirical evidence and experiment are different. *Empirical* means1, "the historic record of holistic expression. It is not the expression of a few variables. *Experiment* means the artificial closed system within which some parameters are changed with the purpose of assessing their causal impact.

Science is about understanding change as it takes place in the broad context of life. Since change in the living is never complete, and always contradictory, it cannot be subject to experiment. In an experiment, a stone remains the same; a cell is in continuous change. Indeed, "about 330 billion cells are replaced daily. In 8 to 100 days, 30 trillion will have replenished—the equivalent of a new you" (Fischetti and Christiansen 2021; Sender and Milo 2021). This remaking of living cells—which has no equivalent in non-living matter—is the expression of creativity: the making and continuous remaking of something that previously did not exist. The purpose: to maintain life.

The distinction between what is alive and what is not aroused controversy over time (Wolfe 2011; for positions along the same line of thought, see the pertinent works of John Bernal, Erwin Schrödinger, Eugene Wigner, John Avery, Robert Rosen, Stuart Kauffman, Giuseppe Longo, Maël Montévil). The reductionist-deterministic view of the world cavalierly dismissed the distinction. However, never proved it that it can explain the living as effectively as it describes the non-living. The extreme endurance flights of all kinds of birds, and of monarch butterflies, and the sophisticated navigation choices of ocean-dwelling animals correspond to a knowledge domain different from that of physics or of chemistry, genetics included. Still, in the spirit of the Cartesian Revolution and of the doctrine of materialism, physics ascertains a false axiom. Since everything is matter, the axiom states that physics can deal with emotions, choice, life's creative nature, etc. with the same ease as if deals with, for instance, earthquakes, the movement of stars, and the making and functioning of machines. Even learning is subsumed in this ascertainment. The fact that machine learning (incapable of doing more than inferring from the past) is fundamentally different from living learning—integrating the past and possible future—is ignored. Subject to gravity, a stone falls the same way. A person can train not to fall, or to control his or her fall (see Figs. 3 and 4 in "Numbers and Meaning"). Some animals move in ways that seem to defy gravity. Life at regular atmospheric pressure is, of course, different from life at pressure 800 times higher. Adaptivity, not only to pressure, but also to temperature, salinity, and toxicity was never explained by researchers guided by reductionist-determinism. In recent years, more convincing arguments concerning how the mind can affect matter (Ellis 2005; Noble 2008, 2012; Noble and Noble 2018) were submitted to the scientific community. Those who reject the distinction life/non-life gave scant attention to such arguments. They are, at most, tolerated (in the oceans of publications), but not pursued as alternatives.

CoVID, the crisis of vision triggered by the dominant science, is not a metaphor conveniently inspired by the name of a pandemic. It is a condition, the consequence of a limited and limiting perspective of reality. Breakdowns prior to Covid-19 and the aggregate of breakdowns experienced during the pandemic are associated not with bridges or nuclear reactions, but with knowledge, with health and behavior, politics and the economy, with the future of humankind. Therefore, it justifies reconsideration of the perspective from which humankind approaches its own changes, vulnerabilities, in particular.

A Conceptual Scaffold

The following definitions are part of the conceptual scaffolding for supporting the quest for integrating the reactive and anticipatory perspectives in a holistic view of the world.

Life Postulate (LP): Life is G-complex

In order to make available an actionable definition, we adopt the distinction originally made by Gödel (1931) in respect to mathematical descriptions in his two incompleteness theorems.

Definition 1 A description of an object or process is said to be decidable if it is complete and consistent.

Definition 2 A description of an object or process is said to be undecidable if it cannot at the same time be complete and consistent.

Axiom of Non-Living Matter Change in non-living matter (physio-chemical reality) is decidable. If this were not the case, we would not have any laws of physics. The phase-state is constant.

Corollary *Physics and chemistry descriptions reflecting the decidable nature of physical phenomena are decidable.*

Definition 3 The dynamics of living matter is the aggregate of physical interaction and anticipation expression.

Based on the above definitions, and on the descriptions of the dynamics of the living, the following axiom will be advanced:

Axiom of Living Matter: A material entity qualifies as living if the description of its dynamics is undecidable.

Let us first acknowledge the empirical background:

- To describe a physical entity is to consider its changes over time. These are described in the laws of physics. They are goal-free.
- To describe a living entity is to capture its uniqueness expressed in its change as it adapts—in an anticipatory manner—to the world in which maintaining life is the purpose.

From these considerations it follows that:

> The unfolding of life, between conception and death, cannot be fully and consistently described.
> The number of variables describing life are part of an open-ended set (the phase-state is variable).

Before returning to the CoVID aggregate breakdown, let us understand the consequences, i.e., the corollaries, of the Life Postulate.

Corollary 1 *The undecidable does not holistically map into any finite representation.*

Corollary 2 *Partial mappings return knowledge of the physics and chemistry of change in biological matter.*

Corollary 3 *Since life is not algorithmic, the dynamics of living processes cannot be reduced to the dynamics of Turing machine-based computation.*

It remains open whether the dynamics of the living might eventually be computable in biological matter-based computations (following Feynman's suggestion to compute in the medium itself). Such computations, if at all possible, would have to be non-deterministic.

Rosen's theorem regarding the non-simulable model of a living system states that such a model is effective, but not computable (Rosen 1991). The path to this conclusion (the same as the one in Corollary 3) is different from that of ascertaining the G-complexity as a threshold for the living.

"Life Wilst to Live"

"Reverence for life" (Schweitzer 2020) affirms Albert Schweitzer's realization that life is the expression of the "wilst to live." Evidently, fighting, preying upon each other, the various forms of competition are "wilst" in action. They reflect the specific expression of intelligence of each and every living organism. Based on this premise, it becomes possible to identify the interrogations through which a pertinent understanding of living matter can be achieved.

A goal such as to make the matter embodied in the artificial realize the *Why?* of bird (Gould and Grant Gould 2012; Barrie 2020) and of ocean animal migrations transcends the causality of the reductionist-deterministic perspective. Birds are not meteorological services; their lives depend on weather, and they are attuned to it. But there are so many other factors at work. The fact that sometimes the processes underlying the unfolding of their lives goes wrong is yet another proof of their non-deterministic nature, but not an argument for replacing them with deterministic processes. The cause of life is not external to life. To prepare, to initiate, and to carry out migrations and, when necessary, to adopt other ways for self-preservation, presupposes a different premise, Understanding the *Why?* of sexuality, at all levels of life, as part of reproduction and survival, is not optional. Or, to refer to plants, the *Why?* of climbing vines: the manner in which plants "hold fast" (through adhesive disks) evolved over time. No doubt, the mechanics of adhesion can be achieved with materials programmed to adhere to some surfaces. But the anticipatory action of climbing is more than sticking to a pole or a wall. Its meaning transcends the *How?*

of the action in order to gain conditions for the *wilst to live* expressed as survival in a competitive environment. Yet another example: the great variety of creatures (some bioluminescent, Widder 2021) living and multiplying at the dark depths of the ocean floor, or in proximity to hydrothermal vents (Scales 2021). The tube worm has neither mouth nor gut, surviving on self-oxidizing bacteria—yet another expression of the *wilst to live.*

A rich variety of anticipatory actions for survival and reproduction under extreme conditions multiplies evolutionary chances. Anticipatory behavior explains why some species of birds, or of climbing plants, or of ocean animals have fared better than others. It explains why anticipatory defense mechanisms, involving physics or chemistry, but not reducible to them, can achieve that. Try to fathom only the anticipatory action of vent-dwelling animals sending their larvae to newly emerging hydrothermal funnels as their own vent dies down. At over 26,000 feet deep in the ocean, the pressure is such (800 times greater than the pressure at the ocean surface) that matter is literally crushed. The extremely low temperatures of such environments prompt even more questions: all beyond the *How?* and all in the *Why?* domain of *"What does it mean?"* Snailfish, with bendable skeletons made of cartilage, survive where physics would expect the matter from which they are made to crack and crumble. Of course, this is no environment for experiments, considered within reductionist-determinism as the final test of *all* science. Any measuring device necessary for carrying out such an experiment would disintegrate.

In all these examples, there are, evidently, cause-and-effect instances. But neither Galileo—first in seeing mechanical principles at work in life—nor Newton—who unveiled the laws of the universe—not even Descartes, would rush to conclude their assumptions are correct. What defines life under the extreme conditions at deepest ocean level is mostly anticipatory action of the *wilst to live*—for which the mechanistic view of the world had no quarter. The adaptive characteristics of the living, in its entire variety of expression, are not in reaction to, but in anticipation of the ever-changing parameters of a challenging environment in which survival is achieved through living for life. This observation is by no means a negation of the relevance of reactive science, but rather an attempt to highlight that change in living matter exemplifies that reaction and anticipation are intertwined, even more, that anticipation cannot be reduced to reaction. To reduce life dynamics to reaction is to ignore its characteristic quality of being alive. Just to illustrate the thought: The condition of gravity, independent of awareness of it or even knowledge about it, is different from that of urges. The living has no control over gravity: awareness of it is at work in many instances. However, when it comes to urges, controlling them confers evolutionary advantages.

The understanding that anticipation underlies evolution is the premise for arguing in favor of disrupting reactive science. The fragmented understanding of life it facilitates guides an unsustainable course for humanity. The aim is to provide a possible sustainable alternative corresponding to awareness of consequences. In all the above-given examples, preadaptation—in anticipation of change, not in reaction to it—explains long timeline life processes. Anticipation action is implicit: Within the entire spectrum of living matter embodied in organisms, it is expressed as autonomic functions. Sexual drive, motoric expression (movement), and nourishment transcend cause-and-effect phenomena. The bar-tailed godwit, the bogong moth, the humpback whale, and plankton, to name a few long-distance migrants, do not live in a stimulus–response behavioral cycle, but rather in open ended anticipatory action. Neither do the Japanese eels that swim over 1500 miles to reach the optimal spawning environment. Their eggs hatch into leaf-shaped larvae that ride the North Equatorial current towards the rivers of Asia, where they grow into eels. The survival instinct—a particular form of anticipatory expression—prompts their choices. The behavioral ecology of climbing plants (Gianoli 2015) is of a similar dynamic. Their living matter makes up a whole—the organism—different from non-living matter, which is devoid of any awareness of self and of others. When Leibniz (1710) wrote that the present is filled with future, he might have intuited the kinds of actions (e.g., sexual expression, migration, active search for alternatives) associated with life. Although not disinclined to entertain the machine model, Leibniz observed that the living scales down to "living machines," i.e., to wholes, not to the elements. Harvey, who echoed Leibniz's view ("the subtlety of its artifice goes to infinity") described the heart as a hydraulic pump. But he knew better than the mechanicists: the parts expand and contract in response to heat and cold, but also to imagination. There is "harmony and rhythm" and there are qualitative changes, "in hardness, softness, colour" (Harvey 1651; 1847, p. 417).

Being alive and trying to stay alive, moreover, to multiply, is a function of higher-level matter organization in which matter itself is continuously refreshed. A stone remains passively the same. Even the melting polar ice cap (so much in the news of our time) has a permanence to it. For centuries it looked the same and changed according to relatively stable patterns. In a living organism, each cell is in perpetual renewal. Holistic anticipatory processes, engaging the entire being, and awareness of the relation between goals and the means for attaining them, are intertwined. Reaction is local by necessity. The dead bark of a tree peels off without affecting the life of the whole. Anticipatory action is integrative. This holds true for societies of honeybees, ants, and zooplankton: They evolved through rich interactions that gave them the evolutionary advantage that leads to their survival—*wilst to live* in its concreteness. Within human society, individual behavior, but also various aggregate manifestations—political, economic (exchange), artistic (emotional)—conjure interaction. These various forms of expression result in beneficial behaviors contributing to the continuous remaking of the human being, or in detrimental, self-destructive behaviors. Human interaction is "embodied" in culture—i.e., the living repository of everything, from tools to theories, pertinent to human activity. Treating the interaction characteristic of social life with the means used to describe physical and chemical interaction brushes aside what distinguishes matter and living matter. The Mariane snailfish lives in a toxic environment. The toxicity is of such high levels that chemistry, in disregard of what makes living matter what it is, cannot explain how life is possible, and even less how come the snailfish successfully reproduces. Empirical evidence reveals that the toxins are exactly what protects the snails from predators. Matter, the abstraction behind which we have the concreteness of compounds and of elements, has no self-reflective capability. In contradistinction to lifeless matter, living matter appears to the observer as endowed with awareness, moreover, capable of influencing even its material make-up. The two different conditions of mater and living matter not reducible one to another.

Civilizations are not Constellations

The universe—in whose dynamics Newton was interested—exemplifies the largest scale dynamics of action-reaction. Its exploration is based on the understanding that matter, in which planets and stars are embodied, change in ways humans can explain and describe. The movement of planets is a convincing example. Observations of the changing positions in the vast space of the universe, in conjunction with hypotheses regarding the forces at work made precise calculations possible. This is the largest machine we are aware of, and knowledge about it forms a large part of what is known as physics.

If all that happens in the world—at micro-level or at the scale of the universe—would be of the nature of physical phenomena, humankind would not experience breakdowns. The dynamics of lifeless matter, i.e., how it changes in time, does not exclude extreme events, such as earthquakes, volcanic eruptions, extreme solar activity. But even these, while mostly unavoidable, can be accounted for. They are observed, measured, described in the language of mathematics, and expressed as physical laws. Changes in living matter, including how and why organisms move in space, are, however, not only the outcome of physical forces at work. The principle of preservation of life goes one step further than Schweitzer's will to live. It states that life is neither created nor destroyed. In some ways the preservation of life is similar to the preservation of energy.

Deterministic science, convincingly providing quantitative descriptions of the vast machinery of the universe, makes possible predictions regarding its change. The universe we are still in awe of (scale, precision) is way simpler than any known form of life. The non-living is subject to change only through cause-and-effect relations. The non-deterministic reality of interactions involving the living (from the cell level to that of communities and civilizations) explains the open-ended variety of expressions of the dynamics of life preservation. *Multicausality*, variable timeframes, and actions guided by possible futures transcend the reductionist-deterministic framework.

Awareness of these distinctions suggests an explanation of the breakdown we are examining. The crisis of vision—CoVID—within which a pandemic—Covid-19—was triggered is the unavoidable outcome of the attempt to reduce life to the non-living. In particular, to postulate the machine model of the living and to practice it made the crisis unavoidable. In this sense, CoVID might be referenced as far back in time as to the Cartesian Revolution. But there is a difference. Explaining the living in terms of machine functioning—e.g., the model of the clock, the heart as a pump, the human being as a hydraulic or pneumatic machine—and actually conditioning the living to behave like a machine—like computers in particular—have different consequences. Until recently, the consequences of this radical reductionism had little (if at all) impact on the future of life in Earth. Therefore, the CoVID reference to the Cartesian Revolution itself while not wrong is a bit unfocused. Closer to our time the practical consequences of the Cartesian Revolution pertain to the global scale, in particular the consequences of expecting the cost of human endeavors to be carried by others. More to the point: to be charged to the future. In the global economy of dramatic disparities, even the demagoguery of ecological awareness is charged to those who cannot afford clean water, waste disposal, sewage systems, hygiene, medical care, and education.

Before globality, as skewed as it is in current practice, became the unavoidable scale of human activity, it sufficed to adapt to change. Or only to react to it. After all, greenhouse gas emissions, and thus temperature increase (with all its consequences) are on record since industrialization gained speed (ca. 1880). Previous pandemics—actually epidemics, not affecting the entire world—decimated large segments of the population. Half the population of Europe died during the Black Plague. But they did not go global, and they were not triggered by self-inflicted activities. In the last 40-plus years, more precisely with the outbreak of the HIV/AIDS global epidemic, Ebola, SARS, MERS, but also with the conflicts that fused into the war on terror (Afghanistan invasion, the Iraq war, the Syria and Yemen civil wars, etc.), the situation changed. The dot-com bubble, the great recession, etc.—against the background of the integrated global economy—illustrated the Upside (benefits)/ Downside (losses) Ratio flip. As desirable as the green energy alternative is, it depends on fossil fuels for processing raw materials, construction, delivery, operation, and maintenance. The benefits fall below the heavy investment and the expected reduction of greenhouse gases. The state of the scaled-up world of moral disparities, in the embrace of determinism, became more and more unstable. In this context, the cost of reducing life to the non-living is comparable to, if not greater than, the benefits of conceiving and deploying technology. The obsession with progress-driven human activity is based on reductionist-deterministic premises: "We need to produce more energy"—instead of using it more responsibly. This obsession exacts high costs in the long term. As a consequence, nothing less than even the possibility of life on the planet Earth is now at stake.

Paradoxically, the anthropomorphizing of everything—granting human identity to phenomena—was replaced by "machinism," i.e., attributing machine identity to phenomena. The initial metaphor—God is like us, for example, or the brain is like a computer, or reality is the outcome of computation—is eventually taken literally. Instead of "like"—all the mechanical contraptions of Galileo's time, and Descartes's exposure to machines (especially clocks) that imitated the human or some animal—it was asserted that all there is alive *is* a machine. The consequence of anthropomorphizing is the abdication of verifiable knowledge—you only see yourself in the mirror. The cost of mechanism: it taxes all resources, including those essential to maintaining life. In fact, it is a negation of the creative nature of the living.

To govern over machines is easier than to discharge political duties within a society consisting of endlessly different individuals. It is, quite seductive to generalize from the knowledge of how lifeless matter changes to the knowledge of how living matter, at all scales—from the mono-cell to the individual and to life of societies—changes. Simplifications come at a cost: explaining away has consequences. Yet, as expensive and fruitless as it is, this simplification has been tried again and again. And ultimately adopted as the premise for science, and as the requirement for social acceptance.

The current crisis, experienced in the sequence of breakdowns for the last 40-plus years, is not reducible to a pandemic—an accident that could have been prevented. Rather, it made obvious the consequences of gambling the future of life on Earth. None of the heralded efforts of governments to get the situation under some control proved to be an investment in the future. Not even the much-celebrated new vaccination methods and boosters, into the making of which so much was invested. The breakdowns—the HIV/AIDS global epidemic, but also the war on terror, the "Great Recession," etc.—did not prompt questions about what made them possible. Nobody asked what should have been done to prevent them. Reactions are nothing more than the endless patching of a leaky ship. One headline (from among many similar) makes the point: "Governments World-Wide Gorge on Record Debt" (*Wall Street Journal* online, July 12, 2021). In the USA, the national debt increases steadily as the "chickens" of bygone years "come home to roost (i.e., the huge student loan debt), and new emergencies arise (e.g., the war in Ukraine). Reactions—to failed educational endeavors, to pursuing misleading strategies—are costly. And in the absence of an anticipatory perspective, what seems to be the only way out of rapidly succeeding crises is what fires the next crisis. The debt against the future, to get the money needed to pay for reactions, is justified by those controlling the public purse as the only way to have a future.

When the current generation lives at the expense of the future, the sense of human continuity is compromised (Garrett et al. 2020). Under the banner of progress, humankind gave up sustainability, that is, its responsibility for the future—its own and of the living environment. The reactive path of more and more technology built on the same misleading scientific premise simultaneously undermines the future. The absence of this awareness undermines science. Reducing the role of the human in order to maintain the stability of societies in favor of control through technology is celebrated instead of being questioned.

This chapter began with simple examples of animal and human activity, selected with the purpose of illustrating what distinguishes matter from living mater. The predictable functioning of constellations—the machine of the universe—and the unpredictable dynamics of civilizations illustrate the difference between deterministic and non-deterministic processes. Documenting the difference of perspective of reaction versus anticipation, even through anecdotal evidence, should suffice in building the case for making us aware of the practical implications of the distinction. The starting point on the journey to this realization is the mapping of the dynamics of reality under the magnifying glass of an extreme situation—i.e., the Covid-19 pandemic. It was brought about by the encompassing crisis of vision that resulted from a science and technology built upon the assumption that there is no difference between matter and living matter.

Questioning a Skewed Premise

Physics and chemistry brought humankind to today's advanced stage—Post-Industrial Society—but at a huge cost. Even the emblematic navigation capabilities of the living (illustrated by the examples given above) and thus their existence are, at this moment, at risk. Birds fly among greater numbers of "flying" machines of all kinds in the skies; fish and other ocean creatures are subjected to changes in their living habitats that have already contributed to extinction of many species. The human being's increased mobility afforded new possibilities, but also augmented the risk to health, life, and the environment. The survival of species, in their amazing variety, moreover the preservation of life on Earth, cannot be taken for granted on the glorious path of progress-at-any-price.

An increasing number of scientists, disabused of their hubris for physics, are arriving at the realization that effective answers to the many challenges of our time cannot be derived from the Cartesian perspective. Consequently, the observation that the survival of the living is becoming problematic is, finally, at least discussed and no longer discarded as an apocalyptic aberration. Hotter summers, more destructive storms, higher ocean levels, more frequent fires, and all kinds of epidemics are on record. Nuclear wars, biological warfare (some connect the pandemic to this possibility), ecological disasters, and the effective impossibility of cybersecurity are concrete threats. Against this background, the empirically grounded realization that life is not reducible to the non-living should no longer be dismissed. Obviously, the various kinds of so-called viruses affecting the functioning of machines, and the SARS-CoV-2 virus (one from among many coronaviruses) are fundamentally different. Appropriate knowledge of what distinguishes them—a matter of perspective regarding the living—can guide preventive action. Unfortunately, this is the time of pseudo-knowledge on behalf of a blind faith in skewed science.

Undoubtedly, measurements describing reality are a source of useful data. But relying on quantitative assessments can be misleading. Numbers are important, but they will not compensate for the missing focus on the significance of quantitative and qualitative aspects. Indeed, meaning is the one thing that the rich variety of forms in which life is expressed have in common. Navigation integrates the motoric and the cognitive. Navigation, for instance, in all its many forms confirms the view of motoric expression as discovery—not reaction to the world (Bernstein 1967). Whoever sees it from the perspective of stimulus–response behavior (i.e., the reflex model) will not find the answer to the *Why?* question, which is *preservation of life*. Birds cannot afford to measure everything before they move from one location to another; it would cost too much energy. Rather, they integrate into their collective behavior what they learn as they engage in life-preserving activities. The living earns its life in a variety of activities (subsumed as metabolism) and through anticipatory action (including sexual behavior).

In view of this broad understanding of life dynamics, the fact that we have reached the limits of a science reduced to physics- and chemistry-based explanations of everything is an inescapable inference. This broader view justifies generalizing beyond navigation and migration of the species, beyond selection and reproductive processes, beyond human activity (in its variety of forms and expression). In describing this reality, either as science or in other forms of expression (such as art, poetry, philosophy, etc.), the human being, as part of the dynamics of change in the world, expresses awareness of anticipation. The challenge is to make it part of the actions through which everything that is alive becomes what it is during its existence. In particular, how humans become what they are, exploring their environment, interacting, making and using tools, machines, and a variety of means of expression, language in particular.

The pandemic, with its dynamics of global spreading, turned out to be yet another argument in an ever longer chain of events for which science, in its reductionist-deterministic condition can be held responsible. The Cartesian view advanced rationality against blind faith, only to become, over time, on account of impressive accomplishments that it made possible, a new dogma. Consequently, the underlying view of causality, unfortunately cleansed of Cartesian doubt, became arguable. There is nothing wrong (or even mildly disputable) with revealing how causes associated with the past might lead to breakdowns. However, accepting that, for the living, causality lies not only in the past—the footprints—but also in the possible future—choices to be made and their consequences—is rarely, if ever, considered. Within the deterministic postulate, the past—and only the past—is supposed to explain change (desired or not). The language of mathematics (statistics, in particular) is usually at work for this purpose. Often it succeeds: the accomplishments of science and technology, as they apply to the physics of the world, are beyond doubt. But not beyond criticism. Especially at the juncture where humankind pays dearly for the impertinence of scientists who ignore the long-term negative consequences of scientific and technological advances.

Reaction is anchored in a limited view of determinism, which was questioned, from inside, by quantum mechanics, and, from outside, by those areas of biology that refuse reduction to physics. Within this view, cause and effect are clear-cut: the arrow of time points from the past to the present. Act upon the cause and the desired effect will result.

The World as a Factory?

Tools, activated by users, and machines, driven by energy other than that of human beings, were conceived in order to leverage their own abilities. From tools and rudimentary machines, used by individuals to factories using individuals to serve their tools and machines, the change is from a human being's own power to harnessing sources of power beyond the human. For example, gravity "moves" the pendulum; falling water activates the mill; steam provides power to motors. Like everything imagined and eventually made to happen, tools are in anticipation of life-preserving activities; likewise for machines. The deterministic physics embodied and the anticipation—the purpose, the possible future—complement each other. As the goals of activity required more powerful tools and machines, dependence upon energy sources other than those needed to maintain life increases. Eventually, machines driven by energy far beyond the ability of machine-makes to produce, replace those who conceived them.

The educational system aligned itself with the same understanding: shared resources for the task of preparing the new generations for becoming productive themselves. Schools (once extensions of families) and universities (once, extensions of monasteries and churches), ever more specialized, became educational factories themselves. Hospitals—whose beginnings are associated with religious institutions—have become integrated factories for repairing various parts of the human machine. Hospices specialize in supporting they dying. Jails and prisons, meant to isolate criminals from society and to give them rehabilitative tasks, evolved into facilities meant to separate individuals convicted of crimes from their victims. They are the brick-and-mortar-embodied reactive understanding of crime, and how it can be isolated from society. Nursing homes and, more recently, retirement communities represent the industrial economy-of-scale solution to aging and its challenges. They are large servicing facilities that enable a society to move ever faster by delegating its social and moral obligations towards the aging. The machine model—currently the digital machine—is no longer a metaphor. It is actually the embodiment of shared energy for the purpose of accomplishing a desired function. Taken literally, it defines the military, a post-industrial-society production facility whose outcome are wars (hot or cold). You win if your energy resources surpass those of your enemy.

"Deceptive self-sufficiency" defines the machine model. It says: Provided that energy is available, everything can be done using machines. Automation is the next frontier: no need for individual participation in the process. Machines can supervise machines. The process is unfolding in spectacular ways. But more consequential is the fact that the meaning of human life is changing. The arc of history extends from individual effort—survival as a personal matter—to shared energy and a sense of interdependence. The world as a machine, or as an automated factory negates the role of community and thus generates a new human profile agnostic of solidarity. It also affects the cognitive condition of individuals.

Nourishment is a source of energy for maintaining life. Hunting and foraging provided primitive people with calories. Agriculture provided not only food, but also a new understanding of space and time. It is therefore justified to examine how the spirit of the Industrial and the Post-Industrial Revolutions transform agriculture. The reductionist-deterministic conditioning of the metabolism is reduced to one successful formula: large-scale production facilities. They are not dependent only on the sun and organic processes for providing more cheap food for more people. Facilitated by the science and technology of the Industrial Revolution, animal feed, hormones, and antibiotics were among the prerequisites of turning agriculture into the food production industry. One specific detail: with the discovery of vitamin D, it became possible to keep chickens in confinement year-round. With the vitamin provided in the feed, hens no longer needed to be exposed to light. If indeed artificial sources of energy (chemicals, artificial light, a variety of medical interventions) augment production in order to cover food shortages, one would have a hard time in criticizing the methods. In reality, these methods produce abundance—sometimes overabundance—and a human being addicted to prosperity. The activity of such a being is no longer related to maintaining life or simply improving it, but rather to ascertaining the right to consumption. In our days, seven billion male chicks are culled every year—it does not pay to feed them. Many millions of female ducklings and goslings are culled in order to obtain more *foie gras* (males are better for this). When more than 50% of the food available in the USA and in West Europe is wasted, one could question the means for producing an abundance actually dangerous to life.

Large-scale industrial farming involves cause-and-effect procedures. There is demand (cause), hence the goods are supplied based on the industrial model of producing more (effect) to meet demand. Farming turned into engineering life is a worrisome activity. Aside from the cruelty involved, the impact on the environment, as well as on consumers, is intractable. The entire genome—of the animals and plants produced, as well as of the people and other animals—is undergoing change. The same holds true for the meat packing industry—a Covid-19 hotbed in the USA and around the world. Upton Sinclair (*The Jungle*, 1906) made Americans aware of the horrible work conditions in slaughterhouses and food processing factories. Except for a number of regulations, and for automation, nothing has really changed since his book shocked its readers (over 115 years ago), neither in the USA nor in other countries. This is where the broader CoVID context becomes evident. The outcome—abundance in almost decadent forms—is definitely below the long-term price that reductionist-deterministic food production exacts.

Schools were set up in order to share precious resources essential to the survival of communities. They became educational factories. Under the alibi of "equality," the goal was to output standardized youngsters able to fit in the industrial society, not to become independent thinkers. Universities, even the most renowned, process a student body; they do not foster intellect or character, granting the highest grade per student demand. The once Within CoVID, the vision becomes blurred: impart knowledge and a sense of community, or become just a part of the consumer economy in which entertainment is dispensed to a generation not yet clear about what the future holds for it. The institution's performance is measured by the number of diplomas issued and by its ability to raise funds. The meaning of education is traded for machine-like performance.

Similar considerations apply to all other instances—medical care, law and order, retirement, etc.—in which more energy than what individuals can afford is made available through specialized facilities (hospitals, nursing homes, retirement communities, prisons, etc.). The ration between successful performance and undesired consequences flipped in the CoVID context: efficiency overrides any other considerations. To reference this process back to 40 years ago might seem a bit vague. But since the HIV/AIDS pandemic and the crises arising since then, it is obvious that the downside of investing in reductionist-deterministic science exceeds the benefits, i.e., the upside. Instead of higher returns, an ever-growing record of losses undermine sustainability. Why? Because within the model in place in some societies, richer nations can afford living at the expense of other people's resources (Africa in the first place), and on energy used in excess, and for which there is no accountability. The military—the *sui generis* war-making machine—is going through the same reversal of values: from a defensive posture that would lead to peace, to an aggressive posture seeking to exploit the opportunity to justify its existence and funding through never-ending operations.

Ecological awareness itself is now commoditized. Seeing the promising label of "Organic," consumers—more scared than enlightened—are willing to pay a premium price in order to avoid harm to themselves and the environment. The label, applied in the USA and Europe to agricultural products, is meant to ensure that harmful substances and methods—chemical fertilizers, pesticides, antibiotics, genetic modifications—are not used. Speculators turned this identifier into a marketing gimmick—reaping profits at the expense of the health of consumers paying for more illusory protection.

Resilience and Scale

Under natural conditions, living matter is resilient. There is life in ocean thermal vents and at depths where survival (under high pressure and low temperatures) is an immense challenge. Energy needs are met along the food chain—part of the ecosystem. Artificial environments afford higher energy sources. Forced upon the living, including the human being, they come at a cost: the ability of what is alive to survive, i.e., resilience, is affected. The toxins that protect life forms—for example, butterflies, very visible and thus easy targets, are toxic to birds—are not the same as the toxins used to "protect" crops. Insecticides and chemical fertilizers end up undermining health and the environment (Fig. 1). Humans increasingly use more energy than what the MEP describes. This energy is harvested through activities that themselves require more and more resources which, at the same time, increase dependencies and vulnerabilities. The record shows that all industrial agriculture facilities, including the most recent aquaculture (fish, shellfish, mollusks) intended to provide more and cheaper seafood, eventually become breeding grounds for various pathogens. Bacteria and viruses already spread from animals to factory-farm workers, to slaughterhouses and those working there. In the pandemic, they became super-spreaders of the virus to consumers. A study produced in 2011 alerted consumers that 47% of meat is in some way contaminated. Ten years later, the percentage is much higher. Likewise for farmed fish. Animal feed is genetically designed, and so are the levers, i.e., growth hormones and antibiotics very visible on the "menu" of farmed fish. They are released into the waters, polluting them, and thus human beings and other organisms. Physics and chemistry, agnostic of living processes, metabolism in particular, underlie these developments. Generalized into the food chain of living matter, nature was remade into a factory. Reductionist determinism in full view: the nine billion cows, calves, hogs, lambs, turkeys, and chickens slaughtered yearly are engineered products. They are destined to reach, via meatpacking factories and food processing facilities, a literally reshaped human, addicted to their consumption. In the stench of their own excrement, their snouts bloody from rubbing against the metallic cages in which they are confined, the pigs, for instance, are far from becoming the

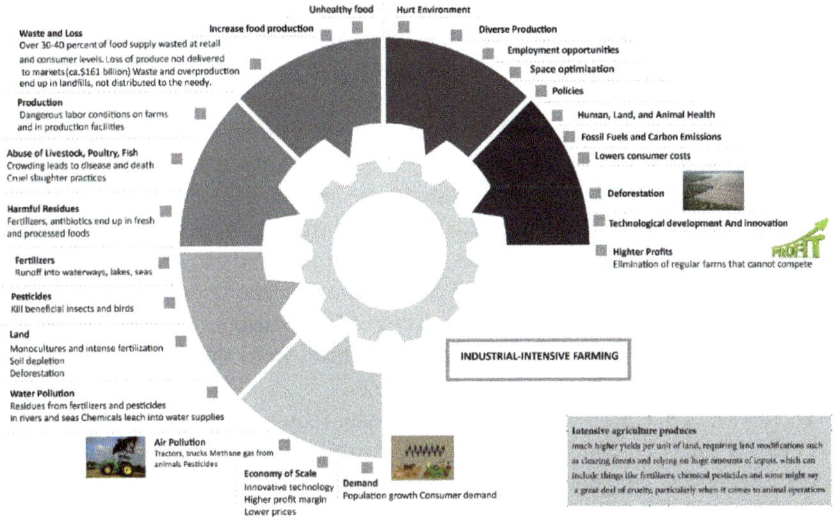

Fig. 1 Transforming nature into food factories. Industrial intensive farming became a high-productivity physics and chemistry process. Natural resources are not renewed but depleted for the sake of prosperity. As enticing as intensive high-yielding industrial farming seems, it is not sustainable

wholesome pork that humans believe they consume. Milk, dairy products, eggs, and processed food contribute not only to overweight, but also to an ever-larger waistline and the illnesses this entails. High cholesterol levels, high blood pressure, apnea, chronic kidney conditions, and liver problems are self-induced conditions. In short, all these lead to large-scale vulnerability.

Covid-19, as research shows, affects exactly those who are captive to machine-based production means. This is not the first corona virus to suggest the connection between the condition it triggers and the reductionist-deterministic view at the foundation of food production. The anticipatory perspective could be translated into practical measures for preventive actions. Such actions could serve as a lever for changed behavior. Of course, this would not automatically translate into getting rid of obesity, or of the broad disease spectrum associated with meat processing facilities, or with sugar-loaded foods and drinks. And it would not automatically create a new ecology. But it will definitely break the vicious cycle of addictions to foods high in sugar content or unhealthy fats, and their long-term consequences. The answer is not a return to the farming of the past—with its own shortcomings, to be sure. Even plant-based meat, which is pushed as a healthy alternative to red meat, is produced at equally disturbing high costs to the environment.

The fact that exactly the obese, the diabetics, and those who eat a big amount of processed foods—plant or animal—become targets of viruses is beyond controversy. Excess energy consumption does not augment resilience; it becomes a liability. The living matter treated like the lifeless matter, processed industrially, is less tasty and less nutritious. Instead of food, with its many characteristics, and which reflects how different human beings are, the consumer is served calories, units of energy—like any machine. Artificial sweeteners, preservatives, and taste (sic!) enhancers turn the living matter of means of subsistence into dead-matter food. It provides calories at the cost of higher and higher energy consumption for their production. The degeneration of the human species through food-related illnesses, in particular binge eating, is a consequence rarely, if ever, discussed. Medicine, practiced in our days as mechanics and chemistry, oversees the degeneration process instead of actively slowing it down through prevention. It offers remedies through surgery and medication—i.e., more physics and more chemistry—instead of awareness of the consequences of indulgence. Prescription and non-prescription drugs also contribute to this degeneration. This is a chemistry (sometimes extremely sophisticated) that makes the abusive lifestyle of consuming processed food tolerable.

There is a lower limit of energy and the data required for survival (the MEP); and there is the open-ended field of ever-higher expectations corresponding to the model of progress congenital with the Cartesian Revolution. At the lower limit, at which nature operates, risks and opportunities are kept in check. The higher the expectations, which eventually are perceived as entitlements, opportunity exacts a higher price on the entire ecosystem. In order to understand what leads to outcomes like those exemplified in the encompassing confluence (Fig. 2), it would be useful to focus on the meaning of the broader context within which coronavirus infections (but also Ebola, West Nile, etc.) occur at alarming rates. Indeed, billions of chickens, turkeys, hogs, and cows and cattle, as well as huge amounts of fish and shellfish, are deemed necessary to meet the expectations of the prosperity driving current capitalist and other types of societies. Therefore, even posing the question of the sustainability of the industrial model is inconsequential to those who understand prosperity as an entitlement—cost whatever it may. Who cares about the future? The fact that 50% of the food goes to waste in the G-20 countries (the most successful economies) conveniently goes unmentioned in the public discourse of sustainability.

The major goals formulated so far in respect to reaching sustainability are, unfortunately, anchored in the same understanding of progress that led to the current crisis. In order to end poverty—a major concern of justified urgency—should access to food that makes people more vulnerable be made available? Usually when help is provided, what is dispensed is processed food originating from advanced industrialized countries that need to keep their production facilities operating. Should the type of public healthcare—i.e., mechanical fixing—that fails so miserably be made available to people who have none? Or should they be empowered to take care of themselves, instead of accepting the false hope of being fixed by others? Awareness of sustainability, i.e., how future states, such as exhausted resources or extreme pollution (never mind cause-and-effect reductionist medicine) should affect our current activity, cannot be relegated to a few social critics. It must become part of the culture—better yet, of conscience, and thus inform actions. No metaphor for consciousness, i.e., awareness, has been found. For sure, it is not a machine outcome, and it is not chemistry in action. The notion was either dropped from scientific discourse or assimilated into the verbiage of a determinism in search of

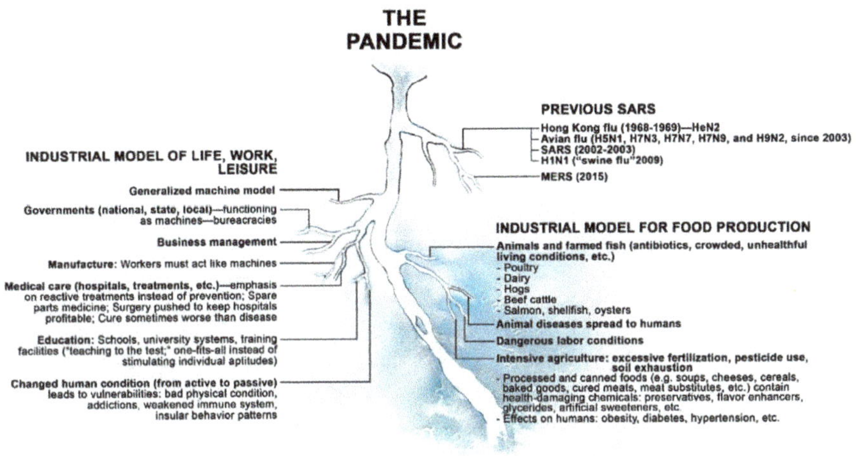

Fig. 2 The pandemic as an encompassing confluence. Reductionist determinism is the origin of its emergence

its mathematical formulation (Grossberg 2017). So was meaning. Anticipation—an expression of this awareness—is once again knocking at the closed doors of the palace within which determinism reigns absolutely. Continuing along the same path, instead of seeking responsible ways to address the future, might be suicidal for humankind. Increased temperatures and higher ocean levels make the headlines. But this is not a matter of "climate change." The seriousness of the current situation is the result of ignorance in respect to what defines life, and why it is not reducible to lifeless matter. This is the overarching subject, unfortunately sacrificed for the pseudo-science of climate change. This speaks in favor of the need to further explain, beyond the definitions already provided, what the anticipatory perspective entails.

Multi-causality

The meaning of the data accumulated during the ongoing Covid-19 crisis aligns with the Sustainability Thesis (ST) of this study: It costs way more to react than to prevent. And reaction does not do away with the danger. The damage associated with this crisis reflects the scale of surrender to a primitive notion of determinism and reductionism—dating back to Descartes, i.e., the seventeenth century. Positivism and the Cartesian Revolution presume that there is no difference in the logic of inquiry across sciences. Matter is matter, and no further distinction is justified. The weakness of the argument's logic is evident: if everything is matter, then there is no matter. To differentiate is the first step in any attempt to explain something. The

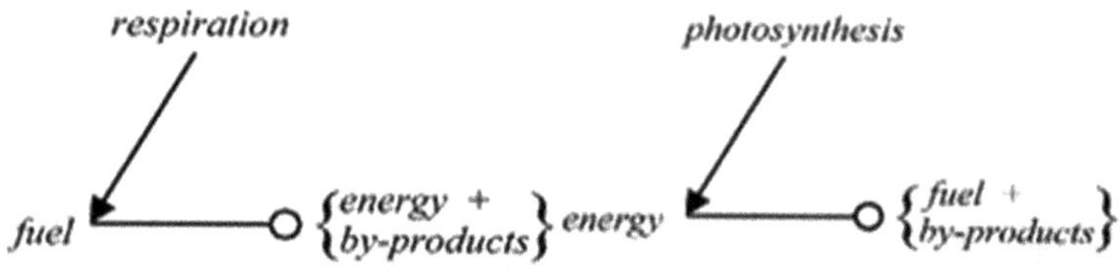

Fig. 3 Life being process, it is obvious that mapping from an input to the output is different from the mapping of non-living entities. Consider only the relation between what life takes in and what the outcome is (cf. Louie 2013)

science and technology of the Industrial Revolution and of our time of post-industrial developments are the outcome of a materialism pushed to the extreme where it loses its logical underpinnings. In order to understand its consequences, let's try to explain what this means.

There is indeed no need for a vital force (a notion inherited from Aristotle's *vital heat* view). Thomas Henry Huxley, the naturalist who focused on the "On the physical basis of life" (1869), identified as protoplasm, tried to identify the components. (The supervenience model probably evolved from Huxley's theory.) Today we can establish, to astonishing precision, the make-up of a cell. But taking all components and trying to re-engineer the cell does not make it living matter. It is the nature of interactions, more than the organization, inside and outside the cell, that explains life. Yes, the role of energy, the inanimate matter, is subject to entropic processes. Living matter uses energy to maintain and enhance life. To use a "like" metaphor: fire springs from fire; the spark mechanically triggered, is fire from which more fire results (Fig. 3). This, of course, does not ascertain that fires are alive. Their light is the light of the sun. And it does not mean that life is fire. Rather: Life requires energy and through interactions learns how to harness it. We infer from this that that causality reduced to cause-and-effect is an incomplete description.

Those who practice the reductionist-deterministic method on the human being know that a "good" pill (for a headache, inflammation, infection, etc.) has a number of possible side effects. Of course, they wish multi-causality—a characteristic of the living—away. A spark in a high-humidity environment will not lead to a fire, even if the spark touches highly flammable material. Wind direction, temperature, the nature of the flammable material—all play an important role. In short: more than one cause. Rather, an aggregate. One more distinction: a fire is an entropic event; life is negentropic: organisms are continuously at work for maintaining, and even improving, their condition. To preserve life is a complex neg-entropic activity. From one generation to another, there is learning—on account of energy. This is how anticipatory abilities are maintained.

With all this in mind, we can understand how the pandemic itself—with the "original" spark in Wuhan (wet market, lab, mine, whatever) or in some other place—can wreak the havoc it does. The initial "spark" led to a world-wide "fire"—globality as the context of the larger crisis of vision that affects our civilization. If we want to understand what is taking place, the multi-causal perspective becomes necessary.

Contamination in areas of centralized activities—large urban settlements, factories, mass transportation, jails and prisons, nursing homes, hospices, among others—was clearly evinced by the numbers of people infected and the number of people who died. The human being, conditioned by the entire experience of a society focused on progress at the cost of high energy, over time became more vulnerable. This vulnerability, at the same scale as the success that the Cartesian mechanistic view has had, has increased over time. The jump from "Humans are like machines" to "Humans are machines" resulted in treating humans as machines. In the name of a rationalism (that is none) the causality characteristic of life was reduces to the causality of machines—easier to describe and control.

Human beings affected by some pathogens are quite different from machines affected by the future of some parts. It is documented that at the beginning of the pandemic patients over 50 years old, with comorbid conditions, were at higher risk of infection. Losing their lives proved to be more probable than that with the younger. Older machines that are not well-maintained break down more frequently than the newer ones. And often are no longer fixed. The skewed perspective led to the hasty formulation: This is a disease of old age. Surprisingly, healthy subjects, even children and adolescents in good shape, and young and middle-aged adults were also among those admitted to intensive care under the Covid-19 rules. Respiratory failure in a "40-year-old man who completed a marathon in October 2019" (Casanova and Su 2020a) is not exactly what physicians trained in a reactive frame of mind would predict. The large number of asymptomatic younger individuals—"machines" not really failing, but with measured parameters that would indicate that they should—also came as a surprise. In the new context of virus variants—especially the VOC (variant of concern), the younger, vaccinated or not, became the new target. At the beginning of August 2021, the CDC's COVID-19 Data Tracker announced: "Child Hospitalizations are at the highest point since the pandemic began." The same assessment followed in the Omicron phase, especially in the winter of 2022. The American Academy of Pediatrics (2022) reported that 1.5 million children tested positive since the onset of the pandemic. In the Omicron phase, in particular the XBB1.5 variant, numbers doubled from one week to the next. All in all, the functioning of the "human machine"—young or old—proves to be anything but the pre-defined functioning of a machine—although many don't give up the machine theology (Nadin 2018a). It became de facto the credo of medical science. Instead of being informed by the physicians facing the pandemic in hospitals and clinics, medical science chose to be informed by machine-generated models.

This is where multi-causality—sacrificed for the simplistic cause-and-effect model embodied in machines—begs for attention. And this is where holism—i.e., the need to understand the integrated wholeness of life—comes into play. The hammer hitting the nail is reduced to a discrete situation: a force applied to an object. However, when a human performs it, the action engages the entire body. There is anticipation in the preparation for the activity—hitting hard or not, maintaining balance, getting the right grip, etc. This understanding of the multitude of parameters involved and of the continuous nature of life processes, is implicit. Adjustments follow as the nailing progresses. Of course, the automated hammer—a machine for driving nails through shingles or wooden boards, etc.—is driven by the physics model: force applied to the target. That's' how guns—hammers with explosive bullets—work. And that's how medical devices, such as those used in the stomach-stapling procedure, work.

Inspired by the same physics model, other procedures (e.g., knee and hip replacements) are performed as cause-and-effect operations. Heart conditions—aggravated by the pandemic—are subjected to repair procedures instead of healing methods. Since Covid-19 evinced early on the fatal role of low levels of oxygen, machines for remedying the situation were deployed. And when the number of such machines did not meet the need, it became a national priority to make more of them. New versions were produced—most of them inadequate, but financed by the government. Empirical evidence shows that intubation and ventilation—techniques inspired by engine maintenance—became almost a death sentence–until proning was rediscovered, and their use was dramatically reduced. The Human Genetic Effort (Casanova and Su 2020b) provided evidence of the role of hereditary factors, as well as other susceptibility factors. Of course, the traditional anamnesis (a person's medical record) could have revealed the same. There is a relation between the genetic description of each person's identity and the biometrics, i.e., the various sequences of data representing biological phenomena such as statistics concerning age, gender, etc., economic status included. But nobody takes time to consider it. Physicians are conditioned to prefer automated processes. Apparently, Covid-19 made, as already discussed, the poor (e.g., Blacks and Hispanics) more of a target than Chinese-Americans or Italian-Americans, German-, Polish-, or Jewish-Americans. In retrospect, the pandemic turned out to be symptomatic of social failure: poverty and the associated lack of adequate medical care, as well as deficient education. Over time, it became clear that the majority of people rejecting vaccination, avoiding tests, or refusing to wear a mask came from the same population segment. The "social machine," with its many moving parts, exacerbates inequality against the background of many shortcomings associated with "America the Economy." The country is so fractured that it could not agree on a coherent course of action. One party blames the other, one tribe demonizes the other. From locality to locality, from school to school, from business to business, each adapted different actions—or no action at all.

Not that Europe did much better. After having broadcasted success in reaching a high degree of hygiene and social cohesion, Germany also succumbed to the fourth wave of the pandemic, And after that, to the fifth. Omicron "came to power" in parallel to a new government—a coalition split on issues regarding what should be done. Bavaria, for example, would not align itself with the states more inclined to the left. Way before that, Italy, a member of the European Union, with more aging citizens than in other countries, first experienced a concentration of cases that few were prepared to understand. Europe gave up being a Union: the old borders were closed. Later, Hungary surpassed Italy; so did the Baltic republics. In France, during the second wave—and wave after wave thereafter—record after record was broken. Because the French are always in a "j'en suis contre" (I'm against it) mode, social unrest escalated as stricter rules (which no one observed) were introduced in order to avoid more harm. Millions of French-people demonstrated against any and all restrictions, while other millions become infected and many died. In the VOC context of the Delta variant (and beyond the Delta variant), India, with extremely high numbers of victims, overcame the crisis; so did South America, with even higher numbers of affected and casualties. Nobody expected that Russia would order a paid vacation—a form of benevolent lockdown—as it faced a new wave of infections. Only 35% of its decreasing population accepted the vaccine. Way ahead of Russia in vaccinations, even Estonia fell prey to the Omicron wave, as did Lithuania, Latvia, and Slovenia—all fully dedicated to preventive measures.

These developments were recorded by statisticians in order to prove that they can explain everything through data alone. Instead of the superstition of reading the future in coffee grounds or tea leaves at the bottom of a cup, the new superstition of Big Data was accepted and blindly practiced. The empirical fact that the living does not rehash its past, as every mechanism embodying determinism does, is ignored. The living, rather, defines its future through choices—some better than others—as is characteristic of non-determinism. This is exactly what science refuses to acknowledge, while at the same time postulating, falsely, that everything is deterministic.

The single patient—in Belgium, Qatar, or wherever—i.e., the individual erased in the methodology of averaging and statistical generalizations, is part of a population, i.e., of a civilization. Healthy, with a solid record of strong immunity, or weakened through one or several maladies, the actual patient to be saved by the physician was not spared reduction to statistics. To understand the selective nature of this pandemic—"Why me?"—is to understand the uniqueness of each person. Through the machine model, reductionism pursues illusory sameness and ignores the dangerous stereotyping that leads to false generalizations. When a certain segment of the population is more affected, and the same group is the target of discrimination—let's consider the elderly—the role of the virus assumes a different identity. Other discrimination targets—the poor, the blacks, the congenitally ill, etc.—are in the same position. The disease can only be understood, and treated, if it is considered in a very large multi-causal chain of interrelated events. It is part of the *whole*: community, country, world, civilization. Unrest and social and political conflict—much of it preceding the virus—are part of this chain. This applies to the USA as it applies to the rest of the world (Turchin 2010; Pinsker 2020; Brzozowski 2020). The guest workers (from Romania and Poland) in Germany, or those in Singapore (Bangladeshis, Chinese, Vietnamese) fared in the pandemic like the Blacks and Hispanics in the USA, or like the Black Africans who made it to England. The poor—white, yellow, or brown—all over the world paid with their lives for the inadequacy of reductionism.

The disastrous failing of the economy—a subject in itself—documents again that human activity cannot be reduced to cause-and-effect interactions and explained in terms of machine functioning. When the living is endangered, machines and everything else that working humans "give life" to are rendered useless. The world economy—with a GDP of ca. 90 trillion US dollars—is a living process. It integrates many types of interactions, involving tools and machines, supply chains, as well as all kinds of transactions. Consumption, as the engine of the US economy—70% depends on it—exemplifies a deterministic form of conditioning. It is as cruel as the consequences of the virus behind the turmoil it entailed. China, the "world's factory," pursued a policy of zero infections, which eventually failed as new virus variants made population control unfeasible. The enlightened consumers of the free world, unwilling to accept any limitation to their so-called liberty, closed their eyes and ears to China's decisive control measures that affect over one billion people. Self-satisfied shoppers continue to ignore the oppression of individuals who are declared social and political threats to the regime. Citizens of the free world bought into the illusion of cheap, disregarding what it entails in terms of human values. This lack of vision—tacit acceptance of the lack of respect for human values—predated the pandemic, but is certainly part of it. Within the Covid-19 debacle, this became evident to even the most fanatic promoters of the global economy. The broken supply chain made it evident that the global economy of plenty was based on cheap labor, cheap energy, and cheap raw materials, not on increased efficiency. A politically and socially fractured world does not hesitate to take advantage of people who can barely survive on their labor. It was less a world stage for bringing the whole globe to a level of decent life, and more the launching pad for fierce competition, regardless of ideology, religion, or ethics. Covid-19 generates a larger market for the machine called the economy, and thus more consumption, with the consequent higher profits. But also more waste, more pollution, more dangers to health of humans and animals, more apathy toward the future. Continuous damage to the environment further undermines sustainability. To consume, as a motivation for maintaining prosperity—the individualism of expectations—is a long-term death sentence. China might have the fastest growing number of millionaires and billionaires in the world. But wealth achieved at the expense of the future is, in China and everywhere, an illusion. Governments pour public money—increasing the national debt—into stimulating consumption. Shortsighted reactions undermine society and whatever is left of an authentic sense of solidarity.

When anticipatory action—pro-active measures, preventive social practices—is sacrificed for the benefit of the immediateness of reaction, society ends up accepting unethical measures. Where else, if not in nursing homes, should the infected elderly be forced to go in order to protect others? It became an acceptable solution when infected aging patients were delivered, by some local government decree, to nursing homes. Others, for instance those who had surgery and could not afford home care, ended up in the same place. Some of them were the ever-glorified but never really respected veterans—good for having sacrificed their lives in senseless wars, but not good enough to receive proper care when in need. "The old would have died even without the virus," was repeated as an excuse. Society felt sorry for each wave of deaths but did close to nothing to help them avoid infection. Multi-causal processes, characteristic of living matter dynamics, were reduced, in this particular take, to the survival of the fittest. Understanding multi-causal processes, such as those defining human behavior during a crisis, deserves attention if indeed the intention is to prepare for future crises of this scale, or even of higher impact. Better yet: to avoid them. In machine-like deterministic systems, breakdowns are unavoidable. Anticipatory actions afford mitigation means and methods of prevention. They are part of the adaptive dynamics of the living.

A Record that Can Never Be Right

Returning to the situation at hand: We do not have, and, given the condition of life, never will have, a full description of all that happens once the SAR-CoV-2 virus—regardless of its variants—interacts with the organism. More precisely, once it encounters the peptidase domain of the angiotensin-converting enzyme 2 (ACE2, for short), a slew of processes become possible: from non-symptomatic (based on currently available data, the majority of the infected never had any symptoms) to terminal condition. We know even less about mutated viruses, e.g., BA.5 and related variations, or the Omicron variants. The focus on the virus's spike and ACE2 (as entry point) returned a partial description. In time, it became known that from the 29 proteins of SARS-CoV-2, five are associated with damage to the blood vessels. The record of Covid-19 shows that, instead of being foremost a respiratory disease, a great number of patients die of strokes and heart attacks. Identification of the proteins expressed by the virus that could lead to damage to endothelial cells (Rauti et al. 2021) is testimony to the efforts to better understand the many dimensions of the pandemic. Nevertheless, even infectivity remains rather ill-defined. For instance: Individuals living in the same household are not equally affected, i.e., infectivity is not uniformly distributed. Some get sick, some don't. And there is the Long Covid, which affects millions. According to some reports (McGraw 2020) two percent ("super-spreaders") affected the 90-plus percent of people infected. This in itself is characteristic of non-deterministic processes. To try to describe the super-spreader distribution or the relation between symptomatic and non-symptomatic, between short-term and long-term conditions, using means of deterministic nature is illusory.

Contingent conjectures as to the variety of causal sequences or to the variety of the virus variants lead to an open-ended image of how Covid-19 is manifested. Clinicians and pathologists took note of patients with seizures, heart attacks, renal failure, cardiovascular damage, encephalitis, endotheliopathy and associated coagulopathy, gastro-intestinal infections, conjunctivitis, liver damage, acute respiratory distress syndrome, Kawasaki disease, "Covid toes," a painful or only itchy skin rash, similar to chilblains, noticed by physicians (Bleicher and Conrad 2020), and thyroid conditions. And this is not all. In children, bulbar conjunctivitis, cervical and mesenteric lymphadenopathies, pericarditis, coronary dilatation were also observed (not to mention red and cracked lips, diarrhea, nausea, skin rashes, etc.). Multiple, non-repeatable intersecting factors, always involving the immune system, suggest that independent processes unfold in a multisystem inflammation. This takes place at various scales and across intervals—testimony to a time structure of extreme variability, and to the holistic nature of anything affecting the living. Together they made up a condition behind which cascading multiple causes are at work.

Reductionism would suggest that if we know the matter (the atoms, the molecules, the cells, the genetic makeup), we could understand the rest. The virus was researched in detail and continues to be examined using everything that Big Science makes available: e.g., new sequencing methods, advanced cameras, detailed microscopy. For instance, everything involved in the ACE2 dynamics became subject to painstaking observations. Yet the various conditions that are triggered are still inadequately understood. Genetic reductionism, stimulated by the availability of sequencing technology developed within the much-glorified genome effort, deserves our attention since Covid-19 was the opportunity not to be missed. In this vein, the medical community took note of the deterministic allegation that "The major genetic risk factor for severe Covid-19 is inherited from Neanderthals" (Zeberg and Pääbo 2020). The press (*New York Times*, no less) presented this as "DNA Linked to Covid-19 Was Inherited From Neanderthals, Study Finds" (Zimmer 2020), as though everyone forgot that an RNA, not a DNA, virus is behind the pandemic. In the final analysis, the obstinate determinism of medical science collided with the findings of physicians trying to help patients. Indeed, single-cell analysis sequencing on nasopharyngeal and bronchial samples (Chua et al. 2020) as a method for assessing the severity of the disease is an impressive technical accomplishment. However, as impressive as sequencing proved to be in identifying the genetic make-up of the virus, it remained more of a "cartography" of the infection. Physicians expect not a navigation guide, but effective treatment methods from which their attempt to help patients could benefit. When such treatment methods (e.g., Redemsivir, monoclonal antibodies, Molnupiravir, and, currently, the very controversial Paxlovid) eventually came out in the market, they were not products of medical science but from the pharmaceutical companies. To demonize pharmaceutical companies only because they are driven by the desire for higher profits is to ignore that they operate in the context of CoVID: no real perspective. They do not understand that living matter changes in a different rhythm from non-living matter. Treating the living entity with the same perspective as you would fill potholes or replace rusting metal parts is the concrete expression of the crisis of vision.

Immediacy reigns—get it done fast and now—and is the mantra for every problem. Within this perspective, Americans, unwilling to assume responsibility for their own well-being, doubled their consumption of prescriptions in the last 40 years (outpacing other nations). This is so because alternatives are almost non-existent. The same Upside/Downside Ration flipping: the outcome hoped for (fixing headaches, inflammation, broken bones, damaged hips, and so many other ailments) is well below the damage that the so-called treatment inflicts. It is an inverted yield curve—everything in the short term. Marketing costs (announcements in medical journals, in door-to-door sales in medical offices), free medication samples and a variety of promotions—there is no lay magazines without several ads for pharmaceuticals—are skyrocketing. The consumer's desire for alleviation often leads to addiction. Covid-19 became a new opportunity: Government-mandated measures and demands fed by misinformation opened the gates to corruption. The public was perversely abused: it had to finance its own exposure to less than validated healing methods.

As a general observation: Covid-19 is multi-causal and multi-phenomenal (i.e., it takes many forms). Severity is a matter that can be understood considering the whole, not just a single cell. There is a spectrum of symptoms, many to be considered as significant in respect to immune response. All these facts were ignored or downplayed (or analyzed in animals, as though ferrets, not human beings, were in danger). Moreover, the sample usually used ignored other factors: age, gender, comorbidities, social identity, patients with mild symptoms who did not need hospital assistance. Breakthrough infections—after vaccination or after having recovered from Covid-19—became a subject very late. Failure was politically unacceptable. If anything, reductionism, involving advanced sequencing technology, undermined its own credibility since the data it generated were inconsequential most of the time.

Spectacular as genetic or molecular structure analysis is today (Zhang and Kutateladze 2020), it does not prove to be the key to our understanding of multi-causal, non-deterministic, holistic phenomena. And it does not really help doctors in treating their patients. If anything can be derived from the crisis of knowledge that the virus made evident, it is the suggestion to examine Covid-19 holistically, rather than in a reductionist manner or exclusively from a deterministic perspective. Covid-19 proved to be an expression of the entire organism undergoing change. It can be slow (in the incubation phase) or extremely rapid (in its various lethal manifestations). Addressing one or another symptom and providing detailed molecular accounts (some being impressive feats of technology) have not helped in saving lives or in treating those stricken. Even more important, it did not prepare the world for the succeeding variants that nobody else but infected human beings produced.

The evidence accumulated in hospitals all over the world shows that when the respirator—a mechanical device for supporting people affected by low oxygen levels—was applied, the results, compared to natural proning (turning a patient on his side) were disconcerting for medical personnel trying to save lives. The majority of those intubated, on account of measured low oxygen levels, died (in New York alone, 88%). Lack of understanding the whole made otherwise justified interventions (such as intubation) dangerous. It is not the purpose of these considerations to suggest or to justify therapies. Rather, society should take note of the fact that the deterministic perspective—reaction within protocols in place—was proven marginally effective at best. Indeed, some patients with lighter symptoms survived. In the end, the lethality of Covid-19—how many of the stricken died—is below three, or even two, percent—way lower than that of MERS or SARS. But for certain categories (e.g., the elderly and those subject to comorbidities), it went as high as fifteen percent. Even these numbers are relative, depending on when and how lethality is established. The variants of the virus are sometimes more infective—take Omicron—but less lethal. This reality will be reflected in the final analysis—although to establish how many people were infected is impossible. Non-symptomatic infections, usually detected through testing, are rarely (if ever) recorded.

One remark from none other than Dr. Rochelle Walensky, Director of CDC, deserves attention: "…nearly every death…due to Covid-19 is at this point preventable," (Holcombe 2021). The notion of prevention is finally making it into the public domain. However, prevention means many things to many people. This in itself invites discussion. A better-educated population, able to understand what is needed to avoid exposure to the virus? An adequate medical care system that serves everyone? A more equitable distribution of wealth? Vaccination? This incomplete list of questions is open to debate.

A Medicine of Prevention

It would be disingenuous to ascertain, without concrete arguments, that anticipatory-based medicine would have spared the world the pandemic. Since Covid-19 was triggered within a broader context, it would have taken more than adequate medical attention to prevent it. Medicine must be understood as part of the encompassing framework within which it became, in recent times, a service provider. It is driven by motivations characteristic of its grounding in the reductionist-deterministic view of life. Specialization (reductionism in action), with its often astounding accomplishments, is its acme. Specialized medicine is extremely profitable, but useless in the context of a pandemic. Not unlike the science and technology to which society owes the level of prosperity and comfort it enjoys—inequalities notwithstanding—specialized medicine has a record of success matched only by the vulnerabilities it prompts. In effect, it provides successful reactions to conditions otherwise difficult to live with: failing organs, damaged body parts, weakened immunity. Health, as a desired state, is traded for survival at all cost—including a medical condition compromised for the long term as a side effect. *Fixing* matter—e.g., the mechanics and chemistry needed to keep a car running—is fundamentally different from *healing* matter. This realization is the premise for contrasting the ambiguity of anticipatory-based medicine to the concreteness of mechanical or chemical interventions. For example, chronic headaches might diminish with proper rest, diet, exercise, mental training (feedback), or by taking a specific pill.

From the history of medicine, we know that so-called anticipatory medicine was practiced more as a form of the *artistry* of medicine. The label "anticipatory" (sometimes used) was not connected to the anticipatory nature of the living, i.e., to how the organism functions. This misleading qualifier gave it a bad name: "Art kills. It was the art that gave us purging, puking, leeches, the gastric freeze," (Zugar 1998). Identifying anticipatory concerns with the "art of healing" practiced by all kinds of imposters did not help. For the wrong reasons, the subject of medicine based on awareness of anticipatory processes in the body was compromised before being understood.

In order to gain credibility, anticipation-based medicine would have to make available effective means for preventive healthcare and for healing. Some of the requirements for achieving such goals deserve attention if for no other reason than because anticipation-based medicine cannot be attained by edict. Or by wishful thinking. The medical community must develop it. For this to come about, medicine will have to free itself from the condition of captivity to physics and chemistry. It will have to realize that it plays an important role in securing humanity's future. On its current mechanistic path, it undermines sustainability as a premise for securing humanity's future. Anticipation-based medicine must be practiced in association with the reactive medicine currently dominating healthcare. Their unity represents the premise for effective prevention.

Among the many consequences of specialization and over-specialization at a global scale, there is one that is impossible to ignore: now everyone is sick, but there is no healing in sight. Sickness is chronic, i.e., life-long. When every detail is under the magnifying glass of an expert (actually, a computer tool) never taking the wholeness of life into consideration, it is impossible for anyone to be assessed as healthy. Under the scrutiny of medical specialization there is always something, i.e., some symptom, that warrants intervention. Often evincing impressive performance, over-specialized physicians bring to mind horse blinders. Focus, or depth, is achieved at the cost of perspective. Not so long ago, internal medicine was complemented by obstetrics and gynecology (women, especially pregnant women, are different from men), pediatrics (children have their own medical profile), ophthalmology, urology, cardiology, among other branches of medicine. And there were surgeons. By now, surgery is specialized to the extent of requiring its own catalog. Orthopedic surgery encompasses several specialties (e.g., shoulder, knee, foot and ankle, hand and even finger). So does pediatrics, extending from the fetus to adolescence. There are specialists in medical genetics, nuclear medicine, as well as in sensor-based wearables and in AI-based medicine. Every type of cancer has its own specialist for detection, treatment, and recovery: colorectal oncology, gynecological oncology, pediatric oncology, bone cancer, cranio-facial cancer, etc. And the trend towards ever-granular specialization continues, with its own language. The use of proprietary tools and methods for keeping the patient in a condition of dependence for as long as possible is expanding. Reductionist specialization is always practiced to the detriment of understanding the wholeness of life. The physician's observation and association skills were overtaken by swift, automated tests—all measurements of physical and chemical parameters, important, of course, but not always significant. As a matter of fact, they were quite useless in diagnosing Covid-19, and even more so in treatment.

Prevention and anticipation-grounded healing are not possible without considering the entirety of one's patient's life, as well as living style. There are many reasons why this path was abandoned: too much to know, too much to understand. In a nutshell: too much to consider. Moreover, "It takes longer to heal than to fix." Working with others and with the patient is difficult: "It does not pay." The alternative, adopted in the vast economic activity called healthcare, is attractive: Can we automate it? Here is the label for the "standardized" condition (documented in a catalog) and the "treatment" (sometimes called protocol): action-reaction, next one please, in a rhythm (seven minutes per patient) corresponding to assembly line activity. In this model there is no place for preventive measures. All this despite the Hippocratic oath—"I will prevent disease whenever I can, for prevention is preferable to cure." Over time, the oath became, within for-profit medicine, more a decorative declaration (like the physician's diplomas hanging on the office walls) than a shared commitment. Payment for prevention is a category not reflected in any insurance policy—although prevention would save money.

Addressing perspective, that is, arguing for a broad view as a premise for anticipatory-based medicine, conjures the placebo effect: What kind of medicine is it when you get a desired result but the "chemical" is not what the "science" says it should be? In testing a new drug for FDA approval, a placebo is always present as a control. The understanding of anticipatory processes that underlie living processes places the placebo effect in a meaningful perspective: the organism is at work, in many instances stimulated more by the thought than by the chemistry. Reports on placebo surgery draw attention not only to the limits of extreme reactive medicine, but also to cases in which reactive medical care proves to be ineffective. In a variety of experiments, it was shown that in some cases, the "sham surgery," i.e., fake operation, followed by make-believe post-surgical procedures (e.g., placebo pain pills, imitation physical therapy) worked as well as the real procedure (Probst et al. 2016). Then there is the vaccine. The statement that almost every death from Covid-19 is preventable is related to the fact that even if a vaccinated person falls sick (the "breakthrough" effect), the disease is milder and only rarely lethal. Indeed, as it will become evident, vaccination is an anticipatory action focused on the immune system. From all that is known, the immune system is the most convincing proof of the anticipatory nature of physiological processes.

From the history of science-based medicine, we can derive an understanding of why specialized reactive medicine and anticipatory medicine went separate ways. The quest for objectivity—the rational explanation—that stimulated medical practice at the end of the nineteenth century explains how medicine constituted not only its own domain of knowledge, but also of authority and economic relevance. Deterministic medicine effectively blocked alternative practices, including those with deep roots, such as Chinese medicine or folk medicine. Most of such practices considered disease as just another state of life, which the body itself would take care of. This was never disproven. Let us remember the memorable credo of Chinese medicine: The best doctor treats illnesses before they manifest (Lihong 2019). Self-healing, a concept transferred to the behavioral practices, is the outcome of engaging the patient. Using one's own resources when facing a medical condition deserves more respect than those who practice reactive procedures are willing to give it. Behavior informed by understanding what can help and what is harmful plays an important rolc.

Why a person develops a certain dominant Covid-19 path (centered on lungs or kidneys or heart, etc.) and not a different one (such as Kawasaki, for the stricken young) is a question that physicians, risking their own lives when they treat the afflicted, had no time to entertain. There is no easy way to determine the path of illness. From no symptoms to fever (in rare cases), body aches, dry cough, fatigue, headache, sore throat, deep thrombosis, hair loss, appetite loss, loss of the senses of smell (the olfactory bulb is affected) and touch, etc.—everything seemed possible. Younger patients who never had any risk of interruption of blood supply to the brain, yet who died from strokes are the vivid example of how difficult it is for the practicing physician to understand such processes. The descriptor "multicausal" can be applied to the variety of processes, some one after another (sequential), others simultaneous, that can result in a blood clot in the lungs (pulmonary embolism) or in the heart, or in the kidneys. The cause-and-effect blinders prevent notice of the multitude of holistically interconnected processes: muscle weakness, tingling in hands or toes, delirium, seizures, stroke. Without the change in perspective that we are arguing for on account of the Covid-19 experience, the next pandemic will prove not only more lethal, but even more expensive.

"The Search for People Who Never Got Covid" (Mallapaty 2021) is an interesting development. The motivation for finding such persons: to develop new ways to treat the disease. The implicit message is congruent with anticipation-based medicine: acknowledge that individuals are unique; acknowledge the non-deterministic nature of biological processes; and overcome the expectations of "one size fits all" characteristic of reductionist-deterministic medicine. Of course, this would contradict the motivation of the entire medical enterprise. It is difficult to free ourselves from the handcuffs of a perspective that claims that the same effect—an infection, in its open-ended possible forms—is due to the same cause, i.e., the virus alone. But no progress in understanding Covid-19 and, in general, what health and disease are, is possible unless a new perspective, within which anticipatory processes are the foundation, is adopted.

The necessary change concerns not only the pandemic, but also the broader commotion: breakdown of the economy, social and political unrest, disregard of values, etc. The deterministic causal vector—from past to present—is screaming for the complementary vector from the possible future to the present. The reductionist path—the focus on the "making" of biological matter (electrons, atoms, molecules, etc.)—is compatible with the complementary focus on behavior. A suggestive image of what distinguishes reaction-driven medicine from anticipatory medicine is the way in which overweight, considered an aggravating factor in Covid-19, is treated. Indeed, a certain way of life (usually qualified as healthy or unhealthy) affects the organism. Comorbidities and Covid-19 further exemplify the thought. What people eat and drink, what they do or don't do, the medical care they get (or don't) are part of a larger image of interdependencies (Fig. 4).

Anticipation means actions. Among them, to vaccinate—but not the one vaccine for all. To stimulate immunity—of the individual, of the community, of society means more than finding the reductionist "miracle" that will take care of everything and everyone in the shortest time. It means to live a "healthy life," i.e., to exercise, eat right, avoid stress, etc. It means to provide access to quality healthcare and education. Unemployment and inadequate medical care associated with Covid-19 contributed to social and political conflicts. These arose not only in the USA, with its unhealed wounds of racism, startling rise in gun sales, and prevailing social inequities, but all over the world. Even in socially engaged Europe, with its more adequate medical care and protection of the employed, conflicts erupted, culminating in reactions opposing government measures supposed to protect the community. The incompetence of those in power, and the fact that they immunized themselves against possible consequences of their actions (in particular, corruption) were deplored in mass demonstrations.

Medical care driven by politics and "controversial treatments" (Zhang et al. 2020) caused friction among physicians. Some drugs were favored over others on account of political fanaticism (the narrative of Remdesivir, Ribavirin, Chloroquine, Ivermectin, etc. will not be rehashed here). In view of these facts, to ask the legitimate question of how to protect the patient and preserve the moral integrity of medicine is imperative. Ethics, which determinism effectively eliminates, must be recovered as an effective standard of quality for medical care and for medical research.

Fig. 4 Reactive medicine versus anticipatory medicine. In restrictive surgery, a surgeon uses one of a variety of techniques to reduce the size of the stomach. After restrictive surgery, one feels full on less food. There is no effort to lose weight. Physics solves the problem through short intervention. The alternative is the long path of proper diet, exercise, maintaining a balanced lifestyle, avoiding stress, and adapting to new circumstances as one advances in age. All these are anticipatory actions

Physics is not an Innocent Culprit

Under the high pressure of the ocean and in the absence of sunlight, there is a lot of life that defies the explanations of physics and chemistry. At 29,000 feet below sea level, the creatures adapted to living under pressure hundreds of times higher than that on Earth and in the absence of sunlight. It is really very cold down there. Such examples invite explanations of life for which reductionist-deterministic science has no credible hypotheses. These living creatures don't fall apart, as other physical objects would. They multiply. Their metabolism is less dependent on sunlight than on other living processes. The explanation for their ability to live an reproduce is different from that describing life on Earth. No experiments with them are possible—to bring specimens to a lab, keeping them alive under conditions different from those where they exist, is out of the question. Extracting them (as some have researchers done) from the very cold environment they live in would "cook" them. Researchers constructed models, not realizing that no model, no matter how detailed, is the same as the modeled living organism. Scientists do the same with insects and birds, unless they end up killing them in order to explain their way of living. As Douglas Adams put it: *If you try to take a cat apart to see how it works, the first thing you have on your hands is a non-working cat.* Amazing technological performance is achieved in examining deep ocean life, viruses and microbes—everything alive. Albeit, technology is useful for illustrating the absurdity of the reductionist model. We know more about the chemistry and genetics of living entities than about their life. Try to explain the *Why*? of sexuality or what animals choose to eat ("have a tooth for") at the level of molecular biology!

Living matter escapes the description that explains the dynamics of stones, water, stars, the universe. It is purposeful, a characteristic that physics discarded as an object of interest. It is creative, in the most profound sense of the word: it remakes itself, it multiplies, it adapts and evolves.

Promotion of physics beyond its legitimate domain by those unaware of the consequences of their choice might justify the qualifier "innocent culprit." But this is not really the case. Atomic bombs were deployed in full awareness of their destructive power. In the logic of wars, this would not be deemed a crime. In the anticipatory perspective, the unethical condition of physics could not be ignored. Likewise, for "change of function," as the re-engineering of viruses is called. Regardless of the motivation, it is questionable. Neither physics nor chemistry has an ethical dimension. Physics hijacked politics in order to make lifeless outer space a priority. One motivation: competing with the enemies. Another motivation: gaining knowledge about the world: Let's find out if there is life on Mars before understanding life on Earth. Let's find out how living matter came into existence eons ago—while at the same time making the claim that there is no difference between matter and living matter—the party line in science. A choice, for which immense resources with a large carbon footprint are consumed, was made. The space program won over the exploration of the life-rich environment in which people live. Oceans are explored mainly in order to find raw materials, not for gaining insight into the many interdependencies of life on our planet.

But this is not the time for incrimination. The Cartesian Revolution—indeed, a revolution—made possible the impressive progress in describing, at all scales, phenomena pertinent to non-living matter. The price of this progress is high. It is reflected in the relatively sad state of understanding the living, to the extent of endangering its survival. Covid-19 became yet another instance for asking whether giving up sustainability is a good bargain compared to landing on Mars or sending a super-powerful telescope to outer space. Producing the most dangerous war machine ever, including bioengineering for developing weapons, did not help when Covid-19 took over the lives of people on planet Earth. Experiencing a pandemic, which ultimately reflects the consequences of the reductionist-deterministic theology in action prompts self-examination. It inevitably triggers questions that might seem unrelated to a discussion on the role of physics in understanding the *Why?* of the terrible price it exacted.

Questioning Positivism

Positivism, an outgrowth of the deterministic reductionism that shaped Western civilization, ascertains the role of measurement as the foundation for experiments. The ruler and the compass (described by Francis Bacon in *Novum Organum*, 1620) are tools that replaced drawing by hand of a line or circle. They were, quite unlike the hammer, the pulley, the lever—extensions of the body—tools of the intellect.

Evidently, in their computational form (often as computer graphics), they are dematerialized, reduced to instructions that electricity-driven machines carry out. There is no reason to object to the rational core of positivism. The development of science and technology, as Western civilization practices them and as they are adopted all over the world, took place on positivist premises. But there is much to object to the arbitrary wholesale generalization of positivist rationality from the physical (i.e., the inanimate) to the biological. The imperfect hand-drawn line or circle are testimony to a process of discovery in which an individual is identified. Living matter causality—moving along a straight line or running in a circle—can be described considering quantitative aspects together with the meaning of the process. The tool does not "know" why we need constructs such as "straight line," "circle," or why they are useful. The causality of drawing versus that of using a ruler or a compass is different. It is the result of quantifying the experience within which the straight line and the circle emerged. In the positivist way of thinking everything can be measured and should be measured if we want to know what it is.

Quantification, originating in positivism, became the new religion. Indeed, the causality of inanimate matter, which can be described fully in quantitative representations (usually through numbers) is reducible to cause and effect. Within the positivist view, complexity is exclusively a matter of data describing quantities. Inferences from the past, described exclusively through data, (representing their quantitative aspect), are the only bridge to the future. In the living, the bridge to the future is the purpose reflected in the meaning of actions that lead to the future. It is in this understanding of time where ethics—distinguishing between the good and the bad—is born. These considerations are not only of theoretic relevance.

The crisis of vision (CoVID), that preoccupies us as the larger context for Covid-19, becomes apparent as the role of meaning in guiding choices humans make comes to light. Tools and machines are instantiations of deterministic processes. They have no awareness of the future, and therefore no built-in ethical dimension. For the living, the future is the continuation of life, or the end of it. The possible future of anticipatory action is a *meaningful future*, not a collection of numbers.

To know is to understand change and its consequences. To realize how different matter and living matter are is the premise for actionable knowledge. Descriptions of reality, where knowledge acquisition starts, integrate the observer, who is unavoidably part of the reality described (Nadin 2018b). Everything involved in generating such descriptions—sensorial and cognitive endowment, genetic and epigenetic conditioning, experience, context—insinuates into the description. The wish for objectivity—independence of the observer—is as understandable as the realization that it cannot be achieved. The living observer can be replaced by non-living surrogates—*unavoidably* measuring devices. Even in this case, the attainable degree of objectivity, i.e., independence of the observer and of the context, is not higher than that embodied in the surrogate. Measuring devices embody the characteristics of those who conceived them and, furthermore, of those who use them. One quote from a book impossible to ignore when we look at how numbers—the data—come into existence: "Our instruments of detection and measurement…are they not like loaded dice, charged as they are with preconceived notions concerning the very things we are seeking to determine?" (Dantzig 1954, 2007). The author goes on to describe measurement as "attempt to counterfeit by number…the world disclosed to our senses." These epistemological considerations might seem alien to the discussion of a virus, the associated pandemic, and the larger confluence into an aggregate of all kinds of breakdowns. But they point to the need to disrupt science in its current embodiment as positivist endeavors.

Most of the effort in containing the pandemic focused on descriptions of an experimental nature and resulted in data. Plenty of it—and more of it daily. They were generated through genetic sequencing, through molecular medicine quantifications, through medical observations, through government-instituted symptom tracing, or through self-reporting (the Weizmann Institute, Halbfinger 2020). Take only the ubiquitous temperature checks, which were ultimately deemed meaningless (as much an expression of opportunistic medical expediency as of ignorance). From China, in the confusing early days of "something" that few people even wanted to acknowledge (the word "pandemic" was not to be used), to the protocols of admitting patients, clients, and travelers only after temperature was checked, we have an example of well-intended but unjustified measurements. According to *Color* (a provider of healthcare and clinical testing), the data show that among a group of over 30,000 individuals (a decent-sized sample), 1.3% tested positive, "and among those, 78% reported only mild symptoms or…were asymptomatic—meaning they displayed no observable symptoms whatsoever. What's more, only 12% had a fever over 100 degrees" (Etherington 2020). Within the theology of measurement, safety measures above and beyond checking one's temperature, became an alibi. This is what is practiced within the dogma of reductionism, in this case measurement of a parameter of low significance. You are not admitted to a medical facility without someone "shooting at you" some temperature gun (as they are called). You could not access your classroom without this. How many thermometers have been sold since the beginning of Covid-19? Never mind how many were calibrated, if indeed those measurements are supposed to mean something? Of course, since individuals are different, it is normal that the description of symptoms will vary. Moreover, all measurements are incomplete, and at times contradictory exactly because the living is above the threshold of the undecidable; that is, it is G-complex (Nadin 2014)—a subject to be discussed in the next section.

But this characteristic, i.e., that we cannot measure everything and expect consistency, is not the outcome of inadequate objective means of measuring, as the fanatics of the experiment would claim. The technology for measuring even what is insignificant is available. Scientists can deal with individual cells and individual molecules, even with electrons and muons. Inadequate knowledge about what is observed and what is measured explains the immense investment in technology—leading to the Big Science model—and the millions of hours of lab work dedicated to a grueling exercise that is meaningless. Covid-19 is not an earthquake, or a volcano eruption, a tornado, or a sandstorm. These can be measured to a level of detail that corresponds to a full description that is consistent. The laws of physics are the proof. Covid-19 is a continuously changing condition of the living triggered by an interaction with a virus that the infected reproduce in mutated forms. The infection can result in death, or in nothing more than mild symptoms—or in none. It is not the state of tectonic plates, subterranean lava flows, weather, or dry plains. It is not a matter of rockets reaching Mars. CoVID, the broader crisis, which integrates the socio-economic and political breakdowns, is about the condition of the living, not of the electrons, muons, atoms, and molecules of organisms. It is the outcome of the obstinate focus on physics, in contrast to neglecting the exploration of life.

References

Alexander R McN (1998) When is migration worthwhile for animals that walk, swim or fly? *Journal of Avian Biology*, 29, 387–394.

Alexander R McN (2002) The merits and implications of travel by swimming, flight and running for animals of different sizes, *Integrative and Comparative Biology*, 42, 1060–1064.

American Academy of Pediatrics (2022) Children and COVID-19: State-Level Data Report. https://www.aap.org/en/pages/2019-novel-coronavirus-covid-19-infections/children-and-covid-19-state-level-data-report/ [retrieved December 19, 2022].

Ayala FJ (2007) Darwin's greatest discovery: Design without designer. Washington DC: *Proceedings of the National Academy of Sciences,* 4:1, pp. 8567–8573 May 15. Published online May 9, 2007. https://doi.org/10.1073/pnas.0701072104 [http://www.pnas.org/content/104/suppl1/8567.abstract].

Barrie D (2020) *Supernavigators: Exploring the Wonders of How Animals Find Their Way.* New York: The Experiment.

Bernstein NA (1967) *The co-ordination and regulation of movements.* Oxford: Pergamon Press.

Bleicher A, Conrad K (2020) We Thought It Was Just a Respiratory Virus, UCSF Magazine, Summer 2020. https://www.ucsf.edu/magazine/covid-body [retrieved July 25 2020].

Bohr N (1932) Light and Life, Opening address, IIe Congrès International de la Lumière, Copenhagen. August 15.

Bohr N (1933) Light and Life, *Nature* 121, pp. 421–423. March 25.

Brzozowski A (2020) Civil unrest has doubled, and COVID-19 could make things worse. *Euractiv.* June 10 2020. https://www.euractiv.com/section/defence-and-security/news/civil-unrest-has-doubled-and-covid-19-could-make-things-worse/

Casanova J-L, Su HC (2020a) A Global Effort to Define the Human Genetics of Protective Immunity to SARS-CoV-2 Infection, *Cell* 181:6, June 11, 1194–1199.

Casanova J-L, Su HC (2020b) Covid Human Genetic Effort. https://www.covidhge.com/ [retrieved June 5, 2020].

Chauvet G (2004) *The Mathematical Nature of the Living World. The Power of Integration.* Singapore: World Scientific.

Chua RL *et al.* (2020) COVID-19 severity correlates with airway epithelium–immune cell interactions identified by single-cell analysis, *Nature Biotechnology*. June 26, 2020.

Dantzig T (1954, 2007) *Number: The Language of Science.* New York: Macmillan/New York: Plume/Penguin.

Dawkins R (1986). *The Blind Watchmaker. Why the Evidence of Evolution Reveals a Universe Without Design.* New York: W. W. Norton.

Dickson B (2020) Energy consumption of AI poses environmental problems, April 17 https://www.pcmag.com/news/ai-could-save-the-world-if-it-doesnt-ruin-the-environment-first [retrieved September 3, 2021].

Ellis GFR (2005) Physics, Complexity, and Causality, *Nature* 435:743.

Etherington D (2020) Color's COVID-19 testing shows most who test positive had either mild or no symptoms, June 15 https://techcrunch.com/2020/06/15/colors-covid-19-testing-shows-most-who-test-positive-had-either-mild-or-no-symptoms/ [retrieved June 15, 2020].

Fischetti M, Christiansen J (2021) Our Bodies Replace Billions of Cells Every Day, *Scientific American*, 324:4, 76. April 1.

Garrett TJ, Grasselli M, Keen S (2020) Past world economic production constrains current energy demands: Persistent scaling with implications for economic growth and climate change mitigation, PLOS ONE, August 27. https://doi.org/10.1371/journal.pone.0237672 [retrieved August 28].

Gianoli E (2015) The behavioral ecology of climbing plants, *AoB Plants* (Special issue: Using Ideas from Behavioural Ecologyto Understand Plants), 7, plv013. March 17. https://doi.org/10.1093/aobpla/plv013 [retrieved July 21, 2021].

Gödel K (1931) Über formal unentscheidbare Sätze der Principia Mathematica und verwandter Systeme, I, *Monatshefte für Mathematik und Physik*, 38, 173–198. (On Formally Undecidable Propositions of Principia Mathematica and Related Systems).

Gould JL, Grant Gould C (2012) *Nature's Compass: The Mystery of Animal Navigation*. Princeton NJ: Princeton University Press.

Grossberg S (2017) Towards solving the hard problem of consciousness: The varieties of brain resonances and the conscious experiences that they support, *Neural Networks*, 877, pages 38–95. March. Amsterdam: Elsevier.

Halbfinger D (2020) How Are You Feeling? Surveys Aim to Detect Covid-19 Hot Spots Early, New York Times, April 2. https://www.weizmann-usa.org/news-media/in-the-news/how-are-you-feeling-surveys-aim-to-detect-covid-19-hot-spots-early [retrieved June 5, 2020].

Harvey WH (1651) *On Animal Generation*. Translated from the Latin by Willis, R. 1847 London: The Sydenham Society.

Hemmen JL van (2021) Mathematization of nature: how it is done, *Biological Cybernetics*, 115, pp. 655–664.

Hofmeyr JHS (2017) Basic Biological Anticipation. In: (Poli, R, ed.) *Handbook of Anticipation*. Springer Reference. Heidelberg/Berlin: Springer.

Holcombe M (2021) Nearly every new Covid-19 death is now entirely preventable, CDC director says, *CNN Health*, June 23. https://www.cnn.com/2021/06/23/health/us-coronavirus-wednesday/index.html [retrieved July 21, 2021].

Kak M, Ulam S (1968) *Mathematics and Logic*. New York: Dover.

Kauffman SA (2014) Foreword: Statable and Non-prestatable Landscapes. In: (Hendrick Richter, Andries P. Engelbrecht, eds.) *Recent Advances in the Theory and Application of Fitness Landscapes*. Heidelberg/Berlin: Springer.

Kurzweil R (2005) *The Singularity Is Near: When Humans Transcend Biology*. New York: Viking.

Labbe M (2021) Energy consumption of AI poses environmental problems. 26 August https://www.techtarget.com/searchenterpriseai/feature/Energy-consumption-of-AI-poses-environmental-problems [retrieved September 3, 2021].

Leibniz GW (1710) *Theodicy: Essays on the Goodness of God, the Freedom of Man, and the Origin of Evil* (Huggard EM, Trans., Farrar A, Ed.). In: C.J. Gerhardt's edition of the *Collected Philosophical Works*, 1875–1890. Routledge 1951. Open Court 1985. Paragraph 360, p. 341.

Lihong L (2019) *Classical Chinese Medicine*. Hong Kong: The Chinese University of Hong Kong Press.

Louie AH (2013) *The Reflection of Life. Functional Entailment and Imminence in Relational Biology*. New York Heidelberg Dordrecht London: Springer.

Mallapaty S (2021) The Search for People Who Never Got Covid, *Nature News*, 29 October https://www.nature.com/articles/d41586-021-02978-6 [retrieved November 16, 2021].

McGlashon A (2019) Are These Birds Better Than Computers at Predicting Hurricane Seasons? Audubon Magazine, https://www.audubon.org/news/are-these-birds-better-computers-predicting-hurricane-seasons [retrieved Oct. 28, 2021].

McGraw E (2020) A few superspreaders transmit the majority of coronavirus cases, *The Conversation*. June 5. https://theconversation.com/a-few-superspreaders-transmit-the-majority-of-coronavirus-cases-139950 [retrieved June 6, 2021].

Nadin M (2014) G-Complexity, Quantum Computation and Anticipatory Processes, *Computer Communication & Collaboration*, 2:1, 2014, 16–34. (DOIC: 2292-1036-2014-01-003-18).

Nadin M (2018a) Redefining medicine from an anticipatory perspective, Progress in Biophysics and Molecular Biology, 140, 21–40. December, Amsterdam: Elsevier. https://doi.org/10.1016/j.pbiomolbio.2018.04.003.

Nadin M (2018b) Rethinking the experiment: a necessary (R)evolution, *AI & Society* 33:4, 467–485. New York/London: Springer.

Newton I (1686) *Philosophiæ Naturalis Principia Mathematica.* London: S. Pepys Imprimatur.

Noble R, Noble D (2018) Harnessing stochasticity: How do organisms make choices? *Chaos* 28, 106309. https://doi.org/10.1063/1.5039668 [retrieved August 12, 2020].

Noble D (2008) *The Music of Life. Biology Beyond Genes.* Oxford: Oxford University Press.

Noble D (2012) A theory of biological relativity: no privileged level of causation, *Interface Focus* 2:1, 55–64. https://pubmed.ncbi.nlm.nih.gov/23386960/ [retrieved August 12, 2020].

Pinsker J (2020) The Pandemic Will Cleave America in Two, *The Atlantic*, April 10 2020 https://www.theatlantic.com/family/archive/2020/04/two-pandemics-us-coronavirus-inequality/609622/ [retrieved April 12, 2020].

Probst P *et al.* (2016) Blinding in randomized controlled trials in general and abdominal surgery: protocol for a systematic review and empirical study, Systematic Reviews, 5, Article #48. March 24.

Rauti R *et al.* (2021) Effect of SARS-CoV-2 proteins on vascular permeability (Research Article), *eLife*, October 25. https://elifesciences.org/articles/69314 [retrieved November 17, 2021].

Rosen R (1991) *Life Itself: A Comprehensive Inquiry into the Nature, Origin, and Fabrication of Life.* New York: Columbia University Press.

Scales, Helen (2021) The brilliant Abyss: Exploring the Majestic Hidden Life of the Deep Ocean, and the Looming Threat That Imperils It. New York: Atlantic Monthly Press.

Schweitzer A (2020) *Ehrfurcht vor dem Leben. Grundtexte aus fünf Jahrzehnten.* Munich: C.H. Beck.

Sender R, Milo R (2021) The Distribution of Cellular Turnover in the Human Body, *Nature Medicine*, 27. January.

Sterling P, Laughlin S (2015) *Principles of Neural Design.* Cambridge MA: MIT Press.

Turchin P (2010) Political instability may be a contributor in the coming decade. *Nature*, 463, 608. February 3. https://www.nature.com/articles/463608a [retrieved July 3, 2020].

Widder E (2021) Below the Edge of Darkness: A Memoir of Exploring Light and Life in the Deep Sea. New York: Random House.

Wikelski M *et al.* (2003). Costs of migration in free-flying songbirds, Nature 423:704, June 12. https://doi.org/10.1038/423704a [retrieved December 31, 2021].

Wolfe CT (2011) Why was there no controversy over life in the Scientific Revolution? In: Dascal M and Boantza VD, Eds.) *Controversies Within the Scientific Revolution*, pp. 187–221, Amsterdam: John Benjamins Publishing Co.

Zeberg H, Pääbo S (2020) The major genetic risk factor for severe COVID-19 is inherited from Neanderthals, *Nature* 587, 610–612.

Zhang Y, Kutateladze TG (2020) Molecular structure analyses suggest strategies to therapeutically target SARS-CoV-2, *Nature Communications* https://www.nature.com/articles/s41467-020-16779-4.pdf?origin=ppub [retrieved June 15, 2020].

Zhang C, Shaoying H, Zheng F, Dai Y (2020) Controversial treatments: An updated understanding of the coronavirus disease 2019, *Journal of Medical Virology*. April 10. https://www.ncbi.nlm.nih.gov/pmc/articles/PMC7228369/ [retrieved May 31, 2021].

Zimmer C (2020) DNA Linked to COVID-19 Was Inherited From Neanderthals, Study Finds (July 3). Updated July 8: DNA Inherited From Neanderthals May Increase Risk of Covid-19 *New York Times*. https://www.nytimes.com/2020/07/04/health/coronavirus-neanderthals.html [retrieved July 16, 2020].

Let Biology Be Biology

Biology is about the meaning of all processes through which life is expressed. The reduction of biology to physics (so-called "biophysics") and chemistry, and more recently to algorithmic computation forced science into a cycle of ever-growing need for more data and energy.

Borrowed against the future, the data and energy characteristic of processes different from those of change in the living fail to represent the complexity of life. Therefore, the knowledge such science makes available is only marginally consequential. Medicine in the form of practices inspired by the reductionist-deterministic machine model fail both the individual and society. (This holds true for education and social activities also.) Justified by the future it makes possible, science must overcome the narrow understanding hoisted upon it by the Cartesian Revolution. "Disrupt Science" means to integrate the anticipatory perspective. In the understanding of a reality in which the living and the non-living are complementary, this integration brings about the possibility and necessity of a holistic view of life.

A Necessary Premise

Even for the most enthusiastic proponents of deterministic science (physics, in particular), it should be clear that neither gravity nor electromagnetism, neither strong nor weak forces, not even genetics explain Covid-19, and even less the aggregate CoVID. To figure out how a virus, with a pedigree of millions of years, interacts differently with its long-term host (the bat) and with the human being requires a different perspective. To understand variants and "vaccine escape" is a matter of biology, not of physics and not of chemistry, even disguised as biophysics, or biochemistry, or computational biology. As a result of interaction with human beings, the virus gets attached to an enzyme or to blood vessels. Metabolism, self-repair, and the immune system, especially of older individuals, are perturbed.

© The Author(s), under exclusive license to Springer Nature Switzerland AG 2023

M. Nadin, *Disrupt Science*, https://doi.org/10.1007/978-3-031-43957-5_5

Unfortunately, the possibility of gaining an anticipatory perspective has been totally blocked by the scientists willfully captive to reductionist determinism. The virus's make-up, analyzed in minute detail—its genome was sequenced over 120 million times as of December 2022—is, no doubt, relevant. It is a particle, visible only via scanning and transmission electron microscopes—of a mixture of 29 different proteins, almost spherical in shape, protected by a fatty membrane. The particle penetrates the cell. This is a physio-chemical process, which sequencing has never explained. The initial recommendations for ways to avoid it—washing hands, not touching the nose, a blocking layer (the better mask became the expectation)—are indicative of the focus on the physics and chemistry involved. But how does the virus spread? That is, what is the pathogenesis? Droplets (the model of flu spreading)? Aerosol? Or who knows what other ways. Science looked for an answer and eventually decided for the airborne path. The World Health Organization (WHO) came late to this determination in—April 2022 (Lewis 2021). Even after extensive research, and excessive cost sequencing, science still does not explain why the virus's viability diminishes over time, or why it varies with the seasons. Or how come animals (cats, dogs, bank voles, ferrets, fruit bats, hamsters, mink, pigs, rabbits, racoon dogs, tree shrews, and white-tailed deer) can be infected with the virus (CDC 2022), or could even become carriers.

Even after more than two years into the pandemic, science cannot define the level at which the virus's presence in the body leads to disease. A twisting strand of RNA—the instructions for transforming genes into proteins—provides what is necessary for the cell to produce similes (Fig. 1). To be clear: virions are not copies of the virus or of some of its proteins; neither are they identical. Cells, each of which is different, are part of a large biological network that generates variation, not identical copies.

The rest is up to how the defensive system—the immune response—is successful or not. An out of-control immune response can be very damaging (Akbar and Gilroy 2020). More essential is understanding what happens in the host. The situation is different from that of bacteria leading to pathologies. The virus triggers, but does not cause, symptoms ranging from asymptomatic to lethal progression. A lot was learned regarding the physics and chemistry of propagation, but close to nothing regarding the anticipatory processes involved in the production of virions by the organism—because the question was not even formulated. The physics or chemistry of the spectrum of conditions associated with the virus—in this case the detailed molecular description or genetic analysis—does not answer the question of what went awry in the organism. And why—especially *Why?* How to explain the "Ocular manifestations of coronavirus disease 2019," (Ceran and Ozates 2020), i.e., hyperemia (red eye in over 20% of patients)? How to explain the photophobia (intolerance of light in over 16% of patients) or the "Symmetric peripheral polyarthritis developed during SARS-CoV-2 infection" (Talarico et al. 2020)? Then there is "The question as to whether SARS-CoV-2 can cause direct myocardial injury" (Bradley et al 2020). These and similar questions indicative of a "non-convergent condition" remain unanswered. The disease is not one, but many: this is why "spectrum of conditions" applies. Major research of data associated with the symptoms focused on the molecular level (Garvin et al. 2020). This is a systems biology approach. Among other findings: the role of the

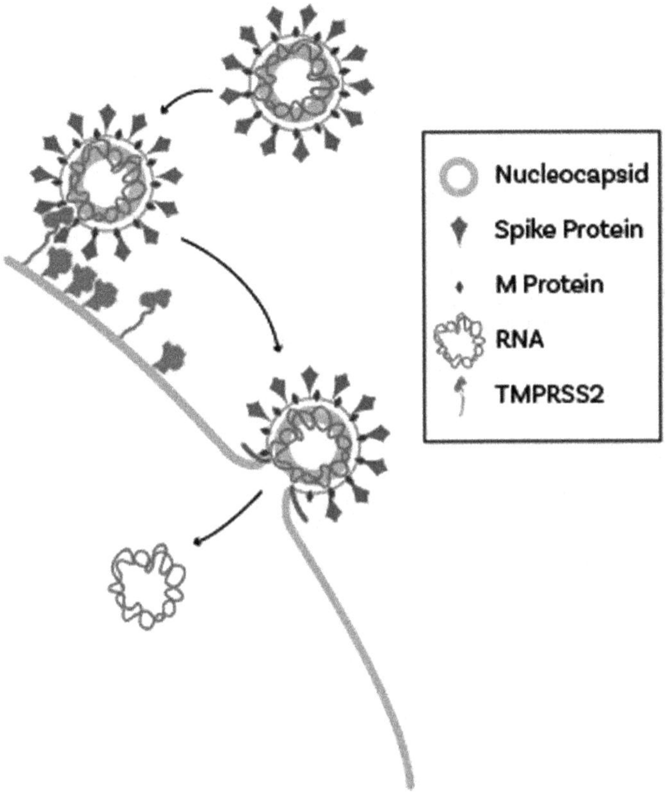

Fig. 1 The biochemical process through which the single stranded RNA virus penetrates the cell. The virus consists of a nucleocapsid, a membrane, an envelope, and the spike proteins. The angiotensin converting enzyme (ACE2), as the entry to the cell, is engaged by the virus through the transmembrane protease (TMPRSS2). Missing is the explicit biology which would describe the larger context in which the genetic processes are usually but necessarily always triggered

circulatory system, in particular the role bradykinin receptors (known to be involved in blood pressure control). A "bradykinin storm—a massive, runaway buildup of bradykinin in the body"—increases vascular permeability. As a result, accelerated heartbeat, dizziness, seizures, delirium, and stroke are present in as much as half of hospitalized Covid-19 patients. Bradykinin can also mess with the thyroid gland, which could produce the thyroid symptoms eventually observed in some patients.

Reductionism, no matter how sophisticated (the findings just mentioned go back to a supercomputer-based study of data), cannot help to guide physicians as they navigate an unprecedented landscape of pathologies. For all practical purposes, from the pandemic's onset, medicine was heroic, but unsuccessful in treating the afflicted. The good news is that, recognizing the situation, some voices proclaimed "The Need for New Ways of Thinking" (Osterholm 2005; see also Osterholm 2020). Just one statement in this respect:

> We need a new balance of values and new ways of thinking and acting. This new thinking must transcend national and institutional boundaries and recognize that, in a globalizing world, health and disease in the most privileged nations is closely linked to health and disease in impoverished countries. Sustainable improvement in health and well-being is a necessity for all, and the value placed on health should permeate every area of social and economic activity (cf. Benatar et al. 2020).

After more than two years of corona, this statement is as current as it was when it was first formulated. During the pandemic, some physicians and scientists eventually arrived at this understanding. But not the medical-industrial complex (as dangerous as the military version). In the USA but also in Europe, recommendations were issued in total disregard of the experience of those who treat patients or advise them on prevention measures. The treatment of children, especially, was a failure. It has not improved over time. Therefore, it comes as no surprise that "Citing botched pandemic response," Dr. Rochelle Walensky, the director of the Centers for Disease Control calls for its reorganization (LaFraniere and Weiland 2022). It is doubtful that a new perspective of biology will guide such a reorganization.

In acknowledging the immense opportunities opened by the Cartesian Revolution, we cannot stop short of noticing how the image of reality is skewed by *machinism*: the body as a machine, and the physician as a better paid mechanic operating on a "car" racing on the highway of life (Fig. 2).

As the pandemic brewed, the data processing machine called *computer* took over almost all attempts to "understand" (or explain) what was going on. The simulation of life, as a non-deterministic process, is not possible within a deterministic medium. Simplified models, such as those used in computational biology, lead to questionable results. Computational biology—flying high on the machinery made available

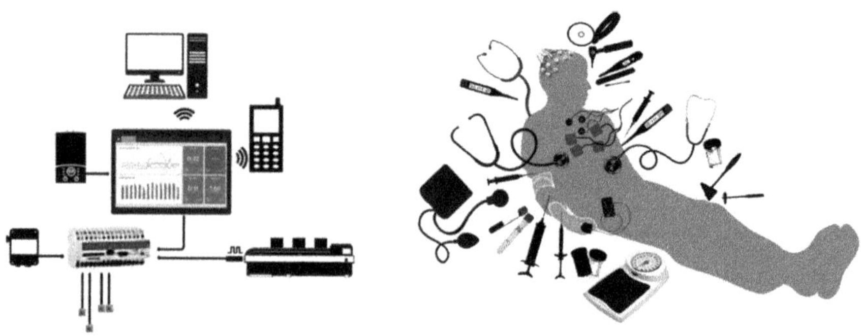

Fig. 2 Machines can be fully monitored. Assuming that the method and means of measurement are appropriate. The more data from measuring (monitoring) their components, the more precise the control process of their functioning. On this path medicine scores high on the use of technology and low on its effectiveness for helping people maintain their health. Excessive monitoring of the human being results in a question medicine should have asked long ago: "In the era of precision medicine and Big Data, who is normal?" (Manrai et al. 2018). The uniqueness of each individual and the obsession with measuring deviations from proscribed values are incompatible. Unfortunately, the premise of the significance of measuring is ignored

through the massive genome project—provided model after model of distressful irrelevance. Genetic analysis—in the form of genetic sequencing—is inspiring, but results only in a multiplication of labels, not in operation able knowledge. It is an impressive analytic effort—what can be called "cartographic"—that does not provide the practicing physician with any actionable knowledge. An analogy: Imagine a ship caught in a disastrous storm. Measuring the coordinates—location at some moment in time—again and again, and with different means, does not make navigation easier. The inventory of the molecules from which the ship is made is of no use. What counts are the forces at work, e.g., a storm has many variables. The crew's competence, engine capabilities, cartography, etc. play their roles. In the pandemic, measuring became an obsession—tests, genetic sequencing, all kinds of statistics. Missing was the understanding of the situation, the knowledge.

The names of a multitude of genetic components identified in the broad spectrum of the Covid-19 are now part of the scientific medical vocabulary (at least in published articles). But as long as the community of physicians does not have an *understanding* of what these identifiers are, they cannot help the patient. The multiplication of names (corresponding to genetic "particles") defies Occam's logical principle of not increasing the number of concepts beyond what it needed to make valid inferences. Indeed, the undecidable cannot be described computationally or otherwise, no matter how many genetic details are generated in the sequencing process. To chase for chemicals that could prevent some of the infection's consequences is to defy the nature of living processes. Focus has to be centered on genetic processes of the host cell, since such processes drive the making of the virions. The anticipation at work in the process results in the multiplication of the pathogens. Questions no one has yet asked: "Why does the cell copy the virus? What is the purpose of the process? What is the evolutionary path in the zoonotic? Due to an enormous effort of scientists in many fields, we know that the cells make virions. We know the syntax. But Why? Well, within genetic determinism the Why? question is left out. Or answered childishly: Because!

For everything that is deterministic in nature, the model guiding the data processing is decidable. In other words, such processes can be described fully and consistently. But life is undecidable. The foundation of a biology no longer conceived as an extension of physics or chemistry is based on this realization.

What is Science?

It is worth repeating: Science is the outcome of a particular form of human activity whose goal is the acquisition and dissemination of knowledge. Science answers questions about change. In a world of sameness, there would be no questions and therefore no need for continuous knowledge acquisition. Life itself is the outcome of change. Agency, as a characteristic of the living, is the outcome of questioning. Questions regarding the past, present, and the future express awareness of change in time. The awareness of a living questioning entity is expressed in questions pertinent to survival: metabolism (acquiring energy) and reproduction in the first place. Empirical evidence, on account of the particular ways awareness of change is expressed, documents that matter (with no awareness) and living matter are different, if not in their making, then in their organization. Therefore, the need for a knowledge perspective adequate for each of them is not an option, but rather a necessity. Ultimately, the subject of inquiry of physics is change in non-living matter. In particular, how elementary particles make change possible, or even necessary. Its accomplishments in this knowledge domain—means and methods for measuring all that there is—are as impressive as its failure to explain life. Chemistry, not unlike physics, is a well-defined measuring-based knowledge domain focused on the elements making up matter. For lifeless matter, it returns actionable knowledge that not only explains the role of the elements in physical change, but also provides tools for engineering matter with predefined characteristics. New materials, with pre-defined characteristics, are made from elements. In respect to the living, chemistry provides knowledge pertinent to the interaction of elements (the chemistry of drug use, or for transgendering, for instance). It also provides knowledge pertinent to what is known as the genetics of life. Inferring from measurements (leading to data) of how matter changes as a result of physical or chemical processes to how life unfolds is seductive, but unavoidably misleading. It cannot succeed because change in the living is different in nature from change in non-living matter.

Biology is specifically about change in living matter. To understand life transcends measuring the matter in which it is embodied, although the role of life-specific components (such as cells and proteins) is relevant in assessing its dynamics. Understanding the meaning of change is made explicit by answering questions such as: Why do they take place? Why in the specific manner in which they are expressed? Such questions are the actual subject matter of biology. Nevertheless, with their different focuses on change, physics, chemistry, and biology are not independent of each other. Fungi and algae on stone surfaces accelerate their erosion in time: in addition to physical interaction (such as the interaction of two masses in Newton's physics), chemical factors are at work. The dynamics of the universe, manifested in cycles of day-and-night, or of seasons, is reflected in the chemistry (reactions triggered, slowed, or accelerated) of plant, animal, bacteria. Nevertheless, the dynamics of life cannot be reduced to chemical processes. Poisons (what viruses are) affect life. The subject of how physio-chemical changes in non-living matter and in living matter are related frequently escapes the interest of scientists. Interdisciplinarity in the age of extreme specialization was rendered impossible.

Dynamics—how fast, how often, how deep, or how broad, how permanent or how accidental, etc.—describes how change occurs in space and time. The dynamics of non-living matter is entropic. Things decay. The dynamics of living matter is negentropic: on account of metabolism, organisms adapt and increase their viability along the cycle birth-maturation-death. Physical change, as well as chemical change, can be fully and consistently described, i.e., it is *decidable*; change in living matter is *undecidable*. In dealing with life, physics or chemistry selects from the whole of the organism a section. Reductionism (to particles or elements) is the methodological premise of the attempt to infer from the determinism of non-living matter to the causal relations of living matter.

Considering the Greek origin of the word, biology is "to speak about life." Or to think about life, since *logos* is the prerequisite for thinking. *Lifelore* was used in the English-speaking world to describe this knowledge domain. It was not until the six-volume treatise published by Treviranus (1802, 1822) that a science pertaining to life phenomena will be labeled "Biology." In the tradition of Aristotle's empirical observations and philosophic focus on life, biology draws much of its knowledge from those who studied plants (botany), animals (zoology), and from medicine. Advances in optics, the microscope in particular, and in dissection and staining, afforded biology some of the earliest tools for discovering life forms and for empirical observations regarding the composition of living matter. Indeed, biology enrolled physics and chemistry in its attempt to understand what life is. In particular, the cell (so named by Robert Hooke) was identified in 1665; but only when Schleiden (1838) and Schwann (1839) described cells as a basic unit of organisms did they become a focus of science. Life unfolding over time—Aristotle's narration of the embryological process—became the dominant perspective. Evolutionary thinking became part of the time perspective of life. Not until the 1940s did the reductionist view of chemistry become dominant in biology. Biological inheritance was the gateway to the genetics revolution through which reductionist-deterministic physics and chemistry disguised as biology is imposed on the life knowledge domain. The life-machine model—in reaction to vitalist explanations of life—preceded the Cartesian Revolution. Machines embody the mechanistic causality supposed to explain the living. In this model, the future, which prompted interrogations leading to anticipatory action, is reduced to an extended past. Anticipation, as a characteristic of life processes complementary to deterministic reaction, is ignored.

For about 400 years, humankind (at least the Western world) has let itself be guided by the foundation set by Descartes and further consolidated by Newton. The functioning of machines was both proof of the adequacy of this explanation of reality and the goal of human activity. The deterministic sequence of *cause* → *effect*, and the reductionist method affirmed a reality so powerful that any alternative was either ignored or deemed unacceptable. Not only was the universe described as a machine, but everything else was considered to be one—the living included. As a description of the inanimate material world, and as an expression of the laws governing its functioning, deterministic-based physics and Cartesian reductionism proved extremely powerful instruments in humankind's overall progress. But neither Descartes nor Newton, nor any of their followers, not to say Einstein and subsequently the quantum mechanics scientists, could have envisioned the revelations associated with research into the complexity of life. Worse yet, it is doubtful that they recognized the potential for breakdowns caused by the science they produced or inspired. The physicist Erwin Schrödinger (usually associated with quantum mechanics) concluded that organisms are subject to a "new physics," (Schrödinger 1951)—referring to the physics that his equation described. Walter Elsasser, close to Schrödinger, eventually tried to provide such an alternative. He undertook an attempt at a new foundation of biology not reduced to reductionist-deterministic physics, but in extension of quantum mechanics (Elsasser 1987). For him, this was a "new physics," and Elsasser understood, indirectly, that it would have to integrate the knowledge of anticipation processes in order to adequately address the living instead of subsuming it to non-living matter. Quantum mechanics, maintaining that phenomena in the world makes possible the understanding of some aspects of anticipation are stochastic in nature. However, there is more to anticipation than what the stochastic view explains.

It should be pointed out that quite often testable descriptions of physical phenomena, in the language of mathematics, have been extended as explanations of the dynamics of life. Fermat's *Principle of Least Time* (1650) from optics (light propagation) or Maupertuis' *Principle of Least Action* (minimum of time translated as in terms of the energy necessary in the motion of a mechanical system, 1774) are examples often given. It is seducing, but ultimately confusing, to take their clear mathematical formulations and generalize to the living. These two examples are mentioned as paradigmatic of how successful mathematical descriptions of physical phenomena inspired attempts at understanding living dynamics. Together with the valid mathematics comes the burden of the deterministic nature of what such theories explain. As we shall see, in the analysis of genetics, biology was taken hostage to the machine model: life was made into another engineering application. Creativity, as an expression of anticipatory processes, was left out, if not openly ostracized. Without introducing the future in the description of change, science remains hostage to the past from which everything seems to originate—including statistical data.

Along this line, it is worth mentioning that in writing about "Simulating Physics with Computers," Feynman (1982) identified possible futures. "The machine can simulate physical processes in nature," but he did not state that the machine is "natural." Feynman goes on: "Just let's think about a more general kind of computer…," and describes the space–time coordinate of the state of such a machine depending on the future and the past (for more details, see Feynman and Hey 1999, Nadin 2012). From all the physicists of recent years, Feynman (like von Neumann, among mathematicians) would be the only one able to open up to an anticipation perspective in science.

On this note, and given the Sustainability Thesis of this book, a distinction begs our attention: preventing versus fixing. Anticipation is always expressed in action. Control procedures evolved over time as ways to correct errors. Aristotle articulated the thought:

> …if every instrument could accomplish its own work, obeying or anticipating the will of others…if the shuttle weaved and the pick touched the lyre without a hand to guide them, chief workmen would not need servants, nor masters slaves (*Politics*, Book 1, Chap. 3)

This describes the functioning of a synchronized whole. And if the organism embodied in living matter is not the example *par excellence*, none better can be found. The interdependence of elements is quite well documented. For the human being, medicine, but not only, is a repository of knowledge about the interdependence in question. Animals and plants are no less integrated entities; their viability depends on each part and on their integration in a dynamic whole. What are the processes through which this integration takes place? Following the arguments advanced so far, this is a question of biology, not of physics, and even less of chemistry. To the extent that mathematics, as a multitude of languages (e.g., of number theory, geometry, trigonometry, integral calculus, differential calculus, category theory, etc.) is the language of physics descriptions, it is useful to examine the construct called control mechanism, and eventually to contrast it to biological processes.

A state equation serves as guidance for understanding how control processes take place:

$$A(y) = f(v) \tag{1}$$

To explain: **y** describes the state of the system to be controlled. It belongs to a vector space $\boldsymbol{Y}$—i.e., one populated by objects (various parts of the body, various organisms, functions, the ecology, etc.). It is in this space where interactions (describable "as forces in action") take place. Sometimes they are not easy to define. The control is described by $\boldsymbol{v}$, which belongs to the set of admissible controls $\boldsymbol{u}_{ad}$. Finally, the function $\boldsymbol{f}$ shows how the control is exercised on the system. For each $\boldsymbol{v} \in \boldsymbol{u}_{ad}$, the state Eq. (1) has as its only solution $\boldsymbol{y} = \boldsymbol{y}(\boldsymbol{v})$ in $\boldsymbol{Y}$. Therefore, to control a deterministic process is actually to find a value in $\boldsymbol{v}$ so that the solution to the equation describing the control system gets the system where it is intended to be. The thermostat, for example, controls the behavior of the many components of a heating system. Further down the line of this description enter the feedback and feed-forward processes, and the attached timing aspects (it takes time to exercise each of them). Steering a car illustrates this thought. The feedback loop provides the input for an action that will change the output. Each step in the functioning of a system associated with a control mechanism is the expression of the causality specific to determinism. The entities controlled remain the same. What is controlled is their interaction. This fact alone indicates that control procedures in the physical of physical entities (artifacts of all kinds) and in the living are fundamentally different. In the first, determinism is unavoidable; in the second, anticipatory processes are at work. The outcome is not pre-determined. In the living, these are non-deterministic processes (which sometimes fail).

Anticipatory action is not of the nature of fixing something (which is characteristic of control systems), but rather pre-emptive. It is holistic, i.e., it engages the entire organism. What informs anticipation is the possible future, not the past of deviations from a given path, or a value of some parameter. Evidently, feedback is of a different nature: it is an inference process, based on the past states of the organism and what it takes to keep it under control. Anticipation is the path towards the future, and as such it might as well deviate from given or expected parameters. Adaptivity is the most suggestive example. Of course, being of a non-deterministic nature, the process can fail—mal-adaptivity, documented in evolution, is a possible outcome.

If all of this seems too abstract in the argument for anticipatory action, let us examine a concrete example pertinent to the issue of sustainability. Way back in time, somebody described the frozen tundra of Siberia as "eternal frost" (*vechnaya merzlota* in Russian) (Yaffa 2022). The same is true for a large territory in Canada and Alaska. When nine million square miles of Earth's surface is now subject to the thawing of the permafrost—not as permanent as the name suggests—it is impossible to ignore the effect of greenhouse gases. Some see the thawing as a problem for engineers: how to get the system back to its dormant state. It's tough to live over a capricious iced surface. Build on cement piles (driven 40 feet into the permafrost) is one method. Another way is to somehow maintain the cold air so that the melting is stopped. Albeit, the system to be controlled is not at all easy to manipulate. Each intervention is very costly. Alternatively, the anticipatory nature of the ecosystem comes to expression: new vegetation sequesters the carbon released from the melting permafrost. Shrubs and trees benefit from the thaw. Photosynthesis removes greenhouse gases, and a new equilibrium is established. In addition, animals that used to live in these habitats return. Some scientists are even considering resuscitation of grasslands through facilitation of large herbivore presence. Natural processes have a rhythm—much slower—different from those corresponding to artificial control mechanisms—and the reactions these lead to. The timeframe for control mechanisms of machines and of processes resulting in a new equilibrium are different.

The realization, more implicit than explicit in the example given, that the dynamics of the world is the expression of the unity of reaction and anticipation is not new. While in this study they are treated as distinguishable from each other, in reality they are interwoven along the line of the notion of complementarity that Niels Bohr advanced. Given the sustainability implications of activities based on faulty science, it becomes critical to transcend the deterministic view. This can come about if the description of the dynamics of change integrates anticipatory processes. The urgency of recognizing the need for such an understanding is reflected in the attempts to learn from the life sciences new ways of approaching challenges for which physics alone, or in combination with chemistry, does not suffice. Living phenomena, such as neuronal activity, inspired "deep learning." But even these benefits, corresponding to biological awareness, are ultimately spoiled when they are used for advancing machine theology. The mathematical construct called *artificial neuron* replaced the living neuron as a "building block" of so-called artificial brains—which are nothing more than yet another type of data processing machines. From the infinite diversity of neurons to the artificial neuron, as the "atom" of every possible neural network, all that is achieved is the need for more data and more energy to process it. In other words, the imitation of what the brain is capable of, at a much lower use of energy and data is compromised by equating the brain with machines. A child learns how to distinguish a cat, a ball, a family member's face, on a relatively limited sample, using a fraction of what artificial neural networks require for their training. The biology of learning processes is different in nature from the physics of machine learning. A lot of progress is made in conceiving machines that imitate what the living achieves at a much lower cost. But no progress whatsoever is on record for understanding how and why the living succeeds (or fails) in surviving and reproducing itself. With this in mind, let us take note of the nature of the distinctions between living and non-living entities.

Gaining Perspective

The perspective of the world that anticipation opens justifies the qualifier "second Cartesian Revolution." The knowledge to be gained by adopting this perspective must be meaningful, i.e., must support consequential action. In this vein, but without an explicit claim to constituting a complete theory in these pages, let us suggest distinctions, informed by empirical evidence, that could qualify as the premise for a Second Cartesian Revolution, and thus be of practical significance.

The living	The non-living (the physical)
Heterogenous	Homogenous
Non-deterministic	Deterministic
Top-down and bottom-up	Bottom-up
Idiographic	Nomothetic
Repetition without repetition	Repetitive
Undecidable/G-complex	Decidable—complicated or simple
Change in the living is:	Change in the non-living is:
– purposeful	– purpose-free
– evolutionary	– entropic
– anticipatory	– reactive
Holistic	Reductionist
Creative/reproductive	Not reproductive
Self-renewal	Sameness

The future state of a physical non-living entity—e.g., a stone, a star, a satellite, a weather system, a ray of light, a neural network—is predetermined. In Laplace's view (1796), we could describe how, inexorably, a stone will turn into sand, and at which rate the process occurs. If factors other than those reflected in the laws of physics are part of the process (e.g., chemical reactions, interactions with living matter, etc.), the rate of change is affected. The future state of a non-living entity, in particular, of a machine is predefined. In a hypothetical world void of life there is no room for ambiguity. The future state of a living entity is in anticipation of possible changes, but not pre-determined. The future state of a particular organism (the individual) is different from, but not independent of, the future of the species it belongs to. Definitory is the reproduction of life, in a process in which difference, not sameness, is the decisive aspect. In evolution, relatively stable continuity characterizes the dynamics of species.

Most important: The living is capable of learning. Among those who took note of this was none other than John von Neumann, the maverick mathematician, trying to understand why the living is not subject to the laws of thermodynamics. This is the explanation for functions acquired over time, as well as for increased performance. Anticipatory abilities, upon which higher performance depends, are acquired through learning. The aggregate behavior (such as that of a swarm, of an ant colony, of communities, etc.) reflects the interdependence of organisms. If nothing else, the preparation for long migrations illustrates the nature of learning. So does the behavior

Fig. 3 The nomothetic and the idiographic illustrated: rocket science and art. Knowledge of physics and chemistry is the foundation of rocket technology. The art of movement conjures meaning

of a beehive, and of those that are part of it, including its biome. And so do various forms of interaction that result in families, tribes, clans, communities, etc.

The list of complementary characteristics of the living and the non-living should be considered as suggestive of a theoretic framework, not as a substitute for coherent conceptions. Awareness of such distinctions should help in understanding, for example, why a physics- or chemistry-inspired approach could not yield consequential knowledge about the change that viruses (such as SARS-CoV-2) might prompt. Unfortunately, the Covid-19 pandemic made it quite evident that such an understanding is missing. An observer of a world devoid of life (seemingly, some planets in our universe are such places) could describe what happens there on account of laws of physics, i.e., laws pertaining to matter. In such a world, the entities making it up are all the same: electrons, protons, neutrinos, etc. The qualifier *nomothetic* (corresponding to law) describes the situation. In a petri dish, every living entity, regardless of size and kind, is unique (labelled *idiographic*). There is no way to describe the dynamics of this small world, except by reducing it to the particles making up the matter or the elements from which its population is made. But once this reduction is carried out, life is left out. Literally for the sake of illustration, let us see how the distinction between nomothetic (pertinent to physics) and idiographic (pertinent to the living) translates into actionable concepts (Fig. 3).

Rocket science and technology demonstrate how knowledge of the laws of physics and chemistry guides the design, the testing, making, and deploying of new means of exploring the outer space. Previously, it was the physics of light propagation (including reflection and refraction), which Fermat tried to describe, and the physics of lens-making that informed the exploration of the outer space by using telescopes. What was not directly observable (the reality hidden to the naked eye) was observed with the aid of instruments. Today such instruments are replaced by digital means and supported by large arrays of data collection and data processing. The laws of physics capture the nomothetic nature of the phenomena observed, i.e., the way in which they are determined. Measurement, resulting in data, provides the description.

The uniqueness of aesthetic expression is not describable through laws. High performance sports men and women, etc. could be featured instead of dancers, musicians, or works of art). The expression is acknowledged in its originality—and valued as unique. It can be imitated (to a certain degree), but the creative process is not reducible to its outcome, as it is not the automatic outcome of new means of expression. The piano became a necessity as the art of music evolved. The photo cameras expanded the aesthetic realm. Art conjures meaning in the act of interpretation. There are no instruments (e.g., microscopes, motion or sound capture devices, etc.) for revealing the meaning. An artistic expression can be (and often is) measured. Numbers can be attached to it. However, what is measured is its physics at work; or its chemistry: How did the masters create those pigments that give paintings or tapestries or sculptures the uniqueness associated with art? But the "art" as such, a fusion of all means used, conjures meaning in the act of its interpretation. By necessity, interpretation is one of never-ending discovery. Over the life of such aesthetic expressions there is continuity—corresponding to what human beings of the past, of the present, and of those to come have in common—and discontinuity, corresponding to new contexts of life, as well as to how everything alive changes.

This is the most significant characteristic of life: a creative process in which uniqueness is continuously achieved. There are no two identical organisms, as there are no two identical cells, and for that matter, no identical DNA sequences. von Neumann (1966) was probably right in identifying reproduction as the concept best suited to informing the interest of mathematician in describing life. This is yet another example of how the dynamics of life inspires many scientists in their attempts to describe reality, in particular, research of phenomena that are classified under the label *Physics*.

As a starting point for discussing how the deterministic perspective undermined von Neumann's theory, let us take note of the following: Life is not self-reproduction in the sense of mechanical production of identical copies of the self. If it were, then evolution could not take place. Moreover: *identical* would mean that at all levels of the living, we would notice sameness. But there are no identical organisms, as there are no identical "parts" of the "whole." The interaction of infinitely diverse cells, making up a meaningful holistic reality is all that there is. John von Neumann ignored this diversity. An automaton—the simplest mathematical machine—sufficed for building a world capable of self-reproduction. It all boiled down to numbers, and to describing, in computational form, a space–time location that is alive: the world as a space–time, 3-dimensional canvas where here and there is a star (easy to describe), a mountain, a river, a living creature (or many of them). The specific dynamic of living matter, originating in anticipatory processes through which life is perceived, was either ignored or discarded because it did not fit the deterministic premise. Life is not a location, rather a continuous exploration, an active presence, a making and remaking of reality.

The fact that the various languages of mathematics undergird the successes of deterministic science does not necessarily mean that only mathematically described phenomena give these descriptions the status of science. Chaitin (1970) seems to echo the expectation of mathematization: If Darwin cannot be expressed mathematically, "...then Darwin must be wrong." Those who hold this view implicitly state: Until we have the data from measurements generating numbers, we cannot validate our hypotheses. Within this view, measurements reduce causality to its deterministic aspect. Given that life is of a more encompassing causality than that describing lifeless matter—the act of describing itself being part of the definition of life—mathematization (not mathematics though) is yet another form of reductionism. The need for languages different from those currently used in mathematics, conceived for biological phenomena, was repeatedly made explicit—most directly by Gelfand (Gelfand and Tsetlin 1971, pp. 1–22) and Bohr (1933). In his lecture at the II Congrès Internationale de la Lumière, he argued for a study of life not limited "to the properties of inanimate bodies." Meaning—in other words, the *Why?* of phenomena associated with purpose—is not captured in mathematical descriptions "of "inanimate bodies." The attempt to reduce biology to either biochemistry—the entire genetic program—or to computational biology, again, leaves out meaning. Since anticipatory processes are not driven by numbers—a construct that does not exist in living nature—but by meaning, it is plausible that semiotic descriptions are a necessary complement to mathematical descriptions. Semiotics, as Peirce's "logic of vagueness," (Nadin 1983) even inspired what is called biosemiotics.

In the spirit in which von Neumann addressed the living, focusing on self-production, those trying to explain reproduction and heredity develop theories that go back to "the stuff life is made from." Gregor Mendel's experiments in hybridization (1806) and the realization that characteristics of the living are transmitted from one generation to another point to processes different from self-reproduction. Friedrich Miescher, examining white blood cells in 1896, identified *nuclein*—what today s called DNA. William Bateson, in reference to a particular recessive disorder, defined (in 1902) the "chemical individuality," which Oswald Avery identified as an acid—deoxyribonucleic acid, which DNA stands for. The reason for recalling this history is that biology is now becoming chemistry. Chargaff, taking note of the difference in DNA composition among species, opened the path to what eventually (1953) was identification of the double helix structure of this acid and thus the description of a language made up of four letters. In 1967, Harvard University decided to split the Biology Department into the Department of Biochemistry and the Department of Molecular Biology. The focus on the unfolding over time of the meaning of life as a creative process (Fig. 4) was traded for the seductive promise of explaining, and eventually making, life by numbers.

From this point, the machine model took over. Machines for sequencing (Sheila Nirenberg, who founded Nirenberg Neuroscience; Feldman and Sanger 2006, among others) and the assumption—not different from the one that von Neumann promoted—that there is a genome, i.e., a genetic description that covers all (at least for a given species) are the outcome of adopting this perspective. What is generated is a statistical aggregate, a generic representation. A criminal brought to justice is identified by his or her specific genetic signature, not by the so-called human genome. Even the famous Watson and Crick had their "individual genome" identified. They did not adopt the infamously expensive human genome in order to look into their genetic make-up. Hopes were high for medicine: identify disease before it affects the individual; treat and cure by genetic means; create a new medicine, precise and capable of erasing the qualifier "untreatable" from the vocabulary. While statements continue to be issued regarding yet another genetic switch (turned on or off) that prevents some condition or cured it, in reality, there is more hope (justifying the acceptance of *hype*) than an effective genetics-based medicine. Francis Collins, former leader of the Human Genome Project, and in charge of the NIH (national

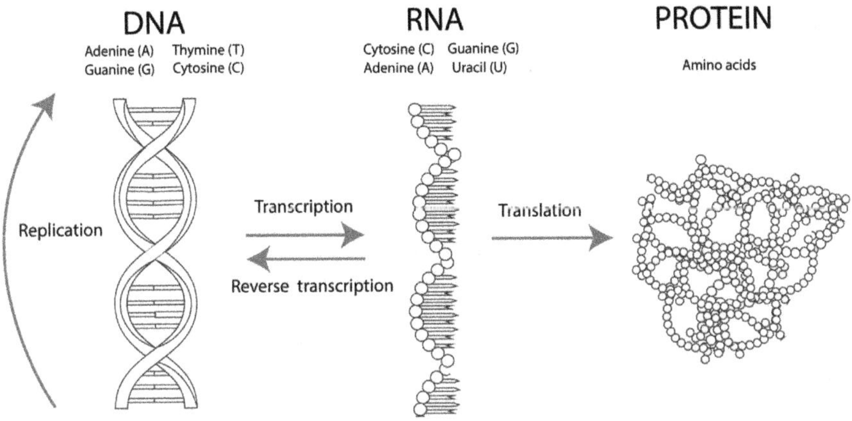

Fig. 4 The dynamics of genetic description is supposed to explain the unfolding of life. The reductionist model is relevant at the syntactic level, but it fails to explain the dynamics of life processes

Institutes of Health) for many years, objectively described the situation: "The genetic architecture of common diseases is turning out to be more elaborate than we might have guessed" (Dockser 2020). At costs running into the hundreds of billions of dollars (ever higher once genome editing, i.e., CRISPR-Cas9 came onto the world stage), the results (cystic fibrosis, hemophilia, and eye disorder called Leber congenital amanrosis, and not much more) are far from what society was promised. The Upside/Downside Ratio characteristic of CoVID—investments exceeding the benefits by many orders of magnitude—is illustrated by the multibillion-dollar genome project touted as the key to ending disease in human beings.

The fact that DNA is individualized—or, better yet, it contributes to individualization—is ignored. For the sake of clarity: there are similarities (estimated in the high 90 percentile, even to 99.9 percent according to some)) among individuals. Still, each individual is unique. Genotypic differences underlie phenotypic differences—but we don't know how, and even less, why. Due to the genetic reductionism of sequencing, not the semantics of DNA "words" (the succession of letters standing for yet other acids) is lost, but the pragmatics: the *Why?* of each genetic sequence. Since science eventually discovered that cells can be engaged in producing proteins, from the DNA through the RNA, this path was naturally pursued in the conception of therapies. But again: if the genetic processes were mechanical—i.e., the generation of identical entities—success would be guaranteed. In the reality of life, however, it is not. To cure or to stimulate the anticipatory immune system is possible only through personalized interventions. Using the data processing performance of computers, genetics abandoned itself to the embrace of algorithmic thinking: There must be a recipe for … (e.g., identifying a disease, a cure, for augmenting functions). This approach generates enormous amounts of data. The fact that the living succeeds, or fails, on "little" data, but on massively distributed and integrated functions, is ignored. Data processing is similar to strafing: indiscriminate machine gun firing. To go back to

Chaitin: If you cut away a leg of a table, balance is lost. There is no compensation from the other parts of the table. Many animals are capable of functioning after the loss of a leg, a wing, an ear, or an eye, since the organism's holistic nature contributes to their viability, i.e., conservation of life. Moreover, there is learning. By ignoring the role of learning in life processes, the data processing model—genetic sequencing generates data—does not account for the meaning of processes.

However, in the reality of life, gene expression is in anticipation of change, not in reaction to it. In very recent years, the focus has shifted to the cell as the "elementary particle" of life. This has led to attempts at rebuilding, and eventually restarting, the process of making life from lifeless matter. According to the hypothesis, if a minimal cell can be defined (i.e., by removing everything that does not destroy it as a living entity), creating life from scratch would take not biology, but chemistry, physics, or computer science. So far, the experiments carried out have convinced the researchers that this form of reductionism is as infertile as all the others performed in the spirit of the Cartesian Revolution (Somers 2022). No surprise then that Elsasser (1981) bluntly stated, "*Biology is non-Cartesian science*" [italics mine]. Furthermore, Elsasser considered actual physics and chemistry as parts of an authentic holistic biology. The consequences of reducing biological phenomena to particles and elements through data processing should actually be met with *awareness* of those consequences.

On the path of synthetic biology, a deep and far-reaching form of "genetic engineering" is making progress. It is impossible to underestimate the upshot of synthetic biology. The repercussions become clear as we examine the implications of understanding the definitory nature of anticipatory processes for the realm of life. Szybalski (1974) coined the term "genetic engineering," in the attempt to describe how genetics could modify the condition of organisms. Recombinant DNA—i.e., gene alteration—could lead to artificial organisms. Bacteria and microbes that had never existed could claim new names in the inventory of species. Over 30 years ago, the city of Cambridge, Massachusetts banned genetic engineering. In our days a great number of scientists is involved in making new molecules (such as those used in the mRNA vaccines), or at least in simulating biochemistry in silico (via computer models). The genome—actually a large computer graphics illustration of how atoms of carbon, oxygen, hydrogen, nitrogen, and phosphorus are arranged in the double helix—is the outcome of a huge analytical effort. In terms of a sustainable (i.e., not at the expense of the future) science of the living, the accomplishments are not inspiring. "GMO" is not only a label appearing on food containers (actually, "non-GMO," to warn consumers), but a factor impossible to ignore in view of the increasing dependence of human health on means of cultivation and production of questionable conditions. Synthetic biology might contribute means for addressing the covid pandemic, but it probably contributed willy-nilly to its development.

Is All Biology Computational Biology?

Computational means and methods, ranging from data processing to computer graphics (used in almost all visualization techniques, from medical applications to genomics), to database management, AI, and Machine Learning have been adopted in every aspect of biological research. Some biologists are enthusiastic; so much of their work is automated, so much transitioned from in vitro (sometimes messy, since life is messy) to in silico (no blood, no fecal matter in computers). Some "wet science" scholars realize that what they never cared for—dealing with numbers or with mathematics—is now handled by others. They become increasingly dependent on competences outside their field. A few of them deplore the fact that biology has become subservient and disappeared behind the numbers. It's no surprise that the Editors-in-Chief of the *New England Journal of Medicine* qualified computational biologists as "research parasites" (Longo and Drazen 2016). They actually meant that biologists acquire the data but depend on the "parasites" to process and make sense of them.

There are two misunderstandings that led to the current situation: (1) What is biology? (2) What is computation? Tso far, the questions are not properly answered. In the arguments presented so far, biology is defined in relation to change in the living. Measuring change in the living is different in its nature from measuring change in "inanimate" matter. Data describes completely and consistently how matter changes; it fails to do the same in respect to living matter. The missing aspect is the meaning of change, i.e., what defines the purposefulness of life at each level. The answer to "What is computation?" is of no less relevance to its adoption in biological studies. Moreover, given the nature of algorithmic computation—the dominant form of computation in our days—it should become apparent that there are clear benefits associated with using computers. But there are also intrinsic limits.

It is significant that the pandemic elicited attention to the process through which the virus is multiplied by the cells it gets attached to. Given the computational reductionism of our age, it is important to understand why the description "multiplication machine" or "copying machine," in reference to the unintentional host of the virus is an expression of ignorance. It was justified when, as already discussed, von Neumann (1966 *op. cit.*) conceived of machine-making of machines—inspired by self-reproduction characteristic of living entities, but quite out of place in respect to the living cell. The error harks back to Hilbert and Ackerman (1928) who were hoping that a mechanistic method could show whether a mathematical statement is provable within the system. More precisely, whether statements within arithmetic are true or false. Human beings are the originators of mathematical statements, including arithmetic, and some mathematicians are pretty convincing in showing whether they are true or false. Turing shattered the hope that such a mechanistic procedure is possible (1937), but in the process conceived a machine that can make any recipe into a machine. This is how typewriters, calculators, ledgers, etc., etc. are embodied in the algorithmic Turing machine—the "mother" of all recipe-driven machines. Computation, in its Turing machine embodiment, is the concrete expression of a negative result: No, there is no mechanistic procedure for proving, in a limited number of steps, mathematical statements. In his famous paper (*On Computable Numbers, with an application to the Entscheidungsproblem*, 1937), Turing clearly demonstrated (in the spirit of the works of Gödel and Church) that everything that is reducible to an algorithm (which means a "recipe") can be computed. This is different from saying "Everything is reducible to an algorithm," and even more different from "Everything is computable"—the theology of our days, and the definition of computational reductionism (or the so-called pan-computationalism). The self-reproduction model of automata that von Neumann produced is algorithmic. The description of what happens when an RNA virus—like the SARS-CoV-2 reaches a certain cell—it inspired is misleading. The virions are the outcome of biological processes not reducible to a recipe. Like all proofs of the impossible (e.g., quadrature/squaring of the circle), this conclusion is also independent of means and methods. No faster computation, and no additional memory can change it. In its Turing-machine embodiment, quantum computation is subject to the same limitation. So are the AI attempts, extending to deep learning and similar techniques.

It is fundamental to understand that the determinism implicit in the Turing machine is expressed in all outcomes of computation. The genome project automated gene sequencing with the use of computers. In other words, it automated the work of chemists dealing with enormous amounts of data. Its enormously successful extensions in the field of molecular biology led to a disputable conclusion (equivalent to an axiom): *Genetic processes are algorithmic*. The epistemological consequences of this false statement could be understood by providing an analogy. Euclidean geometry (of the parallels that do not meet) provided an adequate scaffolding for Newton's view of the universe. Non-Euclidean geometry (of parallels that meet, so to say) is necessary in the articulation of Einstein's relativity perspective of space. Genetic processes are non-deterministic would be the necessary equivalent of non-Euclidean geometry. For the benefit of biologists interested in computation, and for computer scientists captive to the algorithmic, it is worth pointing out, for the benefit of biologists interested in computation, as well as for computer scientists captive to the algorithmic, that Turing himself distinguished between a-machines (automatic, in fact algorithmic, that carry out any computation based on complete instructions), and c-machines (choice, made by an external operator), and o-machines ("o" stands for oracle, and recalls Boltzmann's oracle that can prevent the increase of entropy in a system). Algorithmic computation is, by definition, closed to meaning: it operates exclusively at the syntactic level; that is, it does not understand anything, just like the beads on an abacus have no meaning.

The living is unique: It is not the outcome of some recipe—and even less of some physical or chemical computation. The characteristic of everything that is alive: no two mono-cells are identical, no plants, no animals, no bacteria, etc. are the same. Cells are not the same; proteins and neurons are always unique. There were human beings before us, and more will succeed the current generations—but they are non-identical. Lack of knowledge of biology leads to the claim that the human functions like a computer, or that a cell performs like a machine, that RNA carries instructions, and the subsequent reductionist practice of modeling life in computational forms (e.g., cellular automata, models, simulations, neural network representations). Von Neumann's mathematics of cell reproduction assumes the identity of cells. Living cells are far from being identical. McCullough and Pitts (1990) assume that all neurons are the same. Well, they are not. All of these—Turing's mathematical proof for the *Entscheidugsproblem*, von Neumann's self-reproduction of machines, the entire neural network concept—lead to the misunderstanding of what algorithmic computation is. The claim that—"all biology is computational" is the result of ignoring the science behind algorithmic computation. Science can model gravitational phenomena—from how things fall to rocket launchings—on computers because such phenomena are deterministic, and therefore their mathematical description is decidable. There are "recipes" (i.e., algorithms) for making such models. That is what engineers do. All phenomena and practical activities pertinent to such phenomena under the threshold of G-complexity can be described through algorithmic computation.

Those who ascertain that life itself is the outcome of computation—yet another level of misunderstanding—discard the fundamental science on whose basis the Turing machine (and all its embodiments in algorithmic computation) came into existence. This is not only of theoretic consequence and of philosophic consequence. More than that, it is of practical significance. Nothing from the immense computational output related to Covid-19 proved to be even marginally significant in helping those who are affected, or in preventing or containing the disease. The Summit supercomputer (second fastest in the world, for the time being) at Oak Ridge National Lab in Tennessee set about crunching data on more than 40,000 genes from 17,000 genetic samples in an effort to better understand Covid-19. The data processing resulted in a new hypothesis (the already mentioned bradykinin storm), and 10-plus potential treatments—from vitamin D to hymecromone and timbetasin, many of which are already FDA approved. With one proviso: all of these potential treatments are speculative, of course, and would need to be studied in a rigorous, controlled environment before their effectiveness could be determined. In other words, going in circles—physicians already know a lot about the role of vitamin D, and are aware of what certain drugs can affect. Repurposing (use of drugs developed for medical conditions other than Covid-19) does not need the supercomputer data as a validation tool. It relies on accumulated experience and on sharing it in the medical community. A 60-cent pill used in treating obsessive–compulsive disorder turned out to effectively prevent the vast inflammation resulting from the runaway immune response. Fluvoxamine—a psychiatric drug safe to use and with quick results—has been used for the last 40 years. By some coincidence, the antidepressant was proven useful in preventing clinical deterioration of Covid-infected patients.

Computational biology—an oxymoron *par excellence*—informs attempts that ordinary medical practice had long provided or had deemed inappropriate. It is like saying "But airplanes defy gravity," without understanding that a flying airplane (or for that matter a stone we throw at some distance) is rather the expression of what is needed in terms of forces and energy to fly under the accepted reality of gravity. Action-reaction was well-defined in Newton's laws. In the realm of the inanimate, homogeneity reigns. All machines function on one and the same basis: repetition (as described in physical law). Covid-19 convincingly proved how each patient is different—from the very young to the elderly—and how each death associated (sometimes arbitrarily) with it is different. Vaccines are not uniformly successful. The individual history (*anamnesis*) plays an important role. Anticipation awareness prepares us for dealing with the breakthrough incidents, as well as with what is described as the waning of their protective function.

A Concluding Hypothesis

A computational lemma: In the spirit of Elsasser's attempts at a new logical foundation of biology, it is tempting to prove that algorithmic computation is a subset of complex natural computations yet to be discovered. Such computations are not limited to Boole's logic; their alphabet extends well beyond two letters; their outcomes are not numbers, but meaningful processes.

When Science Becomes Theology

The arguments spelled out so far in support of the idea that anticipatory processes are definitory of life inspire the rhetorical call: "Let biology be biology." Arguing for a science of biology that is not an extension of physics, of chemistry, or of computer science, but rather including them as particular cases, is consequential in the sense in which *Disrupt Science* is. Physics and chemistry are exceptionally successful in taking things apart to see how they work. However, life is the integrated whole. Biology does not ignore the findings of reductionism, but it cannot justify its own legitimacy through the answers provided by physics and chemistry. "What is life?" and "Why change in the living?" are not independent of the make-up, but extend beyond past electrons, protons, atoms, and elements. In contrast to physics and chemistry, biology cannot avoid questions of morality. Crushing a boulder into sand is of no moral concern. Killing of human beings, animals, and plants have an unavoidable moral impact: life is taken. Moreover, taking things apart for the sake of finding out how they work entails critical questioning of legitimacy regarding the future. To find out what a stone is or what it is made of justifies subjecting it to various operations. This is physics and chemistry in action with the purpose of measuring and quantifying. To find out what the slimy mold, or some *archaed*, a plant, a fish, a bacterium, or an organism is entails recording its change over time, not eliminating life from it. When it sleeps, the glass frog hides its red blood cells in its liver. It becomes transparent. This protects it from predators. There is no circulating oxygen, but life is maintained. Killing life in order to establish the make-up, as the frog does, is not an answer to "What is life?" even less to "How and why does it change?" This is an anticipatory process that involves the whole body.

Sustainability—i.e., maintaining life in its diversity—is not a parameter of physics or chemistry. Models of sustainability are meant to highlight activities and patterns of behavior that might endanger life. It is not only about predators, but also about what it takes to survive. On a smaller scale of human activity, the return on the effort of acquiring knowledge has minimal consequences for life on Earth. With globality, the dangers scale up. Opportunity and risk are twins. What the science and technology revolution afforded was by many orders of magnitude greater than the effort. By now, this is no longer the case. Almost all large-scale projects consume more resources than what they make available. As a consequence of the inverted yield of scientific endeavors, humankind borrows from the future—nature is sacrificed, energy resources are depleted. At stake is sustainability, to the extreme—almost impossible to imagine—of extinction of life on planet Earth. In view of such a conclusion, which many scientists study from a variety of perspectives—climate change, pollution, overpopulation, inequities, etc.—it becomes impossible to avoid the question: How did we arrive at such a critical juncture, in which there is more danger than opportunity? The answer (or answers) might guide the efforts alluded to as the disruption of science.

Tools, made of lifeless matter (e.g., stones, branches, pieces of wood), "extended" arms and legs—and gave their makers the illusion of themselves becoming more powerful. Using a lever, a person could move an obstacle that not even several individuals could get out of the way. Same energy, more power. Hammers, pulleys, levers, wedges, the wheel made labor easier and more productive. However, in the process of using tools, those who conceived them became dependent on them. Thus did the makers end up being the "servants" of what they conceived in order to ease their efforts. Language, the most consequential of all human accomplishments, corresponds to a humankind whose abilities—cognitive performance, in particular—increased through ever more challenging activities. Through language, new tasks and new questions emerged. Such tasks and questions transcended the immediacy of challenges that made tools necessary: who or what moves the sun, the moon, the stars; who or what causes earthquakes, lightening, rain, floods, disease, plague; who or what increases fertility (of humans, animals, crops). There had to be something more powerful than they themselves were, even when using tools. The sun, as the source of energy for life, was a good candidate. The highest evolutionary accomplishment within the broader ecology of nature is language. Many other forms of expression involved in the pragmatics of human activity, such as images or sounds, evolved as the interaction among individuals and with the environment became richer. Language assisted in formulating and preserving such questions from one generation to another, and in formulating new goals. But it was not sufficient for describing what they saw high in the sky (day and night), or for imagining what was hidden deep down.

The unknown conjured words, images, body language. It could not be fathomed directly, but only through its reflection on water or, later, in a mirror. This is the origin of the word *idol*. Reflection is a re-presentation, a renewed presentation of what is reflected. Lightening and its reflection on a lake's surface or in a mirror are not the same thing. Likewise for a tree hit by lightning—a reminiscence of what it was before being struck. Awesome—in many ways. Even more so when a sapling starts to grow from the trunk of the stricken tree. Or when eggs turn into fish or birds. Millenia went by before heredity was defined. The intuition of continuity was unavoidable. The magic of words, of images, of sounds, of gestures, of things seen or imagined, of carved wood or stone—they seemed endowed with life. Anthropomorphizing started: like us humans, but more powerful, and immortal. The past lives in them. But they are all pregnant with the future—real or imagined. Pregnant with future is to be alive!

Like the tools—re-presentations of their own limbs—the idols humans made were reflections that conjured the invisible, the unknown. The culture of shared rhythms (gestures or sounds), of words, of images, of movements, and of tools bonded successive generations. So did the re-presentations of everything they kept track of: sheep, members of the family or tribe, stars. Everything, to which they attached numbers. Knots or beads on a *quipu* were not the real sheep or family members of starts. The people who made such knots or beads did so in durable and endurable matter as embodied memory. A cognitive jump from concreteness to the abstract facilitated the path from the embodied representations (knots, beads) to numbers. The process implies learning, and thus leads to anticipatory actions—mainly for avoiding danger. The words of their incantations and what those words evoked were different. Stepping away from the immediate—the specific danger—led to re-presenting it—the words and the intonation. In doing so, humans acquired a sense of possible future: the lurking dangers. The present and what comes next were connected through representations—at the beginning, concrete (e.g., imitating the predator's howling or other sounds), and in time, more and more abstract, i.e., giving it a name. Freedom from immediacy has its price. Humans grew ever more dependent on the representations they themselves created, to the extent of attaching magic powers to them. Expecting something—rain from clouds, fish in a river, berries in the autumn—can easily become "delusional." Idols and rituals—concrete as they were—changed over time, settling in the culture, i.e., the repository of their experiences. It was no longer up to an individual to change them at will. Those who shared, i.e., the family, the tribe, the community, practiced them as part of the culture. Many stones together, i.e., physical entities void of life, are nothing but the matter in which they are embodied and the interactions that shaped them. Such interactions were eventually described in the laws of physics. They are expressed in numbers that result from measuring forces, distances, and speeds. Many individuals are, like it or not, part of a network of interdependencies: they acquire energy (metabolism), they share. Preservation of life through meaningful actions—such a building edifices of stone for shelter from the elements and from enemies—defines such interdependencies. With the purpose of preserving and multiplying life anticipatory actions—sexuality, among them—are concrete expressions of such interdependencies. Family relations transcend description through numbers; they are purposeful. And purpose is anticipatory in nature.

Indeed, living matter, in which organisms are embodied, is always in a state of changing present with an inevitable future (death) vector attached to it. It is a process shared here and now. The historic record documents that the shared re-presentation of the unknown collided with other shared re-presentations belonging to other human networks of interdependencies. It became the path to what would become religion—way of connecting, of binding together (the etymology of *religare* points to this meaning). What defined religion was the pragmatic framework: shared views allowed the community accepting those views to outperform those outside. It transcended the more personal cult of idols, becoming a way of living. In a manner similar to dependence on words, images, sounds, numbers, and tools, people sharing in explanations involving deities (abstractions of idols) became subservient to their own religion.

There is no need, of course, for biology to justify religion. But there is a definite need to transcend deterministic explanations and associate religion to the process of the emergence of biological evolution, even if religion is unwilling to acknowledge the process. Anticipation is at work in the process through which religion becomes an instance of learning, and a premise for successful activity.

The syncretism of the beginnings of "like us" gave way to the role of "professionals" of religion. The process is from "We made them [tools and idols] 'like us'," to "They [gods] made us in their image." Thus, theology was born—the normed practice of religious beliefs, the institutionalized creed. Theology justified and regulated sacrifices, rituals, trances, meditation, singing, dancing. Its role in family life, society, and education, as well as in helping the needy (the poor, the orphaned, the diseased), and promoting ethical values was as important as what tools and the knowledge of nature afforded. From the Plague of Athens (probably typhus) to the Antonine Plague, and from there to all pandemics on record, people believed that their deities (that they themselves conceived of) gave up on them, or decided to punish them. These included the deities of Greece and Rome, the god of the Hebrews, of the Christians, of the Moslems, or the many gods of the Chinese and Indians who could safeguard the afflicted, if they deserved it. Religious cults—monotheistic and polytheistic—attributed miraculous powers to their deities. Worshiping ensued and became part of the religious experience. Theology defined the rules and justified practices meant to place those following them on the good side of the higher powers.

After more than one thousand years of Church-dominated (i.e., faith-based, institutional) explanations of *everything*, a different understanding of their reality emerged. It would give society a better chance in facing the many challenges of a new age in the history of humankind. Enlightenment marked the Renaissance (which literally means *rebirth*) that challenged the power of the Church within the Holy Roman Empire. Descartes (among others) revolutionized thinking (albeit without renouncing God): faith-based explanations were not discarded but rather subjected to reason and scientific method. The Cartesian "Revolution" took the scientific community by storm. It might not apply to biology, but it affected life on earth. Men (yes, men) who already questioned explanations for everything based on religious dogma felt encouraged to exercise reason based on empirical evidence. Many spectacular discoveries, theories, methodologies, and machines resulted. "Like us"—the mimetic (imitation in its many forms) extended to the machines people made. And here the second ironic twist: over time, the creators of gods, and later of machines, became the servants of their creations. This is the start of subservience to science, increasing as science becomes more successful.

The subject deserves closer attention if an adequate notion of biology is of real interest. This condition of subservience, and how humans arrive at it, is of interest in examining the role of knowledge—in particular, science—in society. Knowledge is accumulated through practical experiences. Humans continuously redefine themselves through their activities. This explains why knowledge seems to acquire the appearance of only waiting to be discovered. Thus, the laws of science, formulated by those who, based on their measurements, describe the phenomena of reality, seem to have an existence of their own, independent of those who discovered them. When the sentence "Science proved…" or "Science says…" is used, the humans who constructed science are relegated to the role of submitting to science, not unlike submitting to deities. As a result, instead of a unified description of reality and how it changes, society ended up with many sciences, expressed in many dialects. Subservience to science understood as the science of matter, in disregard of the characteristics of living matter, is probably the most intriguing consequence of the Cartesian Revolution and of positivism. The undifferentiated notion of matter became the new god (or the only god). Physics dedicated to this god is the new theology. Like a curious child, this god knows how to "undo" things: find the components, look at what makes the toy work, and eventually try to remake it—an excellent training tool not just for future mechanics. This theology offers a perspective of accounting: attach a number to each piece and make sure you keep a clean record of the parts. There is no need to know what they are or how they work.

Abandoning doubt as the underlying attitude of any thinking process and hypostatizing the experiment—declared equally valid in testing knowledge about lifeless matter and living matter—are the origins of the theology of science. The universe—what Newton's mechanics described—was declared, for good reason, comprehensible (Weinberg 1977)—but also pointless. Indeed, the movement of stars within the vast "machinery" of the universe *is* pointless. In contrast, the change in each and every cell in biological entities—sometimes not as easily traceable as the movement of stars—is purposeful. There is a huge difference between the unavoidable—predetermined—and the context dependent choices through which life is maintained. The moon does not choose its place in the universe; electrons have no agency. Cells make choices. The specific folding of a protein qualifies as unique, but not as necessary.

Knowledge about organisms was subjected to experiments because they were declared physical or chemical entities—often killed in the measurement process. They became, *volens-nolens*, a particular form of physics (or chemistry, i.e., genetics). This particular knowledge unfortunately never acknowledged purpose, that is, *the meaning of change*—the answer to the *Why?* of it. The medicine of that time, like the incipient sciences, focused on finding causes for all that was. Physics and chemistry were adopted in the process. This incipient medicine—actually a practice of healing more than a subject of study—used the tools it adopted. In the process, medicine entered into the long captivity to them, and thus to what became known as *reductionism and determinism. Reductionism*, the method of reducing things to parts, easier to understand and handle, was congenital. Physics successfully demonstrated that if you understand the parts, which means the simple causes, you are guaranteed to understand the more complicated causality of the entire process. This was very enticing—but misleading. The path from the physics of lifeless matter to the organism seemed promising. Knowing what atoms are made of helps explain how stones turn into sand. Obviously, there is no purpose to account for. But knowing what cells are made of does not explain their life, and even less their "expiration date." Knowing what cells of a bird's egg are made of does not explain how it might turn into a chick—or not. Knowing what cells are made of explains only part of what makes each living entity different and how this results in their being alive. The composition of the glass frog does not account for tis anticipatory behavior. For the biological, everything is purposeful. Becoming transparent during its sleep, the frog avoids falling prey. Meaning captures the motivation of the unfolding over time. Leaving out meaning, or relegating it to religion or philosophy, is a reduction one step too far.

If the sin of the beginnings, i.e., confusing the real with its representation—which is the business of theology—had been eliminated by the Cartesian perspective, its own role in the history of knowledge would be different. If doubt had been anchored in the Method, we would not be facing the consequences of "certitudes" close to those of theology. But unfortunately, despite the progress based on the premise of rationality, one theology—the religious method—was replaced by another method claiming universality. And it persists to this day. The idols and the gods of the time when the unknown was addressed through its representations were replaced by numbers and the mathematical abstraction that became the language of physics. The theology of a primitive understanding of causality took over science. Tools were automated and became machines, using energy—e.g., from waterfalls, steam, electricity—to facilitate work. Given the appearance that machines could do things on their own—you press a button or issue a command and they perform—they became the new idols. Humans conceived them, only in order to declare themselves to be machines (*L'homme machine*, 1747). Subject and object in inversed roles—exactly like those who made gods of clay and stone and claimed to be themselves made by those gods.[1] Through a mental inversion, the makers become the outcome of what they made.

It is a strange but explicable course of events: human abilities are attributed to entities (machines, processes) which humans themselves conceived and made. The current obsession with Artificial Intelligence (AI), together with its Machine Learning (ML) procedures, is the most recent example of this tendency. In this respect, the living is no longer acknowledged through what characterizes life, but rather through what describes the non-living matter in which machines are embodied. That idols are nothing more than human constructs did not stop theology from explaining everything in terms of faith. Words are part of the language developed over centuries of human activity that benefited from the use of language. Numbers are constructs to re-present quantities, independent of what is measured. This does not come to the mind of those practicing the theologies of words and of numbers. Even in

[1] Book of Isaiah, the Prophet, Chap. 44:10–20. "Who but a fool would make his own god—an idol that cannot help him one bit? …the wood-carver...uses part of the wood to make a fire. With it he warms himself and bakes his bread. Then… he takes what's left and makes his god: a carved idol! He falls down in front of it, worshiping and praying to it. 'Rescue me!' he says. 'You are my god!' Such stupidity and ignorance! Their eyes are closed, and they cannot see. Their minds are shut, and they cannot think. The poor, deluded fool … trusts something that can't help him at all. Yet he cannot bring himself to ask, 'Is this idol that I'm holding in my hand a lie?'".

our days, some see them as god-made or god-given, or simply existent like trees and stones, but not as the outcome of the activity of a thinking *homo sapiens*. The theology of determinism and reductionism, in particular that of the machine, is embodied in the belief that everything, living or not, is reducible to the abstraction called matter. Explainable, but pointless! Ontology engineers translate dictionaries—repositories of what words stand for—into the idiom of the almost all-encompassing computation. Of course, computers, as syntactic machines, still don't "understand" words, but at least we can get *Siri* to connect our intentions to some computer processes, or to navigate to wherever other mechanics will take us. Or we can chat GPT into producing texts that read like the scholarly articles used to train the artificial neural networks on which open AI is based. Ontology engineering means making the dictionary, as it stands at some moment in time, accessible to digital processing. In other words, transforming what we know into sequences of zeros and ones—i.e., meaning reduced to syntax in disregard of context. The effectiveness of mathematics, of physics, and more recently of chemistry (disguised as genetics) is so impressive that each has become a theology in its own right, or a "sect" in the larger theology of determinism. Questioning it—the Cartesian *dubito*—is deemed unscientific and treated as such.

The alternative to the theology of matter and machines is the awareness of the creative nature of life itself: its purposefulness. This defines biology as the science of generic creation. In particular, the creative nature of human beings: making and remaking the world, including their own making and remaking, is an expression of anticipation. Creativity is always the outcome of anticipatory actions reflecting awareness of the possible future.

Creativity, just as in the uniqueness of each living expression, is by necessity the knowledge domain of biology. Therefore, knowledge regarding anticipatory processes that undergird creativity is not an option, but a necessary focus. The biological time sequence characteristic of the deterministic view—usually represented as a line from past to present—includes possible futures.

The Mirage of Testable Predictions

One of the most prominent recent proponents of the anticipatory perspective was Rosen (1985). A biological mathematician from Rachevsky's *Relational Biology School* (Rachevsky 1960, 1972), Rosen contributed decisively to the understanding of what life is Rosen (1991). In a discussion of his major ideas, somebody (evidently attracted by the perspective Rosen advanced) articulated what could be an explanation for the low acceptance of his work: "It looks like his ideas did not yield any testable predictions…." This was followed by a challenge: "Anybody who knows of a fruitful line of experimental or theoretical research that was suggested by Rosen's work …." (in the words of Matthew Healy, who has studied biology for over 30 years, *Quora*, October 1, 2016). Both of these remarks (selected from a number of similar takes) are testimony to a major misunderstanding of biology as the physics and chemistry of life. The time has come to advance from the black box of the experiment as proof of ideas concerning physics to the open system of empirical knowledge acquisition. This is what actually defines the method of biology. Expressed otherwise: Let biology—the knowledge domain of what is alive and how it changes over time—be biology, not an extension of the sciences of the inanimate world. More important is contrasting the reductionist Cartesian reductionist-determinism and the holistic non-deterministic anticipatory expression characteristic of life is the grounding of biology.

Neither Rosen nor other scientists advancing the anticipatory perspective propose a standard theory in the sense in which physics does. In rejecting speculation, they placed themselves in a situation similar to those in physics who realized that the world is not flat. Observing the living, and engaging in vast experiments (such as farming and animal husbandry), those many inquiring minds realized that the attributes that the living, in all their embodiments, have in common are expressed in actions that preserve their lives. Biology qualifies as "the encyclopedia" of life, written by life itself in the course of evolution. This explains why knowledge about life is different in nature than knowledge about physio-chemical entities (Nadin 2018a, b). Therefore, it is expressed in descriptions that are different. The language of physics is mathematics. Israel Gelfand, one of the most respected mathematicians of our time (Kyoto Prize, 1989), was aware of Wigner's (1960) essay on "the unreasonable effectiveness of mathematics in physics." He came up with a paraphrase: "There is only one thing which is more unreasonable…and this is the unreasonable ineffectiveness of mathematics in biology." In Gelfand's Kyoto Prize address (and on many other occasions), he argued for the need for a different language for describing life from the cell level, (which he famously studied) to proteins and neurons (which he also studied), to organs and organisms, to motoric expression. Knowledge about life can be formulated only by means congruent with the specific dynamics of the living. Gregory J. Chaitin, discussing *The Limits of Mathematics* (1998) argues in favor of "new principles to understand and explain what is going on" (p. 23).

The transcription of RNA and the synthesis of proteins (not to forget their anticipation-driven folding) entered the limelight as a new Covid-19 vaccine was developed. These are processes that resulted in the course of evolution. Adaptivity documents the particular nature of biological processes. In contradistinction to physical systems, biological dynamics is not locally, but globally, optimal. This means that an unlimited variety of local optima eventually make possible the emergence of global adaptation of a particular species. Evolutionary phylogeny transcends a cause-and-effect interaction. It can be better explained as an adaption to the possible future. Therefore, nothing about life is testable. The expectation of confirmation through experiment (replication) does not correspond to the uniqueness of life expression. The reasons for this drastic statement are:

(a) Experiments are closed systems. The living, at all its levels (cell, tissue, organ, etc.) is an open system part of a larger open system.
(b) Once a certain aspect of life is selected for inquiry (measurement), the holistic nature of the living is contradicted. Experiments are conceived with the purpose of measuring particular aspects of the subject. Whatever data the measurement

returns is significant to the physio-chemical aspect selected, but not to the whole that the living embodies.

(c) Deterministic entities (such as physical objects in the material world) will return the same values of measurement (i.e., the experiment replicates) while non-deterministic entities—everything alive—will continuously falsify the experiment, returning different data.

There are different understandings of what anticipation is, and probably of what it is not. I had the privilege of initiating a dialog among the experts from around the world, in consecutive small-scale conferences. What resulted was not a settled science of anticipatory processes, even less a biology not captive to physics, but a multiplicity of perspectives. They are documented in *Learning from the Past. Early Soviet/Russian contributions to a science of anticipation* (Nadin 2015a); and in the research they inspired: *Anticipation Across Disciplines* (Nadin 2015b); *Anticipation and Medicine* (Nadin 2016). Some participants took the challenge of empirical evidence at heart, but also understood that meaning is as significant as quantitative descriptions are. During the span of their work (from 1930 through 1960s), the forerunners of a modern science of anticipation—the Russian/Soviet School (e.g., Bernstein, Beritashvili, Ukhtomsky, Bechtereva, Anokhin, among others)—provided empirical evidence how and why anticipatory processes take place (Nadin 2015a). Their work still awaits a long overdue in-depth interpretation so that further attempts at understanding anticipatory processes can benefit from it.

The Copernican Revolution (preceding the Cartesian) ascertained the positivist experimental proof of descriptions of reality consonant with Francis Bacon's (1610) attitude regarding the progress of science. The expectation of law, as a description of a deterministic world, is congruent with the notion of proof by experiment. Laws as descriptions of causality eventually morphed into the design of machines embodying regularity and uniformity. The pendulum exemplifies the law of gravity in action. Other machines (with gears engaging each other) followed. More mechanics, more hydraulic power, and electricity led to a variety of automated tools. Their use is an ongoing experiment validating the knowledge on whose basis such machines are built. If they mimicked the way humans operate, they did so in a repetitive manner. Repetition, which experiments are supposed to test and confirm, is the hallmark of machine functioning. Bernstein (1967; see also Nadin 2020) considering the motoric system, discovered that movements are foremost means for acquiring knowledge of the world in which they are performed. They are an expression of activeness—goal-directed learning—not of reactiveness—stimuli triggered. This is what distinguishes the organism from mechanism. Bernstein debunked Pavlov's theory, but the Nobel-prized reflex theory still dominates, only because it fits within the dogma of determinism.

Machines do not learn. For each expression of life—as movement or as any form of pursuing a goal—the neuronal pathways, muscle, ligament, joint motions, etc. will be different each time. Motoric expression is guided by the possible future—stepping into a hole, or stumbling against an obstacle, or hitting the hand against an object. This understanding stands in contrast to the image of the reflected reality that Descartes used, and which Pavlov illustrated with his experiments. A stone rolling down a hill is impervious of any obstacle. Its trajectory is the outcome of gravity and interactions (such as friction). There is no learning in physical processes. In the living, actions, whether thinking, moving, breathing, capture the world. Actions are expression of the implicit knowledge about the world. As one action succeeds another, they are part of the learning continuum immanent in life. Through actions, purposeful expression is concretized. A stone is picked up and eventually thrown against some enemy. Or a branch is taken in order to clear a pathway or to chase an animal away. In this respect "repetition without repetition" (Bernstein 1967 *op. cit.*) becomes a criterion for distinguishing the living from the non-living. A robot could pick up a stone and throw it successfully at some target, or it could manipulate a tree branch. But in the robotic action there is no possibility of changing one's "mind": maybe I want to avoid hurting someone; maybe I give in to a call for mercy; maybe I find a better way to deal with the enemy. The human being is in a *sui generis* learning condition, generating possible futures and evaluating—not always successfully—the consequences. The basketball robot at the Olympics (Tokyo, 2020)—an impressive technological accomplishment—demonstrated to millions of viewers around the globe how well a machine can score. If instead of being handed a standardized basketball it had been given some rudimentary contraption—e.g., children make balls from all kinds of stuff—it would have been less successful, if at all. In this formulation—which can be extended beyond the example given—the implicit idea is that the living is in a state of continuous learning, and thus of adapting to new situations. Tested in each expression of life. This applies to everything alive: plants, birds, bacteria, ocean animals, etc.

Parallel to phenomena described in terms of laws (e.g., gravity, as in Newton's *Principia*) there are singular phenomena of a causality that is not of a necessary nature. The subject was discarded. Windelband (1894) distinguished two "knowledge goals" corresponding to the distinction between the general and the particular. *Nomothetic* sciences aim for the general, expressed as law. *Idiographic* sciences are aimed at describing uniqueness. This is exactly what defines the living at each level of its existence: from the uniqueness of cells to that of organisms and of societies.

For the record, Leibniz (1663) tinkered with the *singular* in analyzing logic and its expression in language, on account of individuation (which he defines as a "physical principle"). It is clear that neither Leibniz nor Windelband would ask for experimental proof of a theory (such as Rosen's theory of life) that ascertains individuation, i.e., the idiographic nature of the living. This note on the broader historic context of the awareness of what is subject to law and what is unique is relevant in view of the implications for understanding Covid-19, and by extension in the CoVID crisis, from an anticipatory perspective. To expect Rosen's, and for that matter anyone's, elaboration on anticipatory processes to fall into the predicated expectation of experimental confirmation is actually denying that there is more to causality than cause-and-effect. It also means the rejection of the pragmatic evidence that there is more to time than the interval between past and present. Reaction, as an expression of deterministic causality, is of nomothetic condition, i.e., it can be described, and is successfully described, through laws that generalize from particular phenomena. Anticipatory processes, as an expression of the multi-causal nature of living processes, are non-deterministic, more precisely, idiographic in nature. They do not generalize because each organism is different. Moreover, there are no two identical cells. Anticipation awareness prepares us for dealing with the breakthrough incidents, as well as with what is described as the waning of their protective function. During Covid-19, as during the entire CoVID period, legitimate and less legitimate treatment methods were practiced. The various drugs that made for headlines cannot be properly tested because while the chemistry might be the same, the organisms treated are not.

Anticipatory expression makes explicit what is meaningful for preserving life. Migration of birds and fish, and of plankton, are unique experiences in preserving the value of life. The nomothetic is agnostic of value. Things fall because there are forces at work. Therefore, the expectation of the testable is perfectly justified for gravity—but not for the glass frog. Experimental replication is more a proof of dogmatic resistance than an informed attempt to understand what characterizes anticipatory expression as a definitory characteristic of life. In reductionist experiments the living is killed in order to prove that some characteristic of life is reducible to electrons, atoms or molecules. Reducing the living to dead matter in order to ascertain that it is not different from the non-living is not experimental validation, but circular thinking: If it does not naturally behave as the laws of physics ascertain, we will change its behavior. The open-endedness of life is thus brought to an end. This means, in practical terms, that the living cell is killed in order to find out what it is: matter without life. Logic makes evident that the holistic nature of anticipatory processes cannot be subject to experiment because the behavior of the whole is sacrificed for the sake of segmenting and separating, as is carried out in reductionism.

Experiments—which are measuring activities within closed systems—return numbers: quantitative descriptions of physio-chemical phenomena. Anticipatory processes take place in the open system of life unfolding. While dependent upon the underlying physics and chemistry of biological matter, such processes are defined through qualitative as well as quantitative aspects. Meaning and value are significant. Of course, an integrated view could facilitate a more adequate understanding of reality. This means that numbers, describing quantities pertinent to biological matter, and their meaning, for understanding their significance for maintaining life are to be considered together. Many reputable physicists (Erwin Schrödinger, Niels Bohr, Walter Elsasser) have acknowledged that the dynamics of the living—how it changes, and how changes in its environment affect it—is different from the dynamics of physical, inert matter. Von Neumann joined them while trying to provide the mathematics describing their difference.

It is time for a role reversal. Instead of justifying themselves for questioning the view that biology must be an accessory of physics, scientists advancing alternative views should submit physics to scrutiny. The discipline claiming to explain everything should be held responsible for its many shortcomings. This is not the place for revisiting atomic bombs, the chemistry of gas chambers used to exterminate millions of human beings, function gain in genetics, and the modern digital technology of wars. But their extension into engineered genetic wars cannot be swept under the rug. It is tempting to assume that life is related to the emergence of properties not existing at the "Big Bang (or whatever other references to beginnings). But there is no proof for such an assumption. Empirically evident is that there was a period without life in the planet's history, and that since life emerged, Earth's dynamics changed. In the billions of years since life is documented, there were many instances of change, some captured in evolution theory, for which physics comes up short in explaining them. The nature of viruses, for instance, eludes physics and chemistry. If we consider only pandemics (by no means the only events of extreme consequences for the living on planet Earth), we have enough evidence for what physics could not, and still cannot, explain. Furthermore, if we consider the broader image of progress—or what is termed as such—it might be the time for a critical reevaluation of physics, chemistry, and—why not?—of the biology they subsumed.

A Biological Law?

The realization that biology does not qualify as a noematic domain—of change described in laws—but rather as idiographic—describing uniqueness—makes even the thought of law in biology qualify as inconsistent with the distinctions already spelled out. Bernstein (1967) produced evidence regarding motoric expression, i.e., how the living moves around. His is an empirical observation: No motoric expression in living entities endowed with a motoric system is a perfect repetition. In his word: What seems a repeat movement (even on purpose, such as in showing a pupil how to move the bow on violin strings) is actually "repetition without repetition." Let us generalize this observation (see also Nadin 2015c) in the form of another thesis (since claiming the status of law would contradict what it states).

Irrepeatability Thesis (IT): The living is the domain of repetition without repetition.

This applies to each level of life: from the cell and its dynamics to the organs and organism, to species, to everything alive.

Because of this characteristic of the living, it is counterproductive to seek a common denominator where none is possible. Those who attempt to create life would probably realize, at some moment in time, the impossibility of their endeavor. For this realization to occur, they would have to become aware of the anticipatory processes involved in how life is expressed: always in uniqueness. This is not a statistical feature, but an expression of anticipatory processes that guide, for instance, at the unfolding of life from the stem cell. Differentiation is a process irreducible to the mechanics of copying.

In virtue of the impossibility to reduce life to the functioning of a machine, the expression "a life-making machine" qualifies as the mother of all oxymorons. Uniqueness and sameness correspond to aspects of reality impossible to reduce to one another. Conceptually, it is easy to understand that the impossibility (of certain mathematical operations, of certain physical or chemical processes) corresponds to the nature of processes, not to inadequate means of evaluation. Measuring means and methods, appropriate to describing in nomothetic aspects of reality—such as those expressed in the laws of physics—return no knowledge when applied to ideographic aspects of reality. The nature of a work of art is by no means adequately captured in its weight or in its geometric data, or in the composition of the pigments. The human being, and for that matter every living entity, is not meaningfully described through physio-chemical measurements, but rather through the meaning of their activity. This observation, inspired by the Irrepeatability Thesis (that points to the uniqueness of living expression), can be extended to the concrete aspects of the Covid-19 pandemic, including the use of surrogates (mice, rates, bats, etc.) to legitimize treatments or vaccines. Everything has been measured: from the distance that a droplet of saliva travels, to the degree of hair loss, to the change of toenail color. Missing is the understanding that each patient is different. This resulted in reductionist expectations pushed to the extreme in the uncritical practice of vaccination. If indeed a vaccine prompts immunological processes, why insist that those who already went through Covid-19 be vaccinated? Wasn't it stated that if you had the flu, you don't need to vaccinate? (Dr. Fauci on C-Span interview 2004). And why insist on everyone being boostered? Why state that a booster might save your life? Ignorance and incompetence march together, one behind the other. Medical interventions, which members of society want to assume are always well intended, did not align with the expectations trumpeted by politicians, scientists, and the powerful pharmaceutical companies.

The community of medical practitioners, whose members are dedicated to their profession, cannot be blamed for being conditioned to seek the help of machines in treating patients, or even for adopting the convenient model of the human being as a machine. The call: "Med school needs an overhaul" (Goldfarb 2020; many others stated the same) is indicative of a painful realization prompted by Covid-19. Indeed, continuing and even refining the training of physicians in the machine repair mode is where this all originates. The watering down of higher education all over the world affected medical education as well. In many medical schools, training on a trade school level rather reflects market expectations than higher standards of a profession dedicated to maintaining life. It costs less to train than to educate. For medical doctors who also pursue a PhD, the education costs more and more, but the degree earned is not a convincing testimony to independent research and expertise. In over-specialized training, breadth is eliminated in favor of deceitful depth. Eliminating culture, instead of enlisting it in medical education, is indicative of historic myopia. Medicine ended up creating patients instead of helping those needing effective healing methods.

The other realization concerns the inadequate system for validating and sharing knowledge pertinent to medical care. In particular, in the case of Covid-19, society was bombarded with opportunistic publications rarely meeting the standards, as disputable as they are, of integrity associated with the experimental method. From almost 700,000 rushed papers published, less than one percent formulated testable hypotheses. Surprisingly (or not) "Non-replicable publications are cited more than replicable ones (Serra-Garcia and Gneezy 2021). A headline such as "Coronavirus Tests Science's Need for Speed Limits" (Yan 2020) is probably suggestive of the dangers of making an opportunity (for recognition, funding, tenure, etc.) from a crisis. "Against pandemic research exceptionalism" (London and Kimmelman 2020) is yet another example of the awareness of the pitfalls in "expeditions" research in a crisis situation. Efforts were made to facilitate the "datamining" of the huge volume of publications (Lever and Altman 2021), as well as the structuring of this production (therapeutics, forecasting, vaccination, "long Covid," etc.). Evidently, where there is "gold," i.e., precious findings, not even machine learning will miss it. Fake articles were generated and, to the dismay of the community of researchers, were published—even in prestigious publications. A more rigorous review process, especially more transparent evaluation, would have spared many trees and prevented waste of precious time. Never mind science's loss of credibility. The energy consumption in medical research facilities—especially those in league with Big Science—is only one aspect of the unsustainable path chosen by medical science captive to reductionism-determinism.

To a great extent, machines for measuring have replaced the human (nurse, physician's assistant, physician). There are measurements—diagnostic imaging such as X-ray, MRI (magnetic resonance imaging), CT scans (computed tomography), etc.—of such a level of sophistication that they require a specialized technician in order to perform them. Practicing physicians do not have such competence. Frequently, they end up treating tests instead of patients. The results of measurements, which are prescribed for a variety of reasons (not the least because of defensive legal considerations), is often beyond their ability to interpret. X-rays, films or scans, or computer tomography images (such as urography), not to mention MRI's, etc. are delivered with the interpretations of the radiologist (the best-paid physicians), who examine not patients, but all kinds of visualizations. This is consequential: more people die under medical care or end up dependent for life on it.

Medicine moved its attention from the patient, as a subject of medical care, to representations of the patient, and to further representations of representations. The next step is automation: mechanical evaluations of such representations, and the generation of a database of average values—artificial thresholds of disease. Enormous amounts of data are used to train all kinds of neural network diagnostic procedures. Nobody seems to realize that captivity to such data effectively means that nothing out of what the data describe can be detected. Such methods are blind to the dynamics of the living. When biology is reduced to physics and chemistry, medicine increasingly turns into a specialized form of statistics and thus falls into the open arms of computer technology. This is particularly evident in the case of genetic evaluations. To identify changes in chromosomes, genes, and proteins, geneticists operate in a knowledge domain specialized to the extreme. Some perform molecular study of genes; other chromosomal tests; and yet others, biochemical tests. It is not always clear what the benefits are, and even less, the risks inherent in generalizing from genetic data to the treatment of a patient. There are no proven paths from such measurements to helping patients, or to generating protocols for preventive medicine. Often all of this creates data for the sake of data. Once again: the uniqueness of living entities (expressed in the Irrepeatability Theses) renders the use of surrogates—mice or computer models—absurd.

The extent to which such measurements are justified should be of concern to everyone involved in medicine. Automating medicine is a goal nobody explicitly supports, but everyone contributes, sometimes without understanding what it means, to achieve it. Furthermore, why machines are increasingly deployed despite the limited benefits of using them, and despite their huge energy consumption and the consequent ecological damage (caused through the use of extremely toxic materials, for instance) should have prompted critical evaluation a long time ago. Reductionism is expensive, especially computational reductionism, but not necessarily effective. Just one simple observation: the machinery of gene sequencing and the tools of computational biology have so far produced very modest consequential knowledge pertinent to the pandemic. This became a multi-pronged crisis, which the excessive deployment of technology neither predicted nor appropriately explained—never mind helped to cure!

Here, too, the background is the larger crisis of vision, within which goals other than healing or disregard of ethical standards sanctify the means. As a result, "Global Sentiments Surrounding the COVID-19 Pandemic on Twitter: Analysis of Twitter Trends," (Lwin et al. 2020) changed rapidly. Trust in science was lost with each new politically motivated deployment of deterministic means and methods: the lockdowns, practiced in many countries, exacerbated the situation; travel restrictions were imposed only after spreading began. From fear, sadness, and anger, everything was expressed. Xenophobia found a niche, as did social and political criticism. Societal concerns of all kind made it into the CoVID confluence, i.e., the aggregate crises faced in a variety of manifestations as the pandemic evolved into its next phase. Despite the publicly expressed confidence in data—the core of the Cartesian view of science—advice from public health officials was often informed more by ideology than by biology. The larger framework of corruption, undermining the health of societies, explains why the public lost trust in science. When a leading medical journal (*The Lancet*) gets involved in politics, it ends up publishing articles that do not withstand peer review (and need to be retracted). Such publications lobbied for one party or the other, instead of focusing on medicine as a practice independent of politics. Formulated behind the scenes, letters of support were generated for self-promotion. They follow the pattern of funding procedures, which most of the time endorse whatever has become established instead of encouraging the exploration of alternatives. Journal editors block submissions before peer review can even begin. Science shielded from responsibility became part of political campaigns. (National leaders ignorant of science spout their support for the science *du jour*, such as biophysics). "Vote for our candidate and you will get a viable medical system;" or, as *Scientific American* put it: "You will get more support for science." This is indicative of yielding its creative, exploratory function in favor of becoming part of the capitalist profit economy and the government machine. There was never in history more money spent for science, despite its miserable record. In situations such as pandemics, when it is supposed to prove its worth, science, medicine in particular, became more business opportunities than pursuit of knowledge. The medical-industrial complex monetized the Covid-19 misery by socializing inadequacy. In the broad CoVID crisis of vision, medicine became a culprit whose hands cannot be washed clean, neither by journals, nor self-congratulatory conferences (also paid by the public), nor all kinds of prizes and medals.

Confined to their homes, many individuals suffered from a variety of conditions that have nothing directly to do with the virus, but which were aggravated by the pandemic and by confusing messages from experts. The aches and pains of the labor market cannot be causally retraced to the bat virus, but neither can anyone ignore the multi-causal aspect of the breakdown within which the pandemic played a major role. Which measurements, if any, should be considered if the well-being of patients seeking help from the medical system currently in place is indeed the focus? The misguided focus on matter—genetic expression for instance—literally eliminated the understanding of the human behavioral component. Testing became a money-printing machine for those who produce less than reliable testing technology. That it is as good as stepping on the scale every minute to see if you lost weight, instead of changing your behavior (rational diet, exercise, etc.) is cavalierly lef aside by those who promote testing. The anticipatory perspective, which integrates the awareness of long-term consequences of deterministic medicine, is not the panacea, but rather the challenge for a more comprehensive understanding of the dynamics of life, which is the expression of matter dynamics and of behavior.

Disrupting Science

The notion of disruption (actually creative disruption, Schumpeter 1942) cannot be separated from its description as a way to make capitalism more efficient. Therefore, to avoid any misunderstanding, a straightforward clarification is necessary: "Disrupt science" is not about increasing the productivity of science in order to make it more competitive, and even less about monetizing it even more than is already the case. Science as a whore could please those who control the purse, but cannot save the future. Disruption calls for reinventing science on a broader foundation than that provided by reductionist-determinism. In order to stimulate life, instead of undermining it, those involved in scientific activities need to rediscover and to understand what it is. The crisis of vision—corresponding to a progressively more limited view of reality—could turn into an opportunity. If humanity is to have a future, those who are part of it should better realize how it can and should inform the present.

Science frequently fails to provide guidance in critical situations. Among the increasing expressions of concern regarding this fact are those informed by analyzing the ever-growing number of scientific publications. A recent study reveals that no less than 60 years of data pertinent to 45 million articles, and almost 4 million patents were considered (Park et al. 2023). Science is "becoming less and less disruptive" means nothing other than regurgitating instead of opening new paths of inquiry. "Low-hanging fruit" is neither confirmation of old results nor a way "propelling science and technology in new directions." Benefits to individual careers (the self-destructive "publish of perish") are different in nature from creative research. Aligning woth the prevalent perspective is easier than challenging the premise upon which institutionalized science (politically manipulated) operates.

To overcome the limitations of the attempts at reducing biology to physics, chemistry, computational science, or any other particular reductionist-deterministic discipline in no way implies discarding them. Diminishing support for their further development is not a wise strategy. The past is relevant. However, to limit science to analyzing how the past affects the present and even the future, turns those dedicated to this understanding into agents of predestination. Only on the basis of an adequate knowledge of reality can science meet the challenges of our time. The change from a reaction-grounded course of action to a prevention-based behavior is decisive in respect to sustainability.

Determined to understand what life is, von Neumann focused on complexity as a distinguishing characteristic. He advanced the idea that living matter is substantially prevalent in the world because it evolves autonomously. In his view, biological growth, development, and learning reflect the complexity of life. Below the threshold of life, matter decays. Above the threshold, which von Neumann did not identify, matter evolves. So far, evolution theory is the most convincing description of the process—but probably not the final one—if there is such a thing as final in science. The development of language, the diversification of cognitive activity, the ever-growing number of forms of interaction are empirical arguments in favor of a new perspective of science. The variety of ways in which living performance is expressed further consolidates the idea of a threshold of complexity above which life unfolds. Science will not survive if it does not develop means and methods for accounting for the reality above the threshold of complexity.

The notion of *threshold* suggests not only levels quantitatively described, as when it is reached, something else takes place. It also suggests horizons: an opening, a door, new starting points. For von Neumann, the threshold of complexity, which he contemplated, meant that life can be described as a process of self-reproduction. In the spirit of mathematical reasoning, he omitted the detail that there is no sameness in the realm of life. The focus on scale, i.e., measurable quantity, obstructed his understanding of life as an expression of self-awareness. Stones know nothing about their existence or composition. Cells, on the other hand, are involved in interactions impossible in the absence of self-awareness. This is why anticipatory processes are consubstantial with life. The preservation of life, moreover its self-reproduction, is not in reaction to, but in anticipation of change. Von Neumann wondered how come living matter is prevalent once life emerges. In order to explain this, he argued from a perspective of thermodynamics. The same conclusion, but grounded in the dynamics of life—i.e., how it changes over time, including the sequence birth-philogeny-death—can be reached by identifying anticipatory processes. Such processes integrate quantitative and qualitative aspect, on particular the realization of meaning in a context of generic ambiguity. In the living, the same elements mean different things in different contexts. Life is not possible in the absence of carbon, but life can come to an end when carbon is present in some combinations (think CO_2). The living, which continuously creates itself, does not start in sameness, but actually ends in it. In the interval between procreation and death, life never repeats itself. This is why it cannot be fully measured. Moreover, its dynamics is the outcome of contradictory forces at work.

With these observations in mind, we realize that the condition of being alive meets the definition of what Gödel described as undecidable (Davis 1965). The focus of his reasoning was not the description of the living, but rather on thinking itself as an outcome of life.

The consequences of properly defining the threshold of complexity for disrupting science will soon become apparent. Gödel articulated a logical principle. In his two *Incompleteness Theorems* (which belong to the knowledge domain known as mathematical logic), he dealt with the notion of the decidable as it applies to formal axiomatic theories. In the spirit of his definition, we take note of the fact that in the description of reality a distinction can be made regarding entities that can be fully and consistently described, and those which are undecidable. There is no need to generalize from characteristics of a formal system to reality—an epistemological path to be considered carefully. Rather, the concept of decidability can be used as a defining criterion in connection with complexity, and with von Neumann's threshold between what is alive and what is not. Justifying the generalization of the qualifier *decidable* (and its pendant, *undecidabl*e) is the rational premise: If in describing something—an aspect of reality per se, or representations of such an aspect—the purpose is to come to an effective way of handling the reality or the theory about it, decidability becomes a threshold value for a precise expression of complexity. Thus, science can put it to use in some of its ways of questioning reality, i.e., articulating answers to questions pertinent to it.

A science that serves the purpose of sustaining life must be able to provide knowledge about life. The living, which continuously creates itself in its uniqueness, is a process. Therefore, it is undecidable. This might even qualify as an axiom of life. All the data accumulated to date about life unfolding leads to this inference. It is incomplete and contradictory. One practical consequence is at hand: to disrupt science means by necessity to open its perspective as to include the undecidable, instead of reducing it to the easier to handle, more operational decidable. The threshold between the decidable (non-living) and non-decidable defines *G-complexity*. The interactions characteristic of life are of a different order of magnitude from those characteristic of lifeless matter. If undecidability becomes part of the knowledge acquisition and dissemination of reality, science unavoidably will change. Disruption does not do away with science; rather, it augments its role by changing its perspective. Activities such as the practice of medicine, urban development, education ought to account for the whole of life (holistic aspect), not for its reductionist simili. Medicine as effective prevention, and education as a framework for shared knowledge are examples of the possible instantiations of a new science that integrates the future in its perspective. Such activities could adequately meet the practical challenges of addressing the implicit inconsistency of change in the living.

Those inanimate parts of the reality that make up the subject of physics and chemistry evince a dynamic that, for all practical purposes, can be described completely and consistently. Humans build homes, make materials, cars, airplanes, computers, and a variety of soft products (i.e., software performing a desired functions), all of which can be measured in detail. As a matter of fact, the plans (or designs) for making them are decidable descriptions. Engineering does it; technology makes it possible. When they break down, the humans who made them can fix them. They can be repaired. And when this costs too much, they are discarded.

The dynamics of the inanimate is limited. The laws of physics describing the way non-living reality changes are the expression of their decidability. Newton's physics and, for that matter, Einstein's are paradigmatic of this perspective. The living, of unlimited dynamics, can be understood only from within the awareness of its undecidable nature. Empirical evidence contributed to the awareness of anticipation as an expression of this condition. Self-reproduction and preservation of life are the outcome of anticipatory processes. The knowledge about the living has, by necessity, to integrate the undecidable. A knowledge domain appropriate to descriptions of living processes cannot be derived from the mathematics and logic of the inanimate. There are no laws to be discovered. At best, science could formulate theses about life and its unfolding. "Repetition without repetition" (cf. Irrepeatability Thesis) is an example. There can be no complete and consistent description because, as the living interacts with the world in which it unfolds, it changes, it continuously remakes itself, it creates itself. And it also changes the world. This explains why its phase space itself, of variables describing its processes, changes. The laws of physics or of chemistry apply to the matter in which life is embodied, but do not explain the dynamics of life and cannot predict its future state. An open system—which the living is—cannot be fully characterized. This applies to patients, students in a class, drivers on the road, communities, societies.

The consistency clause ascertains that inferences implicit in determinism (same cause → same effect, without any thought left for uncertainty and ambiguity) do not hold for entities that are G-complex. Every medical practitioner exposed to the Covid-19 pandemic has experienced this. Shifting recommendations of all kinds (e.g., mask or not, various medications, distancing, temperature control) succeeded each other, because the science was either absent or actually wrong. Pandemic or not, nothing is cause-free. And nothing is cost-free. The abject reaction to the pandemic became the stage for more expensive and only marginally effective science. Expense of an unusual scale at the cost of the future.

This corresponds to the Crisis of Vision reflected in the obsession with "big science." The tendency began after World War II, and it entailed expanding to a larger scale within the impetus of globality. If this were only a matter of ever larger projects—costing the equivalent of the budgets of some mid-sized states—there would be less reason to worry. Scale in economy is associated with higher efficiency. However, in its "big" condition, science has not become more efficient or more adequate. Evidently, projects such as atomic and nuclear energy research, space exploration, and super-computing are in virtue of the necessary resources, predestined to attain large dimensions. The Large Hadron Collider became, by necessity, an international undertaking. The record shows that such projects grow larger and larger. They depend on more and more energy, and almost invariably have a military dimension. While initially dedicated mainly to physics projects, Big Science eventually took over domains associated with biology. To date, the most impressive endeavor was the genome—the project intended to reveal the blueprint of life. Research of the human brain is of no modest size. A headline such as "Big Science failed to unlock the mysteries of the human brain" (Mullin 2021) cannot arouse celebration. Its pendant, *Models of the Mind*, (Lindsay 2021) has a telling subtitle: "How physics, engineering and mathematics have shaped our understanding of the brain." Nobody wishes science, little or big, to fail. But the opportunity to take note of the reason for such failures cannot go unheeded. Of all sciences, biology, in the proper sense of the concept, was not part of the attempt to understand the brain. In the absence of an appropriate science, able to affirm the complex nature of biological processes, wasting resources in stubbornly reducing life to non-living processes is no longer an acceptable accident.

Big Science's carbon footprint is testimony to demagoguery. It gets funded for promoting sustainability. The big climate change studies projects were supposed to focus on ecological concerns, clean energy resources, waste management, etc. In reality such research undermines the ecology by suggesting a truncated science that reduces everything to physics. In the same vein, the huge investments in genomics, produced impressively detailed analytical artifacts but very little in terms of health maintenance. Instead of healing, the "fixing" means and methods end up transforming patients into lifelong prisoners of the medical establishment. Genetics celebrated the identification of sex identifiers—the x and y chromosomes—to a society in the vortex of transgenderism, i.e., genetic engineering based on the dubious distinction between biological sex and gender identity. Puberty blockers are dispensed as a new entitlement to a generation already confused by the lack of vision of the society it lives in. The blockers are the victory of reductionist-determinism embodied in a science that undermines life. No liberation is achieved; a life-long dependency on chemistry begins. It not only considers the human being to be a machine, but also contributes to making everyone become one. Having facilitated industrial agriculture, Big Science inspired medicine to accept the rules of capitalist economy, in which profit is the only goal. This occurs to the detriment of medicine's anticipatory condition, ethics in the first place. There is no ethics in creating dependencies, be they to opiods or to all kinds of medications. There is no medicine in the sense of healing in the new shops where knees and hips are replaced in the manner in which mechanics replace brakes or tires on cars. There is no education in the real sense of the word in the large online classes (millions of subscribers) on transitory skills peddled to the world. Education decoupled from a sense of the future as expressed in anticipatory actions does not stimulate discovery of one's self and the role of interaction with others. In its current condition, education ends up being either an ideological activity, or a costly form of entertainment in the transition from childhood to maturity.

Disruption of this kind of science becomes necessary if the future means anything to the human being, regardless of one's social, political, economic, religious, etc. identity. Capturing the dynamics of life's physical and chemical substrata, determinism and causality, together with non-determinism and multicausality, could afford a more adequate understanding of how life is expressed. Within a condition of complementarity, in the sense in which Niels Bohr defined it, to react and to prevent, should become a societal goal. The unity between reaction and anticipation makes possible an effective practice of sustainability, instead of just wringing hands over its waning. This understanding, as the foundation of the new science, should inform medical care, for instance, in distinguishing between health and illness, or in characterizing various deviations from averages that make little sense. It should inform education, urban development, social programs, ecological awareness. More sources of energy, for example, which the physics and chemistry of our age are working at, only increase the tendency to use more and more energy without consideration of the future. In the absence of anticipation considerations, the individual is placed into a deeper state of dependency that eventually will be impossible to pay for.

Disruption of science entails broadening its foundation: from the limited reductionist-determinism to the complementary foundation that integrates the holistic understanding of anticipatory expression. Moreover, from the obsession with law to the realization that most processes in nature are idiographic, the road is fraught with dangers. A new way of thinking cannot be imposed from outside science itself. From the machine model of life, embodying laws of physics and/or of chemistry to the awareness of non-deterministic processes underlying change, a lot will have to happen in terms of expectations. Awareness of the price of never-ending progress—the mirage or our times—will not ensue by decree. From the focus on the decidable expressed in the simplicity of physio-chemical interactions to the broader view that integrates the undecidable, there is a lot to be done towards discarding the useless and the obsession with quick fixes.

Science has to re-acquire the sense of doubt in order to free itself from a blind faith theology into which it maneuvered itself by confusing its own predicament with reality. Questioning must become the premise for describing the world, replacing blind submission to a fractured rationality. As a human activity whose outcome is knowledge in the service of society, science and the dissemination of science cannot be limited to what is below the threshold of G-complexity.

The reduction of biology to physics and chemistry forced science into a cycle of ever-growing needs for more data and energy, all borrowed against the future. The living itself, in its amazing variety, could inspire the disruption of science: It survives and continues to be creative within its means, in terms of the energy necessary and the data at work.

Life is about meaning and results in creating more meaning. In the final analysis, science is justified by the future it makes possible. A science that by its limited and limiting condition undermines the future of life ought to be disrupted.

References

Akbar AN, Gilroy DW (2020) Aging immunity might exacerbate COVID-19, *Science* 369:6501, 256–257, 17 July 2020.

Bacon F (1610) *Novum Organum* (New organon, or true directions concerning the interpretation of nature).

Bernstein, NA (1967) *The co-ordination and regulation of movements*. Oxford: Pergamon Press.

Benatar SR, Gill S, Bakker I (2020) Global Health and the Global Economic Crisis, *American Journal of Public Health*, 101:4, 643–653, April 2011.

Bohr N (1933) Light and Life, *Nature* 121, pp. 421–423. March 25.

Bradley BT *et al.* (2020) Histopathology and ultrastructural findings of fatal COVID-19 infections in Washington State: a case series, *The Lancet,* 396:10247, 320–332, August 1.

CDC (2022) Animals and COVID-19. January 5. https://www.cdc.gov/coronavirus/2019-ncov/daily-life-coping/animals.html [retrieved April 5, 2022].

Ceran BB, Ozates S (2020) Ocular manifestations of coronavirus disease 2019, Graefes Archive for Clinical and Experimental Ophthalmology. 2020 June 6. 1–5. https://www.ncbi.nlm.nih.gov/pmc/articles/PMC7274940/ [retrieved July 20, 2020].

Chaitin GJ (1970) To a Mathmatical Definition of Life, *ACM SICACT News*, 4, January, pp. 12–18.

Chaitin GJ (1998) *The Limits of Mathematics*. Singapore: Springer.

Davis M (1965) *The Undecidable*. New York: Raven Press.

Dockser MA (2020) Covid-19 Raises Questions About the Value of Personalized Medicine, *The Wall Street Journal*, May 8. https://www.wsj.com/articles/covid-19-raises-questions-about-the-value-of-personalized-medicine-11588949927 [retrieved March 29, 2022].

Elsasser WM (1981) A form of logic suited for biology. In: Rosen R (Ed.) Progress in Theoretical Biology, Vol. 6, pages 23–62. New York: Academic Press.

Elsasser WM (1987) *Reflections on a Theory of Organisms*. Frelighsburg, Quebec: Orbis Publishers.

Fauci A (2004) Influenza Vaccine, *Washington Journal*. October 11. https://www.c-span.org/video/?183885-2/influenza-vaccine [retrieved November 22, 2022].

Feldman R, Sanger J (2006) *The Text Mining Handbook: Advanced Approaches in Analyzing Unstructured Data*. Oxford: Cambridge University Press.

Feynman RP (1982) Simulating physics with computers. *International Journal of Theoretical Physics* 21, 467–488.

Feynman RP, Hey A (1999) *Feynman Lectures on Computation* (Frontiers in Physics). Boulder Co: Westview Press.

Garvin MR *et al.* (2020) A mechanistic model and therapeutic interventions for COVID-19 involving a RAS-mediated bradykinin storm, *eLife*, July 7. https://elifesciences.org/articles/59177 [retrieved July 9, 2020].

Gelfand IM, Tsetlin ML (1971) Mathematical Modeling of Mechanisms of the Central Nervous system. In: Gelfand IM *et al.*, Eds. (Beard CR and Barlow JS, Trans.) *Models of the Structural-Functional Organization of Certain Biological Systems*. Cambridge: MIT Press. (Originally published in 1966 as a memorial tribute to Nikolai Aleksandrovich Bernstein, the eminent Russian pioneer in cybernetics).

Goldfarb S (2020) Med School Needs an Overhaul, *Wall Street Journal*, April 13 https://www.wsj.com/articles/med-school-needs-an-overhaul-11586818394 [retrieved May 22, 2020].

Healy M (2016) What does the field of theoretical biology think of Robert Rosen's book "Life Itself"? https://www.quora.com/search?q=Anybody%20who%20knows%20of%20a%20fruitful%20line%20of%20experimental%20or%20theoretical%20research%20that%20was%20suggested%20by%20Rosen%E2%80%99s%20work%20 [retrieved May 16, 2020].

Hilbert D and Ackerman W (1928) *Grundzüge der Theoretischen Logik* (Principles of Mathematical Knowledge). Berlin: Springer.

Laplace PS (1796) *Exposition du système du monde* (The System of the World). (Published by Courcier, Paris in 1808.

LaFraniere S, Wiiland N (2022) Walensky, Citing Botched Pandemic Response, Calls for C.D.C. Reorganization, *New York Times*, August 17. https://www.nytimes.com/2022/08/17/us/politics/cdc-rochelle-walensky-covid.html [retrieved December 10, 2022].

Lwin *et al.* (2020) Global Sentiments Surrounding the COVID-19 Pandemic on Twitter: Analysis of Twitter Trends, JMIR Public Health and Surveillance 2020, Apr–Jun, 6:2, May 22 https://publichealth.jmir.org/2020/2/e19447/ [retrieved July 20, 2020].

Leibniz GW (1663) *Disputatio metaphysica de principio individui* (On the principle of the individual). Baccalaureate thesis, University of Leipzig.

Lever J, Altman RB (2021) Analyzing the vast coronavirus literature with CoronaCentral, *PNAS* 118:23, April 5, e2100766118.

Lewis D (2021) Superspreading drives the COVID pandemic—and could help to tame it, Nature 590, 544–546, February 23. https://www.nature.com/articles/d41586-021-00460-x [retrieved March 13, 2021].
Lindsay G (2021). *Models of the Mind: How Physics, Engineering and Mathematics Have Shaped Our Understanding of the Brain.* New York: Bloomsbury Sigma.
London AJ, Kimmelman J (2020) Against pandemic research exceptionalism, *Science*, 368:6490, 476–477, May 1.
Longo DL, Drazen JM (2016) Data Sharing (Editorial), *New England Journal of Medicine* 374, January 21. pp. 276–277.
Manrai AK *et al.* (2018) In the Era of Precision Medicine and Big Data, Who Is Normal? *JAMA*, 319:19, My 15, 1981–1982. https://doi.org/10.1001/jama.2018.2009 [retrieved August 23, 2020].
McCullough WS, Pitts W (1990) A logical calculous of the ideas immanent in nervous activity, *Bulletin of Mathematical Biology*, 52:1/2, pp. 99–115.
Mullin E (2021) Cambridge MA: MIT Technology Review. August 25. https://www.technologyreview.com/2021/08/25/1032133/big-science-human-brain-failure/ [retrieved December 2, 2022].
Nadin M (1983) The logic of vagueness and the category of synechism. In: (Freeman E, Ed.) *The Relevance of Charles Peirce.* La Salle IL: The Monist Library of Philosophy, pp. 154–166.
Nadin M (2012) What speaks in favor of an inquiry into anticipatory processes? Prolegomena to the 2nd edition of Anticipatory Systems, by Robert Rosen, in (George Klir, Ed.) *International Book Series on Systems Science and Systems Engineering*, pp. xv-lx. London/New York: Springer.
Nadin M (Ed.) (2015a) *Learning from the Past. Early Soviet/Russian contributions to a science of anticipation.* Cognitive Science Monographs, 25. Cham, CH: Springer.
Nadin M (Ed.) (2015b) *Anticipation Across Disciplines. Anticipation Across Disciplines.* Cognitive Science Monographs, Vol. 26 (373 pp.). Cham, CH: Springer.
Nadin M (Ed.) (2015c) Variability by Another Name: "Repetition Without Repetition." *Learning from the Past. Early Soviet/Russian contributions to a science of anticipation.* Cognitive Science Monographs. Cham, CH: Springer. Vol. 25, 329–340.
Nadin M (Ed.) (2016) *Anticipation and Medicine.* Cham, CH: Springer International Publishers, December.
Nadin M (2018a) Redefining medicine from an anticipatory perspective, Progress in Biophysics and Molecular Biology, 140, 21–40. December, Amsterdam: Elsevier. https://doi.org/10.1016/j.pbiomolbio.2018.04.003.
Nadin M (2018b) Rethinking the experiment: a necessary (R)evolution, *AI & Society* 33:4, 467–485. New York/London: Springer.
Nadin M (2020) We are our actions. Would N.A. Bernstein Research BCI? In: Latash M.L. (ed.) *Bernstein's Construction of Movement: Original Text and Commentaries.* Routledge: Abingdon UK, 242–264.
Neumann J von (1966) *Theory of Self-Reproducing Automata* (Burke AW, Ed). Champaign: University of Illinois Press.
Osterholm MT (2005) Preparing for the Next Pandemic, *New England Journal of Medicine*, 352:18, 1839–1842.
Osterholm MT (2020) Covid-19 doesn't obey politics, it only obeys the laws of biology, physics and chemistry, NPR interview on Fresh Air, June 17.
Park M, Leahy E, Funk, RJ (2023) Papers and patents are becoming less disruptive over time, *Nature* 613, pp. 138–144.
Rachevsky N (1972) *Organismic Sets: Some Reflections on the Nature of Life and Society.* Mathematical Biology Inc.
Rachevsky N (1960) *Mathematical Biophysics: Physico-Mathematical Foundations of Biology* (3rd edition) Mineola: Dover.
Rosen R (1985) *Anticipatory Systems: Philosophical, Mathematical and Methodological Foundations.* Oxford: Pergamon.

Rosen R (1991) *Life Itself: A Comprehensive Inquiry into the Nature, Origin, and Fabrication of Life*. New York: Columbia University Press.

Schleiden MJ (1838) Beiträge zur Phytogenesis, *Archiv für Anatomie, Physiologie und wissenschaftliche Medicin*, 137–76. Berlin: Müller.

Schrödinger E (1951) *What Is Life? The Physical Aspect of the Living Cell*. Cambridge: Cambridge University Press.

Schumpeter J (1942) *Captialism, Socialism and Democracy*. New York: Harper & Row (1950).

Schwann T (1839) Mikroskopische Untersuchungen über die Uebereinstimmung in der Struktur und dem Wachstum der Thiere und Pflanzen. Berlin: Reimer.

Serra-Garcia M, Gneezy U (2021) Non-replicable publications are cited more than replicable ones, *Science Advances* 7:2: eabd1705.

Somers J (2022) The Final Frontier, *The New Yorker*, March 7, pp. 16–20.

Szybalski W (1974) In vivo and in vitro initiation of transcription. In: (Kohn A, Shatkay A, editors), *Control of Gene Expression*. New York: Plenum Press. pp. 23–24.

Talarico R *et al.* (2020) Symmetric peripheral polyarthritis developed during SARS-CoV-2 infection, Comment, *The Lancet Rheumatology*, July 13. https://www.thelancet.com/pdfs/journals/lanrhe/PIIS2665-9913(20)30216-2.pdf [retrieved July 14, 2020].

Treviranus GR (1802–22) *Biologie; oder die Philosophie der lebenden Natur für Naturforscher und Aerzte*. Göttingen: J.F. Röwer.

Turing AM (1937) On Computable Numbers, with an Application to the Entscheidungsproblem, *Proceedings of the London Mathematical Society*, Volume series 2, 42:1, 230–265.

Weinberg S (1977) *The First Three Minutes: A Modern View of the Origin of the Universe*. New York: Basic Books.

Wigner EP (1960) The Unreasonable Effectiveness of Mathematics in the Natural Sciences, *Communications in Pure and Applied Mathematics*, 13:I (February), 1–14. New York: John Wiley & Sons, Inc.

Windelband W (1894) *Geschichte und Naturwissenschaft* (History and Natural Science, Rectoral Address), reprinted in Praludien, Vol. 2, 136–160 35–379. Tübingen: J.C.B. Mohr. (Translated as "History and Natural Science," G Oakes, Trans. NKR 287–299).

Yaffa J (2022) The Geat Thaw, *The New Yorker*, January 17, pp. 30–39.

Yan W (2020) Coronavirus Tests Science's Need for Speed Limits. *New York Times* April 14. https://www.nytimes.com/2020/04/14/science/coronavirus-disinformation.html [retrieved June 10, 2020].

Vaccines—is This the Happy Ending?

In reaction to the pandemic, the large-scale deployment of a new type of vaccine remains an accomplishment impossible to ignore. Stimulating the organism's immune capabilities is a medical option with a good record. At closer look, the unusual path pursued in respect to the SARS-CoV-2 virus raises many questions, especially since the crisis lingers and more repeatedly vaccinated persons get sick (even more than once). Since survival is at stake what motivates science in the age of CoVID to pursue its goals is not a rhetorical question.

Once the vaccine was released, the number of hospitalizations and deaths through COVID-19 was substantially reduced. The record of this novel genetically engineered vaccine is undisputable. In contrast, the realization that the pandemic is a self-inflicted tragedy is still slow in coming. Confounding societal responsibilities and interests—in particular, economic and military priorities— make for a difficult-to-interpret image.

A false sense of entitlement explains, but does not justify, options—including the vaccine and antiviral drugs used for treating affected persons—which in the long run are unsustainable. Ultimately, neither medicine nor pharmacology can substitute for behavioral choices: individual responsibility steadily exercised. This could have spared the world the agony of a pandemic that should not have happened.

An Unfinished Saga

No doubt, books will be written and movies made. Since billions of dollars are at stake, expensive litigation will keep the justice system busy. This has already begun: issues of patent infringement and shared ownership are on various court dockets in the USA and abroad. Katalin Karikó (the American biochemist of Hungarian origin), working with Drew Weissman, succeeded against the odds in suggesting a less toxic mRNA solution. She might receive the most desired prize that scientists aspire to. Or maybe it will be Robert Malone, who claims that he discovered the new vaccine (Malone 1989), but who, after questioning the mRNA-based vaccines, became *persona non grata.*

Another story from among the many behind the vaccine revolution prompted by the pandemic: "She is a genius!" This is how Nita Patel, a scientist, originally from

© The Author(s), under exclusive license to Springer Nature Switzerland AG 2023
M. Nadin, *Disrupt Science*, https://doi.org/10.1007/978-3-031-43957-5_6

a farming village in India's Gujaparat state, was described. She is credited with the idea for successfully testing the protein behind the vaccine that everyone was talking about, but nobody used—until July 2022. The Novavax, a subunit protein vaccine developed by the company Patel works for, is a more traditional concoction. It uses moth cells (for making the protein) and tree bark (as an adjuvant). Instead of an mRNA vaccine (Pfizer, Moderna) or a viral vector vaccine (Johnson & Johnson), Novavax stimulates the making of antibodies for the spike protein. Once authorized, it became yet another alternative among the almost 300 products, still not approved, prepared in reaction to the pandemic. Almost 200 more are still in different stages of testing. The most recent push is Operation Nasal Vaccine for population-wide respiratory mucosa immunity. With a world population of almost eight billion, there is good money to be made with vaccines since national governments heavily subsidized and paid for them. Europe, with its almost 500 million citizens, even considered a mandatory vaccination program.

There are plenty of characters—including impersonators—in the COVID-19 vaccine drama. Many actors in the social media sideshow, and many experts, famous for claiming an expertise they actually don't have, took center stage in the unfinished saga. And there is, or should be, a Hollywood "happy ending"—exactly what everyone expects after the ordeal of the pandemic. But, after three years, is there really a happy ending in sight? Or will humanity live through the misery of the pandemic (morphed into an endemic state) forever? The Long COVID still torments millions of victims—China reports (or actually does not) millions of infections and thousands of deaths—but there is no solid science to help effectively those affected (Taquet et al. 2022). The vaccines and the boosters have become a fixation in the COVID-19 saga. All kinds of speculators turned it into an opportunity. The winning numbers are in the "lottery" of COVID vaccines and treatments.

For all practical purposes, vaccines seem to be the turning point in the COVID-19 narration. In all fairness, if vaccination were what everyone expected and their promoters claimed, books and movies would be justified. Also prizes—some already conferred. Is the name "Breakthrough Prize" some kind of irony? Not intentional, of course, since breakthrough infections, which means getting COVID-19 after vaccinations, or even after recovery, are nothing to sneeze at. The numbers of breakthrough infections (infection and re-infection) after vaccination do not look good. New research is trying to find out what happens to the immune system of people repeatedly vaccinated (Irrgang et al. 2022). Generously funded by the entrepreneurs at Google and Facebook (among others), with the aim of recognizing work that opens new horizons, the Breakthrough Prize acquired almost the status of the Nobel Prize. Through such forms of recognition, the public was pushed into believing that with mRNA vaccination a new age started: flu will disappear! Even cancer. To the list of miracle drugs produced recently (although long in the pipeline), Paxlovid and Molnupiravir—antiviral medications meant to treat COVID-19—were added in a rush, side-effects and relapse notwithstanding. The role of Paxlovid, an anti-viral that is supposed to prevent hospitalization (Owen et al. 2021) and death, is still unclear. For those older than 65, it barely helped (if at all). Relapse after treatment seems a major side effect. People on medications for high blood pressure went through life-threatening moments.

With all this in mind, let us consider vaccines and antiviral drugs in a broader perspective. In COVID (Crisis of Vision) times of a skewed perception of reality, there are many characters seeking legitimacy by attaching their names to miraculous cures, or to resistance to vaccination, or to mask wearing, lockdowns, social distancing. In Germany, France, Holland, Romania, Italy, as well as in several states in the USA, people took to the streets to express their views on the matter. Around the world, peddlers of miraculous ways and means to avoid COVID, or to survive it, made money off people's ignorance (and imbecility). Many dark horses are still running in the race to get out of the pandemic darkness, as its fourth year began, or to make the best of the opportunity. To complete the picture: Everyone is against having their lives regulated (access to restaurants, games, church services, etc.), and against lockdowns (called "shelter-in-place" in the USA). Very few, however, are willing to sacrifice anything for the common good. Prosperity, profit, and fun are idols we'd better not anger.

The mRNA vaccines are, in one way or another, game changers. There is a lot to learn from what it took to make them, and what they actually are. As there is a lot to learn from the virus spike protein made by moth cells, and from other vaccination methods. In particular, it is time to ask why means of prevention—because that's what vaccines are supposed to be—are turned into means of reaction: Have it if you prefer a milder infection. At least until their protection decreases—the waning effect. Is it true that people who received a third injection (the booster) are ten times (or more) less in danger than those injected only two times? If yes, then for how long? The second booster (never mind the new third booster) could [sic!] protect those over 65 years of age. Should we understand that vaccines are now means for reducing the risk of hospitalization? "Official" science collided with opinions (between incompetent and highly qualified) during a time when the media were asked to align with authority. A sobering account by David Wallis-Wells (2023) states: "If you get a vaccine that cuts your risk of dying from COVID by 90 %..., but infections grow five times as common, you are only twice as safe as you were before." The Paxlovid pill is supposed to have a similar result. It does not. Rebounds are not rare. Will those who already had COVID (some two or even three times) be even more better off—the so-called "super-immunity"—if vaccinated (Callaway 2021). Are vaccines now necessary for those who already acquired natural immunity? From a logical perspective, this does not make sense. From a medical perspective, it changes the foundations of immunology. The same holds true for the fact that the majority of those dying of COVID are vaccinated. Statistics can backfire: If everyone is vaccinated, then everyone will die of COVID. Of course, these questions and the associated considerations—some absurd—deserve explanation—even justification. After all, the subject concerns human life, not only the commercial success of new products and the credit some governments take for themselves.

Anticipation: The End is Where We Start from

The built-in preventive processes usually associated with the immune system reflect its anticipatory condition. It is a massively distributed system, with a great deal of local activity. The immune system is anticipatory in the sense that regardless of the pathogens, it recognizes danger and engages the entire organism, sometimes to dangerous extremes, in defense. It is the expression of the holistic nature of life. Non-deterministic in nature, the distributed immune system can nevertheless fail—as it does when, for instance, it goes into overdrive. COVID-19 prompted many explanations of the extreme danger posed by exaggerated immune system activity: the so-called cytokine storm.

Let us recall that anticipatory action, as an expression of the immune system, is characteristic of living processes. Behavior that prevents harm—physical injury or emotional disturbance, for instance—is in some ways a "vaccine." The root (*vaccine* means "pertaining to cows") recalls injecting the cowpox virus in humans in order to prevent smallpox. The operative notion is: PREVENT! But if you had small pox, natural immunity is expected. After all, vaccines—get a "smaller" infection—are a path towards natural immunity.

In life, a lot is learned, a lot is discovered: the "vaccine" of prevention includes avoiding certain foods (to prevent sickness, overweight, damage to organs, etc.). Hygiene is a *sui-generis* "vaccine." Let us also remember that anticipatory processes are non-deterministic. This means that the same anticipatory action (such as vaccination) can lead to a variety of possible outcomes, not just the one hoped for. The Johnson & Johnson vaccine—one of the four vaccines approved in the USA—triggered blood clotting. This is not a desired outcome. In other words, vaccination can be successful or not. Reaction is deterministic: the same cause leads to the same effect. Hitting the nail on its head with a hammer will drive it into a piece of wood. In the case of the COVID vaccine, the expectation implied by reaction is: Let's vaccinate and everyone will be just fine. Well, what about countries with a high vaccination rate but still experiencing an increased number of infections? What about a fully vaccinated and boostered cardiologist, in good physical shape and with a healthy immune system, who became the first to be infected with Omicron (in Israel)? The distinction between cause-and-effect and anticipatory action is also in regard to timing: in a reactive system, cause comes before the effect. The pandemic caused the rush to concoct a vaccine (many, actually)—that is, in reaction to COVID-19. Anticipatory action means to *prevent* something—i.e., keep the cause from happening in the first place. In the case of COVID-19, there was no prevention of any kind. There could have been. There should have been. (This assertion will be explained later in the text.) Once the pandemic began, all measures—vaccines and various treatments—were in reaction to it, and to its many consequences.

To react to whatever affects life—virus, bacteria, microbes, food, pollution, etc.—takes place *after*. If a person doesn't act in anticipation—i.e., avoid getting hurt, or getting poisoned, or becoming mentally damaged—but rather in reaction, the result of the effort is different. Giving up preventive action—avoid harm—in favor of reacting to damage—e.g., treat a burn, counteract poison, swallow a pill for emotional distress—is sometimes inevitable. But even treatment, i.e., a reactive practice, can become anticipatory: healing by engaging the organism, and redefining its interactions with the environment (clean air, clean water, proper nutrition, etc.). Treatment in its reactive form is preferred for reasons of convenience. Mechanical interventions (surgery, for instance) or chemical interventions (drugs of all kinds meant to remedy a life-threatening condition), which now dominate medicine, reflect the understanding of the acceptance of the model of the human being as a machine. The physician becomes the "mechanic" of the "human–machine." Such interventions—genetic healing, or surgery performed using sophisticated robotics, for instance—add new means and methods to the toolbox for repairing machines—human or not. As things stand now, instead of stimulating prevention through long-term anticipatory action, society prefers the immediacy of reaction. If you have a headache, just take a pill—no matter how much it may cost, or what the long-term consequences might be. No one has (or wants to take) time or patience. The determinism practiced is a hallucinogenic that leads to the decrease of the cognitive capability of its victims.

To detail what *anticipatory vaccination* is supposed to be is a prerequisite for understanding the dangers implicit in the misappropriation of anticipatory action for reactive purposes. As a prelude: the goal of vaccination is to prevent illness. But in the context of COVID-19, the goal post was moved from prevention to mitigation of disease. In the context of medicine that is increasingly becoming reliant on repair technology, the objective changes. This became an example of science by *fiat*. Data document that instead of effective immunity against SARS-CoV-2, a less symptomatic condition manifests itself. After vaccination there is no need, or only minimal justification, to be treated in a hospital. This helped contain the pandemic. In retrospect, after the "Get vaccinated" edict (close to mandatory in some countries), several vaccines developed specifically for the COVID-19 condition turned out to be antiviral drugs. The difference: vaccination of the healthy vs. antiviral treatment after infection began. Therefore, it is justified to examine how and why the respective vaccines were developed. The focus of the analysis is not on the scientists and their respective sagas—intriguing stories—but rather on the perspective of science that informed their research.

No one should or could dispute the fact that millions of lives were spared and costly hospitalizations avoided. Nevertheless, the underlying science invites questioning. Take only "Why are we boostering kids?" (Zweig 2022). The subtitle says it all: "The CDC and the FDA have ignored other countries' caution, the WHO's chief scientist, leading American experts, and their own data." This is about endangering the lives of the young (12 years of age and older) through a side effect: inflammation of the heart, known as myocarditis. Again, the death toll was not significantly lowered by vaccination: 350,000 deaths in 2020; 475,000 in 2021; and 270,000 in 2022 (Wallace-Wells 2023).

It is obvious that the science involved is connected to politics and market pressure. The selection and promotion of some drugs over others, to the detriment of obtaining valid remedies deserves to be scrutinized. The most flagrant example: an Emergency Use Authorization (EUA) was issued in December 2021 for Paxlovid, a day later, for Molnupiravir. It turns out that these drugs "may ultimately be found to be ineffective in vaccinated people, or almost certainly cost-ineffective" (Milijkovic and Prasad 2023). The purpose is not to hunt for villains, but to finally understand what undermines actions intended to save lives. The perspective of science has to be re-evaluated.

The Immune System is an Anticipatory System

Let us start with a conclusion: If the understanding of the anticipatory nature of life processes had guided the preventive effort as well as the vaccine development effort, the world would have been spared a pandemic. If this sounds ominous, it is because the events leading to the February 25, 2021 *Emergency Use Authorization* (granted in December 2020) of some of the vaccines come close to foreshadowing, if not an evil, at least a misguided course of remedy.

The anticipatory immune system is where it all starts. It is the impressive outcome of an evolutionary path along which anticipatory action driven by the possible future facilitates maintenance of life. From the system science perspective, the possible future means the ever-larger open set of possible states, from incipient life to death). The immune system is not a structural entity, like control systems in machines. And it is not some incidental or even pre-programed biochemical configuration supposed to protect from pathogens. Rather, it is a large whole-body encompassing functional process. For the immune system, the possible future (virus, microbe, poison, wound) is what might affect biological processes that keep the organism alive. Precursors of immune cells are continuously produced from bone marrow. The variety of precursor cell types corresponds to the open-ended nature of living processes. These cells make it into the skin, blood stream, thymus, lymphatic system, spleen, and mucosal tissue. And they are continuously renewed. The knowledge acquired from observing how the process unfolds and changes over time inspired various methods for testing the performance of the immune system and for stimulating it.

The possible future can be an accident: e.g., a wound to the skin from a rusty nail. The immune system as such plays no role in avoiding physical or chemical accidents. It works in preparing the body to defend itself against what such accidents could lead to—infections, for instance. The toxin-producing bacteria (in this example, *Clostridium tetani*) from the environment, escape the immune system's defensive processes. Even the tetanus vaccine (the DPT—diphtheria-pertussis-tetanus vaccine administered shortly after birth) will not *prevent* infection. But it enhances the body's ability to defend itself against a nasty pathogen.

Behavior—training, risk awareness, sense of responsibility—is the premise for prevention There is no vaccine to prevent breaking one's legs in downhill skiing, or PTSD for people exposed to extreme negative events (war-related, in particular, or medicine-related, for those having to deal with COVID-19 in treating patients). But after the accident, instead of anticipation-underlying healing, reactive medicine is usually called for to fix the injuries. The possible future can also refer to exposure to pathogens in general. Behavior helps: e.g., reduce exposure. In respect to COVID, China and New Zealand took this path (successful until the Omicron/BA.2 variant), forcing people to behave according to the rules set forth by the authorities. So did Japan (until BA.5). In Hong Kong, people returning from travel abroad had to remain quarantined (for three weeks), at their own expense. They were continuously tested. This is an authoritarian system for socialized prevention, in which the individual must abdicate individual choice for thc common good. Some find this model difficult to justify. When China finally gave up on it, a contradiction became evident: millions of lives saved, but at what cost? The conflict between exercising freedom of choice and responsibility for the family and society at large is real.

The Vaccination Narrative

There is so much to the history of the immune system that the temptation to rehash it is difficult to resist, even in a day and age of Wikipedia. There is drama: the body fights whatever might cause harm. No surprise that the Babylonian epic *Gilgamesh,* or the written testimony from ancient Egypt, as well as writings from the territory of what are today India, China, the Middle East, Greece, and Rome are always mentioned as early proof of awareness of what today is called immunity. It took a long time until the rather elusive processes of self-defense characteristic of any form of life—plants, animals, bacteria—were analyzed in detail. What seems beyond dispute is the realization that there are innate processes of self-defense—for preserving life—as well as adaptive processes. The two have in common the anticipatory aspect: organisms are not waiting for something to happen. Rather they are continuously preparing a defense against what might prove detrimental to their remaining alive. Several lines of defense are built and rebuilt:

- prevent infection (the skin is characterized as "the first line of defense" producing a variety of antimicrobial proteins);
- contain infection (the mucosal tissues often perform this function);
- eliminate pathogens (e.g., the lymphoid progenitor stem cells mount responses to specific microbes and exterminate them).

Adaptive immune processes emerged and evolved over the lifetime of living entities. The timeline of adaptive processes is different from that of processes deployed by innate immune activity. Innate self-defense is faster; adaptive immunity is achieved over time and reformulated as the context of life changes. Enhancing immune system performance is possible only through understanding its anticipatory condition. Reductionists think otherwise. They chase after molecules and cells, pursuing the physio-chemical path. Empirical evidence testifies to immunity destroyed on account of such interventions. Vaccination itself, properly understood, proves this point. Theories of immunity (e.g., Mechnikov's cellular theory, Ehrlich's "natural self-tolerance," Richet's injection of dead or weakened microbes) led to vaccination, more a practical path than a theory. Authentic vaccination is not based on physics or chemistry, but on biology. Its aim is to mobilize the auto-defense processes, characteristic of the living, in order to save lives. Despite progress in documenting the role of vaccines, the outcome is still poorly explained. Therefore, it comes as no surprise that there are specialized vaccinologists, but no coherent theory of vaccination.

Vaccines are inspired by empirical evidence of how immune processes develop. The milestones in vaccines (cf. Nature Milestones in Vaccines, *Nature*, 28 September 2020) include variolation (practiced for more than 6000 years in India, China, and even in the Ottoman Empire): infect the skin with scab or smallpox lesion cells to provide protection. Edward Jenner made the first known vaccine (1798, *An Inquiry into the Causes and Effects of the variolae Vaccinae*). Immunizing against smallpox (by using the cowpox virus) during an epidemic that killed almost one third of the children in London was a victory of prevention. By 1980, the World Health Assembly declared naturally occurring smallpox eradicated. Take note: no molecular genetics, no reductionism, no inventory of cells, rather, an intuitive approach reflecting the understanding of the underlying meaning. The skin is the entry point. The distributed immune system facilitates interactions that result in prevention. Louis Pasteur used live strains of pathogenic agents (chicken cholera in one case, and human rabies for another vaccine); Daniel Elmer Salmon and Theobald Smith used non-living pathogens for their protection of animals.

The subject is very broad: Why use a live or a non-living pathogen? In the case of the virus, this question is reformulated to account for the fact that the virus is not alive. What are the consequences of choosing one vaccine over another? Moreover: Does the vaccination augment the autonomic function of the immune system? As more vaccines were developed, the physio-chemical path became dominant. Artificially produced pathogens in combination with other substances were increasingly used. Consequently, another question arose: Why make the immune system dependent on engineered procedures? An even more focused question: Why human genome-informed treatment, ignoring the need for individualized action, but not virus genome-informed prevention? After all, a virus can be described in minute detail. It is a physical entity with a limited number of variables. Such interrogations cannot be ignored, but to focus on them is to switch attention from the immune system to ways and means of stimulating or even remaking it. The anticipatory model highlights factors that the reactive model ignores, or has no knowledge of.

Humankind is living under of circumstances of increasing abuse the organism. This is a characteristic of COVID—"Don't worry, everything can be fixed." Stress, processed food and artificial ingredients, polluted air and water, abusive medical treatment, among similar risk factors, make good health more and more the exception. Food produced in all sorts of "factories" (huge farms for the production of meat and milk) are then processed in other factories. They are based on chemistry, not biology. Their production cycle is industrial. The time of the machine, not natural time, i.e. not natural rhythms, defines the process. In addition, air pollution, abusive medical treatments (such as transplants, prostheses), chemotherapy, puberty blockers, etc. result in the fact that health is compromised. New so-called rights, reflecting the entitlement mentality, are ascertained in total disregard of the nature of life. As a consequence, all kinds of vulnerabilities emerge, and once in place continue to increase. The opioid addiction pandemic—still unfolding (over 200,000 victims in 2021)—is a condition society can do without.

Given the nature of contamination perils, society arrived at the juncture at which public health in our days cannot be conceived without vaccination. More important: vaccines have become part of the medical culture. Research into the various aspects of stimulating immunity have afforded results that are of benefit in many areas of concern: immunodeficiency (susceptibility to infection from particular pathogens), diminishing effectiveness of medical interventions, acquiring herd immunity (reducing disease incidence by protecting as many individuals as possible). Vaccines cannot compensate for harmful social behavior, but could help in preventing consequences. Medicine encoded knowledge regarding genetic and environmental factors, aging (and the associated immune-senescence), side effects, and antigenic overload (overwhelming the immune system, especially of the young and very young). In this context, it was unavoidable that after natural means were used in making vaccines, the first genetically engineered concoction—the Hepatitis B surface antigen recombinant DNA—was successfully tested. Most recently, with the advent of sequencing, immunology changed. The victory of reductionism over the broader perspective of a holistic view comes with a high price tag.

Caveat: in our days, the dominant understanding of genetic processes corresponds to capitulation to the machine model of life. This is how the living matter of DNA—a biochemical compound—was associated with the newest machine: the computer. And this is how the metaphor of programming came into place. It comes as no surprise that one headline reads "Vaccines 2020: the era of the digital vaccine is here" (Pizza, Pecetta, Rappuoli 2021). The scientific foundation of using simuli of pathogens in order to stimulate the immune system is pretty clear. Even inactivated bacteria, or viruses multiplied in chicken eggs or all kinds of cell cultures, can be used in making vaccines. Not yet clear is whether genetically engineered vaccines—making antigens based on genetic sequencing descriptions and delivering them to the body—offer the same performance as vaccines based on natural pathogens. Moreover: whether their deployment meets standards such as expectations of efficiency and absence of undesired outcomes continues to preoccupy scientists (Dolgin 2023). The use of the mRNA-based vaccine is spectacular in terms of lives saved, but also in terms of redefining expectations. The choice implies a conceptual change: vaccination is no longer really vaccination. It becomes a method, far from perfect, for reducing consequences of infection. At most, these vaccines, massively deployed (at times proof of vaccination was required in certain situations) triggered passive immunity. There is no prevention. In order to stimulate prevention, active immunity must be triggered. The mRNA method, which is still insufficiently understood in all its possible consequences, does not accomplish this. In view of this reality, affecting a very large section of the world's population, digital vaccine optimism, as an extension of the mRNA vaccine technology, seems at least hurried. The claim that it is attractive to envision "a future scenario whose pandemic vaccine is designed and developed in one week by a single research group and produced globally by a series of robotic stations distributed across different continents" (Pizza et al., p. 3; Venter 2013, p. 236) sounds rather dystopian. This is another instance in which the reactive mode of thinking has the upper hand. Worse is the deterministic take, like fixing computer viruses, producing one version after another in the living, yet again declared to be a machine screaming to be fixed.

To analyze what makes up DNA, RNA, and the ribosomes is relevant to the physics and chemistry underlying life, but not to understanding life itself and its continuous remaking, including its defensive processes. These are acids and enzymes and all kinds of chemicals. But after Crick, they became—by name—genetic information (Watson and Crick 2013). Descriptions generated by genetic methods (molecular biology) are all data extracted through measurements of physical and chemical characteristics. To sequence a virus is justified thusly: get to the blueprint of the poisonous particle via measurements that generate data. Being an RNA virus, it can activate cell reproduction processes: copies of the virus (i.e., protein) are made by the body itself. Thus the edict: We'd better try to understand the physio-chemical process of the copying process. Currently, this promulgated practice is science by faith, of the same nature as any kind of theology.

Data Driven and Knowledge Driven—Two Different Perspectives on Vaccination

The COVID-19 story is actually more about a new epistemological threshold. The focus is neither on living nor non-living pathogens, but rather on descriptions of pathogens. More precisely, descriptions of proteins meant to train the immune system to fight the "real" pathogens—the viral load. The new vaccines are the extreme example of data-driven vs. knowledge-driven science. Instead of understanding the meaning of immune activity, scientists focus on data that describe the matter in detail. The fact that they deal with living matter is cavalierly omitted. No more knowledge-based understanding of change (in a person's health), but data-based predictions for interventions that address the change. Instead of meaning interpretation, statistical inference.

To vaccinate ultimately means to induce antibodies. This was achieved, in the past, through means closer to intuition: fight fire with fire, infection with infection—*un diavolo scaccia l'altro* (One devil chases the other away). The empirical evidence for this practice is relevant only insofar as *knowledge* about living processes is often derived along this line. The term *antibody* described what in the immune serum neutralized the toxins (poisons, or virus infections) and pathogenic bacteria. The body, as living matter, produces antibodies that interact with pathogens. Synthetic vaccines trigger the production of different antibodies. A small wound is a live class in how this takes place. Live replicating strains or killed pathogenic organisms lead to replications that produce antibodies for an immune response, but not for more disease. Enough replication, but within limits—this is a difficult order. To achieve this goal for non-live vaccines, all kinds of adjustments intended for increasing the immune response are used. For instance, toxoid vaccines (such as the one used against diphtheria) are formaldehyde-inactivated protein toxins purified from the infecting agent. Preservatives, emulsifiers, and stabilizers (such as gelatin) are part of the mixture. Allergic reactions of all kinds are the unavoidable consequence of such additions. Such reactions explain, in part, why some people oppose vaccination.

Vaccine-prompted protection is not without controversy, despite evidence from vast amounts of studies of the various aspects of immune activity. New approaches, such as viral vectors and nucleic acid-based RNA and DNA vaccines, prompted even more debates. Robert Malone, the first to use an RNA-based vaccine is convinced that there is a risk of intoxication. He demands that the vaccine makers inform the public about the dangers. Although he himself was vaccinated, he experienced breakthrough infection. The cognitive gap is such that it would take a long time to explain why, instead of the natural pathogen, live or dead —collected from a wound or cultivated in a lab—genetic descriptions, pertinent to the physics and chemistry of the process, are used. Viral vectored vaccines actually imitate natural infection in order to induce humoral and cellular responses. No adjuvant is necessary. The RNA and DNA encode the target antigen using protein nanoparticles. This enhances their efficiency.

The rather abysmal performance of some vaccines, such as the annual flu vaccine, and the swine flu mass vaccination effort of 1976, are examples that cannot be ignored. They cast a shadow on the spectacular record of the role vaccines have played in the eradication of quite a number of life-threatening conditions. It does not help that the inadequate medical care in the USA, focused on reactive means, feeds even more hesitancy among those who could benefit from having their immune system enhanced through proper vaccination. The poor, often uneducated, suffer the most. In Europe, mistrust in medicine has grown not because the physicians are incompetent, but because the government intervenes in decisions concerning individual health. Victims of measles (according to the WHO, 2019 had the highest number since 2006) are a painful reminder of the inadequacy of well-intended programs meant to facilitate access to vaccination.

All this background is justified by the need to explain the extremely sophisticated technology behind modern vaccine design and production. And even to make clear what is meant by "digital vaccine" (a catchy misnomer). But it does not explicitly show what happens when in fact the immune system is conditioned to behave in a deterministic manner, despite the fact that as an anticipatory process it is by its condition non-deterministic. In all instances when the statement was made that the reductionist-deterministic view hijacked anticipatory understanding, what is emphasized is how the living is reduced to the physical and chemical substratum. Or, more to the point, how living matter is reduced to matter devoid of life.

The focus on meaning was abandoned. The smallpox vaccine treated the body with a living pathogen. And the body, in its anticipatory expression, "understood" the meaning of the danger. From 1796, when Jenner discovered that an animal virus could stimulate the immune system to protect against a human virus (smallpox), to 1885, when Pasteur (using desiccated spinal cords from rabbits) saved the life of a nine-year-old boy who had been attacked by a rabid dog, there is a jump from active viruses to inactivated viruses. Some 50 years later (1937), Theiler grew the yellow fever virus in non-human cells, while submitting alterations to it so that it could trigger immune responses instead of causing infections. These are simuli, or surrogates, used in individuals in order to prevent infections. Recombinant DNA (early 1970s) was a further step: the generation of a bacterial protein, still a natural surrogate of biological processes. A fourth breakthrough occurred in 1980: Richard Mulligan and Paul Berg published the findings of their experiments with an *Escherichia coli* gene (Desmond and Offit 2021). The latest step: trigger chemical reactions within the cells for the purpose of making proteins beneficial in immune self-defense. Viral vectors, DNA molecules, or the so-called messenger RNA (mRNA) can do that. In a metaphor used repeatedly during the pandemic, this means to activate the *machine* in the human being, and to *program* it to produce the drugs we need. But the metaphor is by now reified: it is taken literally. And this is where the confusion lies.

Suddenly vaccination becomes a "programming" task—send instructions to the cell. Since messenger ribonucleic acid in interaction with cells lead to making whatever proteins could be beneficial (or proteins undermining life), it can provoke the immune system into fighting for the life of the organism. The thought is: Train the immune system on something that looks like the enemy so that when the virus makes its way into the body, it will be recognized and annihilated. Think about a security system to prevent break-ins: Provide the picture of the criminal known to have robbed households on your street. If the virus did not change in interaction with the body, the dangerous corona protein, or any other protein, could be easily identified. But there are variants. The attacked body makes them. The robber caught *in flagrante* will change: wear a mask, a wig, whatever. Recognition becomes more difficult. Omicron, a more recent but possibly not the last variant, is an example. This is why the next booster—a vaccine to reinforce the effects of vaccination—was developed for the BA.5 variant. As though there is enough knowledge that nothing else might change the infectivity landscape. But since BQ1.1 currently dominates, and XBB1.5 is becoming more prevalent due to very high infectivity, this would mean adapting the vaccine. Like a digital device that prevents break-ins, train your device using faces of known criminals: if somebody looks like the criminal, sound the alarm and get the system to fight it. Sounds simple. The genetic view according to which the metaphor of a four-letter language (A, C, G, T, standing respectively for adenine, cytosine, guanine, thymine) becomes the reality of the body invites the thought of algorithmic computation. The premise is clear: All that is needed to control the situation is to identify the "recipe" of the infection. Cell dynamics, include genetic processes. But they are part of the holistic process of maintaining life. That is their meaning. The organism is not a deterministic machine. What the mRNA molecule carries to the cells are descriptions pf physical and chemical processes. Of course, the variants of SARS-CoV-2 have different genetic make-ups. They affect host cells in different ways. Since no two cells are the same, the replication is also different. Contrary to the dicta of genetic determinism, the cell is not a passive copy-making mechanism. It makes choices.

One more thing: While various researchers have studied the mRNA carrier, it is far from clear how it works. Even more important: there is no basis for evaluating long-term consequences of using it. It was determined that COVID-19 affects the immune system. Evidently, vaccination does the same. Indeed, chirality—which means ultimately how interactions take place—is affected. The BA.5 custom-made booster actually was not clinically tested. The claims associated with its release make reference to how mice reacted to flu vaccines. The chemistry of the mRNA carrier was altered in order to avoid rejection by the body. Does this change make a difference? The same holds true for the mechanism of delivering the mRNA. (A rather rich-in-detail report is "The tangled history of mRNA vaccines, *Nature*, 14 September 2021.) With all this in mind, it is worth checking the timelines—there are several—of how the mRNA vaccination model appropriated what eventually grew into the humongous COVID-19 business. His took place while the government assumed the function of physician to the nation.

A Timeline that Arouses Suspicions

The first timeline: genetic research. André Boivin (1947), in a paper that went almost unnoticed, advanced the idea that the DNA (deoxyribonucleic acid) makes the RNA (ribonucleic acid). Most important: RNA can lead to protein synthesis. In 1948, Boivin visited Romania (where he had worked from 1930–1936). In his presentations, he explained the process through which the body neutralizes pathogens producing antibodies (Mesobreanu 1948) He also distinguished between "active immunity" and "passive immunity." (If there will be a Nobel Prize for the mRNA vaccine, the Committee should call it the Dr. Andre Boivin Prize in order to posthumously honor his contribution.) It took almost 50 years to turn the accumulated knowledge—never rewarded by any prize, and rarely fully acknowledged—into the genetic tool for protein generation. The phases are as follows:

- May 1961 is a landmark: a *Nature* article (1961, 190, pp. 576–581) describes how an "unstable intermediate" makes the path from genes to ribosome for protein synthesis possible.
- Another text (*Journal of Molecular Biology*, 1961, 3, pp. 318–356): a first-known description of the protein synthesis process.
- In parallel, liposomes—drug-delivery media—are conceived and tested (starting in 1965).
- mRNA and liposomes come together some 30 years later, when experiments in mice and rats are carried out. Lipid nanoparticles are approved, while in parallel modified mRNA is experimented with (the goal being the reduction of immunogenicity).
- In his two articles (1989), Robert Malone describes the use of an mRNA-based vaccine on animals.
- Nanoparticles are approved as a transport medium; in parallel, experiments with mRNA or modified mRNA (supposed to reduce toxicity) are carried out. This opens the way to clinical trials (by 2008–2009) for cancer immunotherapies.

For some reason, neither SARS (2003) nor MERS (2012)—corona virus-caused conditions—attracted the attention of researchers who eventually (by 2020) formulated mRNA-1273 (the Moderna vaccine) and BNT 1628 (the Pfizer-Biotech vaccine). These are the vaccines rushed to the public through an emergency approval process. None of the many vaccines developed claimed to prevent the virus from reaching the human organism. Rather, the goal is to enable the immune system to respond to infection and thus prevent the full-fledged disease: recognize the enemy, a 3-dimensional corona-shaped protein (carbohydrates cover the spike's surface) that locks into human cells.

There are explanations for what motivated researchers to take the path of what is called synthetic biology. Mostly, it is the convenience of reductionist biology: If genetic processes can be seen as machine processing, the task is to tame the machine. So far, considering waning and breakthrough, as well as side effects such as myocarditis, it turns out that what was supposed to behave like a machine does not. The cost of the inadequacy of the mRNA to act as a vaccine, in terms of lives lost, suffering (after vaccinated individuals got infected), and stress on the medical system is high. And it will increase, once long-term consequences start to add up.

Abdication

The change in the meaning of vaccination reflects an abdication: from prevention to milder symptoms, avoidance of hospitalization, reduction in number of deaths. Paxlovid and Molnupiravir—the inhibitor of the main protease of the SAR-CoV-2 virus—are effective if used during the first five days of infection. Both perform almost as equally as the vaccines, but they are not dependent of the immune system response. Unfortunately, the side effects are quite intense, and relapse after treatment is not infrequent. Worse yet: they are ineffective for those over age 65.

Since Jenner, the medical community has realized that vaccines against infectious pathogens help to prevent, and, when necessary, to contain epidemics. But they were not perfect. Vaccination is a rather difficult path: Once a pathogen was identified, it takes a long time to "prepare" it for deployment. It is easier to try to contain the danger, i.e., isolate those who might infect others. Of course, controlling human behavior raises many questions. The HIV/AIDS pandemic and the current monkey pox epidemic prove the point. It is more punishing for the affected not to take note of reality. Condescension and political correctness only make the situation worse for them. The more tolerant societies become—this unavoidable process qualifies as an inexorable social law—the more difficult it is to engage individuals in preventive activities. Even the making of vaccines, from the most rudimentary to the most intricate, is not a straightforward process. The Jenner experience cannot be scaled up. And once synthetic vaccines, mimicking natural pathogens (or some parts of such pathogens), are designed, a number of unforeseen factors need to be considered. Even social aspects of a nature different from those affecting behavior come into the picture. To make vaccination a goal (forcing it, or legislating it) transcends political

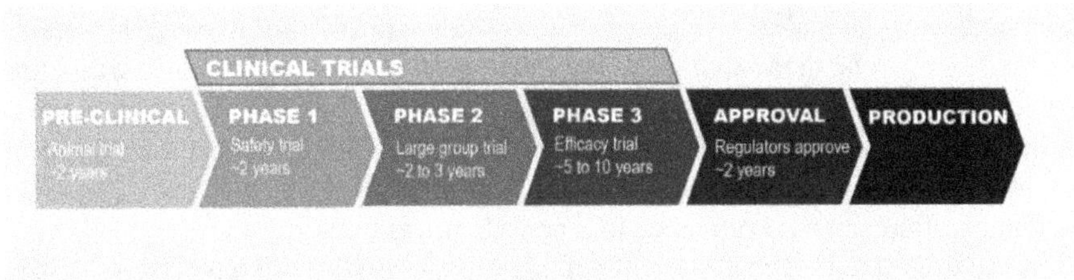

Fig. 1 The long path of certifying vaccines (or any other drug). It reflects the deterministic view of experimental evidence. If anticipation considerations are ever adopted, the process will have to be redefined

and ideological confines. To optimize vaccines, chemistry provided adjuvants, which did not always "behave" as intended. The issue of vaccine side effects, and the issue of waning (decreased protection) are as old as vaccination itself.

Before producing vaccines, approval—a very tedious process, meant to prevent causing harm—is expected. Elaborate means of production had to be designed and tested (Fig. 1). It is a long procedure, unfortunately tainted by political and economic calculations of all kinds. In addition, experimental evidence is an assumption derived from ascertaining that there is no difference between the living and the non-living (for more on this, see Nadin 2018).

All this—a process that can extend over 15 years—means money: to go through all the stages is a logistic nightmare, close to the model of financing a new movie or a skyscraper. Producers of illusion products (as movies are) and the pharma industry share in the experience of an investment that could go wrong. Therefore, a prerequisite came into place: *somebody else's money.* Worse yet: Some viruses, triggering chronic or repeated infection, could not be effectively controlled. HIV-1 was the major example—there is still no vaccine for it. But there are also the syncytial virus that affected the respiratory system (RSV), and the herpes simplex, not to mention the influenza virus, Ebola, Zika, streptococcus, and a whole slew of corona viruses. This in itself was enough not only to spur seeking new ways to address dangers to human (and not only human) health, but also to jump start new businesses focused on the science and technology of synthetic biology. Known forms of vaccination were complemented by genetics-based methods—which after a while replaced the former.

As always, extreme situations lead to extreme solutions. Another pandemic—the 2009 influenza—marked the switch from live or weakened or "killed" (i.e., inactivated) viruses to promoting the making, by the body, of proteins that emulate the target virus. When RNA fragments are injected in order to induce the cells to produce proteins that "look and feel" like the pathogens, the immune system is trained on these synthetically designed proteins that mimic those produced under the virus attack. For this to come about, researchers, such as Katalin Karikó and Drew Weissman, had to find ways to change the mRNA. Under real-life circumstances, the organism would annihilate incoming RNA. To make the organism accept the synthetic mRNA molecule—engineered from four kinds of nucleosides (adenosine, cytidine, uridine, and guanosine)—it had to avoid degradation and the possibility to cause inflammation, and so as to be made to translate efficiently. Inflammatory response was a first obstacle. In order to address it, Karikó and Weissman modified the uridine (using a so-called pseudo-uridine). As a result, the immune system's dendritic cells would not fight the injected mRNA. But the chemistry of the RNA was changed, and thus the human organism was subjected to a new chemical reality.

As the story was told again and again, the researchers did not have it easy. Vaccination in its original straightforward form, as an anticipatory action, succeeded most of the time, as empirical evidence reveals—but not always. Scientists who try to develop new anticipatory actions by using the genetic make-up of the living are kept in check by scientists laboring in the reactive mode. They are, invariably, supported by taxpayer money through the mechanisms of research funding, that is anything but breakthrough. Neither Karikó nor Weissman, and later on the scientists from the J. Craig Venter Institute and from Synthetic Genomics, qualify as anticipatory-driven scientists. In contrast, Derrick Rossi's attempt to use mRNA to encode protein that transformed mature cells into stem cells is closer to an anticipatory path. Stem cells are a promising way to engage the organism in self-repair. But Rossi, who withdrew from the Moderna group, made no progress with his own idea. Almost everyone who developed mRNA-based vaccines adopted, within their respective specialties, the reductionist-deterministic view dependent on the powerful sequencing technology.

The purpose of using the mRNA was to make available new ways of vaccination. New therapeutic avenues for the diseases that plague modern society, cancer in the first place, are worth pursuing. The complementarity of the method—using genetics from a mechanistic perspective—and the implicit anticipatory understanding deserve to be acknowledged. Indeed, the possible future in this particular case is the protein to be generated by the cells according to the mRNA instructions. Let us roll back the movie to the year 1961.

- From the transient RNA molecule—labeled as messenger RNA (mRNA) by Brenner et al. (1961)—proteins of interest (such as the spike protein of COVID-19) are made by ribosomes.
- Robert Malone (1989) injects animals with a vaccine that produces antibodies.
- Conry conceived a vaccine that encoded cancer antigens (Conry et al. 1995).
- Mevix Bioscience is on record of being the first mRNA company (founded in 1997, changing its name to Argos Therapeutics).
- In 2009, Benjamin Weide is credited with the first trial of an mRNA-based vaccine (an attempt at cancer immunotherapy).
- In 2011, a carrier was found—lipid nanoparticles—and a vaccine was prepared for testing. This was a vaccine platform for a possible influenza pandemic. At the time, a company called Novartis specifically addressed the possibility of a pandemic.
- In March 2013, there were three cases of avian influenza each with new characteristics. A synthetic RNA-based vaccine was designed and prototyped, ready for preclinical testing. Very few people are aware that the US government actually

bought the vaccine and prepared a strategic reserve. The avian flu pandemic never materialized—more due to luck than anything else.

The purpose of this revisit to the past can be expressed in one question: Could have a design and prototype for a possible corona virus (one from the many we are aware of) pandemic succeeded before COVID-19 struck?

During testimony before the House Science and Technology Committee (March 2020), Dr. Peter Hotez, Co-Director of the Center for Vaccine Development at Texas Children's Hospital, made public that his team had created a Corbevax vaccine that may have been able to prevent the pandemic outbreak (Hotez and Bottazzi, 2021). "We had the vaccine ready to go…nobody was interested in a corona virus vaccine." The vaccine—almost as effective as the mRNA vaccines—was eventually certified in India. It will remain unpatented in order to make it affordable to those who need it. The same question can be posed in respect to Paxlovid. Merck and Pfizer had some protease-based drug, meant to control viral proteins, since 2003. But the final word was: "No disease, no market for drug." Pharmaceutical companies could not see any profit in preparing for a catastrophic event, and even less in preventing it.

The immense bureaucratic machinery involved in giving the green light to anything of medical interest is not known for being visionary or for understanding the value of new ideas. This is where the anticipation perspective illuminates the Crisis of Vision (COVID). The notion that every avenue for protecting life should be pursued is probably beyond controversy—even in this age of extreme egotisms expressed as entitlement-based credo: My life is more important than anyone else's. In this respect, what counts is not so much the underlying science, but the actual results. Given the fact that quite a number of conditions affecting the human being proved beyond what classic vaccination offers, seeking alternatives is an anticipatory action. The *Coalition for Epidemic Preparedness Innovations* recognized the possibility of using synthetic genes for rapid development of vaccines (and for other applications). However, what is not clear is how prepared society is to invest in prevention, and thus avoid the huge costs of reactions. Corona viruses became a subject of interest more than 60 years ago as "cold viruses," (Kendall et al. 1962). Prior to this, in the 1920s, respiratory infection in chickens, and later—late 1940s—animal coronaviruses associated with brain disease and hepatitis had been discovered. In particular, upper respiratory tract infections in children were traced to them. Since 2003, at least five new coronaviruses affecting humans have been identified. The SARS infection (reported in 29 countries) and MERS prompted a short-lived surge of interest in treatment methods and in preventive vaccination.

The connection between animal and human coronaviruses became a subject of research. The likelihood of spillover from animals, and the possibility of pandemics were spelled out. MERS-CoV became a prototype pathogen. Investments were made into the study of the particular virus associated with MERS. Moreover, a vaccine (MERS S-2P 90 mRNA) was tested in mice. This became the proof-of-concept for the mRNA method of vaccination.

Two observations:

(1) Research in mRNA and lipid and polymer chemistry was focused mainly on cancer treatment. Therefore, the investment is relatively specific. Moderna Therapeutics (founded in 2010) obtained full support from the US Biomedical Advanced Research and Development Authority (BARDA); CureVac AG of Tübingen, Germany, as well as BioNTech (Mainz, Germany) also benefited from public money. Nothing wrong about this; just as there is nothing wrong about focusing on the lucrative market for cancer treatment.
(2) To present "SARS-CoV-2 mRNA Vaccine Development" as "Enabled by Prototype Pathogen Preparedness" (Corbett et al. 2020) is disingenuous. No one disputes the urgency: saving human lives at any price. No one disputes the technological accomplishment: "Within 24 h of the release of the SARS-CoV-2 isolate" and "within 5 days of sequence release," a vaccine was produced: Using "current Good Manufacturing Practice (cGMP) production of the mRNA/LNP expressing SARS-CoV-2 S-2P as a transmembrane-anchored protein with the native furin cleavage site (mRNA-1273) was initiated in parallel with preclinical evaluation" (Corbett, *op. cit.*). Sixty-six days after the release of the virus sequence, a first human Phase 1 clinical trial (March 16, 2020) started, followed 74 days later (May 29, 2020) by a second Phase. Confirmation of expression and antigenicity in vitro preceded the vaccination of the first human subject; immunogenicity was documented in several mouse strains.

Against the background of more than $12 billion (to which more billions were added) made available through Operation Warp Speed, the development of the mRNA vaccines for COVID-19 was qualified as epochal, triumphal, phenomenal. Nine new billionaires, from the COVID-19 vaccination business, joined the list of those who, in many ways, became rich and famous on account of the reactive response to crises (medical, social, political, or economic).

There is nothing to object to enthusiasm, or to getting rich in a decent manner (i.e., not at the expense of others, in particular not of those hoping for help). Saving human lives is worth all the money. But there are questions that just do not go away, especially in the context of wondering how COVID-19 came into our lives: the spillover from bats in a market in Wuhan or in some mine in Southern China; Italy as a source of the pandemic; the engineered virus (by no matter who, the Chinese or the Americans). For the latter possibility, laboratories involved in function-gain ("make them more terrible through genetic engineering") are the suspects. Many are trying to resolve the issue. (Even more make money on conspiracy theories.) The EcoHealth Alliance (with Peter Daszack at its head), the Andersen Lab (with Kristian Andersen in charge), Harrison and Sachs (A Call for an Independent Inquiry into the Origin of the SARS-CoV-2 virus, May 19, 2022, PNAS), DARPA and numerous other actors owe society answers regarding the origin of the pandemic. In an unrelated case, it turns out that the monkey pox antiviral Tpoxx originated as Post-9/11 Bioterrorism Defense product (Walker 2022). It is not inconceivable that in the treasure trove of the military complex there are traces of research and development related to COVID-19, or to some other dangers humankind recently faced.

Regardless of what might be eventually uncovered, one conclusion is beyond dispute: reductionist-determinism, with its unavoidable obsession with progress at all cost, is the culprit. This is COVID in its broad meaning: creating a context for the possibility of disaster. Not yet sufficiently clarified is that a scientifically grounded concept—anticipation-driven immune response—can easily be turned into a reactive procedure when the expected outcome of vaccination is not attained. For the sake of clarity: Whatever the process through which vaccines were conceived, tested, and produced, it is beyond doubt that the chances of getting COVID-19, and eventually of dying, are higher for the unvaccinated. (There are many reports on this—no reason to compare them, more will be produced.) Moreover: should a vaccinated person get infected (the so-called breakthrough), the symptoms are usually milder and, more important, the long-COVID—a miserable condition lasting way past the disease proper—is marginal. Recent research made it clear that the long COVID is where science has failed to come up with valid methods for prevention or treatment (Ledford 2022a). Many individuals choose not to be vaccinated only to find out, the hard way, what might happen. Be this as it may, a vaccine, in the strict sense of the word, is not available. The fact that China, after failing to police the pandemic, reached to the mRNA vaccine makers, confirms that currently there are no better solutions. It is possible that for some conditions vaccination might not work as prevention method.

The Pandemic that Could Have Been Avoided

The Crisis of Vision and the vaccine euphoria, leading to a massive effort to immunize, are related. Indeed, prior to SARS-CoV-2, there was no way to prepare for the specific disease condition associated with the SARS-CoV-2 pathogen. Still, the pandemic could have been avoided. But not in a world whose prosperity depends on high mobility, and even less in a society of increased disregard for others. These are systemic characteristics. However, the vaccine platforms already developed could have been tested. Should have been tested, is the correct statement. High coronavirus infectivity was already documented. Hence: Why not fully develop a SARS or a MERS vaccine? And use the time after SARS of MERS were gone to test and to validate. Why not stimulate alternatives? Why not fund vaccines already at an advanced stage of testing? In a relatively calm environment, questions such as those we face in our days of instability could have been addressed.

As the pandemic becomes an endemic condition, new questions arise. How effective are the vaccines? How stable? Will waning occur? Are there breakthroughs? Boosters or not? What kind? When? What about variants? And what about a sense of interdependence in our integrated world? After the justified rush to adopt vaccines that are not really vaccines, such questions are legitimate. Explanations of all kinds, from the abstruse to the absurd, were considered fit to print, or to be broadcasted. In advance of the release of a vaccine specific to the BA.5 variant, scientists remembered the original antigenic imprint: immunological consequences of being exposed to infection, including vaccination. Evidently, every medical intervention changes the state of the organism. But to evoke immune memory in the decision-making process for changing the vaccine is shortsighted (van Zelm 2022).

The waning of vaccine efficacy, not a new phenomenon, could and should have been researched. The issue of virus variants could have been examined. From an assumed efficiency of 96%, associated with the Emergency Use Approval, to the less than 40% achieved in reality (regarding the Delta variant), there is a great difference. It indicates that currently society does not have the science adequate for justifying the vaccines as more than a pre-therapeutic concoction genetically engineered. Or as an after-infection stimulant of the immune system (the T and B cells). The booster, inspired by the pharmaceutical companies, is considered a smokescreen. The newest, adapted to the BA.5 variant, is probably the last publicly funded. The list of ingredients, made public, does not disclose anything relevant to how immunity is expressed. This is chemistry dressed up as information science. The demagoguery of "a universal coronavirus vaccine" (Koff and Berkeley 2021) also qualifies as a smokescreen. Those not yet able to produce an effective vaccine for COVID-19 are already peddling a new threshold: convergence of more technology and biochemistry for the purpose of submitting the human being to further questionable reactive means. Long-term consequences are of no immediate concern.

The immune system works on meaning—exactly the dimension absent in the thinking of those "married" to "one-size-fits-all" interventions. There is not even an inkling of individualized vaccination protocols. Individuals with a strong immune system and those affected by immunodeficiency (often induced by medical treatment, including excess vaccination) are different. The machine mentality took over. Treating humans as machines in mass vaccination campaigns is a convenient political and economic game. But it is also a failed medical path. Almost 40% of those hospitalized in the USA with the variant of the 2022 spring and summer, were fully vaccinated and twice boostered (Muller 2022). Too many of those in the medical establishment became lobbyists—or even political activists—for the companies producing vaccines. The vast network of enterprises delivering COVID-related products, sometimes under questionable circumstances, exercises undue pressure.

To expect either the vaccines or the various treatments available to be 100% successful is to align with those who advance the deterministic view. Indeed, non-deterministic processes underlying life are such that, as a matter of principle, there is no universal cure for anything. And there cannot be a fully successful vaccine. The living cannot be forced to behave like a machine only because it is convenient for those trying to fix it. Having this understanding in mind, it follows that means and methods for addressing health should be individualized to the extent possible.

It used to be that entrepreneurs risked their well-being in pursuing visionary ideas. The new method is to get the public to pay for someone's success—or, even better, to pay for their failure. Socialization of risk and reward leads to opportunistic reactive interventions. In the 1976 swine flu outbreak, the political campaign to mass vaccinate the population was not based on biological arguments. The vaccine used had many questionable aspects. For instance, as a side effect, Guillain-Barré syndrome affected a number of people, leading to their death. Somebody wanted to be re-elected. Precisely the same happened again in 2020—a "radioactive," aberrant president and the vast government machine would throw everything but the kitchen sink into bending the curve of infection. "Bend the curve" became a political slogan. By 2021, not too much changed (another curve to bend, which only means high infection rates and many deaths), except that all the people might need a booster—or even a second one. But not even this is sure: BA21 became BA2.121 and the game continued in 2022 and even in 2023.

Through a variety of arguments, some governments are trying to make vaccination mandatory. And if not for all, at least for adults over 50 or 60 years of age, or for people serving the public: police, military, medical workers, etc. Germany's new coalition government wanted this measure badly; so did Italy. Some form of vaccination mandate was tried, one way or another, arousing public uproar. Demonstrations, all over the world, testify not so much to resistance to vaccines as to the people's fear that COVID-19 might become another attempt to transform this crisis into a form of social control. For an observer from distance, the entire COVID-19 saga appears as the plot to force society into new forms of submissiveness: the dictatorship of genetics, of the digital, of larger governments, and of the exercise of more executive power. Missing from the public consciousness is the crisis of vision—COVID—context. For the benefits of immediacy, society gave up self-determination and responsibility. Progress as entitlement. Payment due in the future. Even the poor and needy living in capitalist societies have it have it better than ever. Millennials, who grew up in the digital age, and Generation Z, native to the digital, seem to accept the deal. Reality, in this view, is what one experiences on monitors, large or small, not what an individual might have to do for a living. In the USA, there are no less than 170 million Millennials and Generation Z together. They are charging into the future of guaranteed minimal income but not necessarily of a guaranteed sustainable life.

It is clear that the mRNA platform for vaccination could have been used for appropriately testing the viability of the method before the pandemic. The Paxlovid "miracle" drug could have been fully developed instead of being put on hold until the time when it would pay off for the drug industry This would have alerted the public to what was experienced since vaccination began: breakthrough, waning, side effects, long-term consequences (such as a weakening immune system). But Moderna and BioNTech are not the only players tethered to governments. In the European Parliament, a member displayed, to the TV cameras, the agreement between the makers of vaccines and the European Council. Fourteen pages were blacked out. The provisions, which basically relieve the companies from any responsibility (no liability whatsoever) are kept secret because fundamentally they are illegal. As opposed to car manufacturers, to makers of canned food, even to makers of toys and cosmetics, makers of vaccines meant to save lives are given assurances, which nobody else gets, that they will not be held up to norms of quality and expectations of soundness. Someone (Bastian 2021) came up with a suggestive formulation: Pharmaceuticals are generally a gamble. To get a product validated for release, a whole process, involving many people, and requiring quite a bit of money, is initiated. The big players are always ahead. They can bribe the government into releasing public money for their probable success. As the record shows, various drugs, peddled by government agents in guise of experts, were developed as a treatment method—and failed miserably.

In the same article, highlighting the gamble, a major statement was made: "The mRNA Vaccines Are Extraordinary, but Novavax Is Even Better." For their vaccine, Johnson & Johnson engineered a harmless adenovirus as a shell to carry genetic code to the cells. The spike protein is made by the cells themselves. It was supposed to be a one-shot vaccine. Until it became a 2-shot method of vaccination. Finally, the government took note of the fact that it causes frequent clotting, and hence stopped supporting it. To expect the public to trust such ad hoc re-evaluations is at best impertinent. Oxford-AstraZeneca produces a carrier vaccine also. No doubt that Novavax, Johnson & Johnson, and the players behind Oxford-AstraZeneca also could have worked in anticipation of the real SARS-CoV-2 since they focus on spike proteins. Incidentally, so do the Cuban vaccines Abdala, Soberana, and Membisa. It turns out that actually Novavax was active in pursuing vaccination for MERS. The use of synthetic genes, a goal of the Coalition for Epidemic Preparedness Innovations, was already a centerpiece of their activity as they, like everyone else, targeted cancer. Ironically, cancer, as the merciless enemy of the living, itself reflects the pitfalls of life within a civilization increasingly addicted to the industrial making of everything (from food to clothing). No other disease promises higher profit for everyone involved in treating it.

In a different situation are those vaccines coming from China and Russia. They use inactivated virus for triggering production of antibodies. CoronaVac could not have been conceived before the pandemic; neither could have the Russian Gammaleya Institute prepared their "Sputnik," a product of questionable performance offered as a bribe to countries in desperate need of Russian support (or dependent on its sources of energy). But China, devastated by the realization that not even locking up its population can avoid infections, is by now opting for the mRNA path, which it initially dismissed, more on ideological grounds—although the Chinese vaccine is performing as expected (Fig. 2).

As recently as mid-December 2021 (Gordon Joyce et al. 2021), the Walter Reed Army Institute of Research shared the details of a Ferritin nanoparticle COVID-19 vaccine (SpFN). The synthetic protein, to which variable spikes can be attached is based on ferritin nanoparticles (SpFn). It elicits, according to the released information, powerful humoral and cell-mediated responses. As a result, replication of the virus is prevented in upper and lower airways. The hope is that a robust immune response can be achieved over the entire SARS-causing betacoronaviruses. It is fair to assume that it will take a long time until such claims can be tested and validated. In the years to come, other methods will be considered and eventually tested.

The fact that all kinds of means and methods for facilitating preventive actions are considered deserves to be celebrated. For example, nasal vaccines (operation "Nasal Vaccine is already mentioned and lobbyists are at work in tapping some public money for it). David Curiel, the person who in 1995 produced the proof of concept for the mRNA vaccines, is one of those involved in nasal vaccination. Applied to the cells commonly exposed to the virus, such vaccines could be more effective and trigger less resistance to vaccination. However, failure to understand

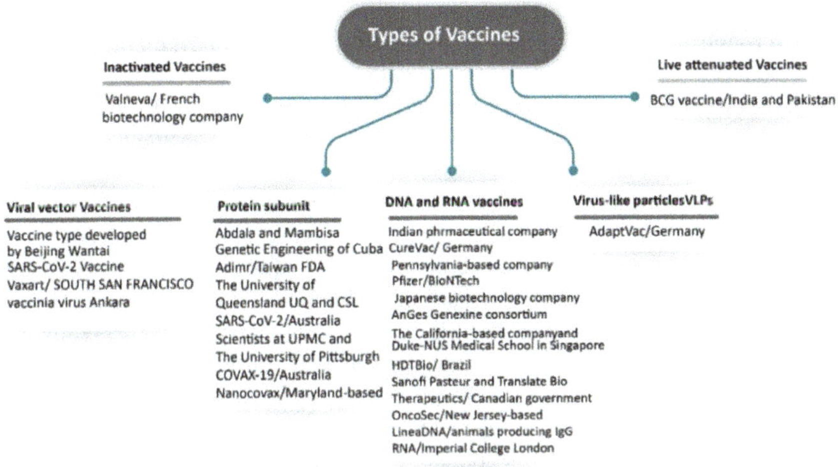

Fig. 2 Types of vaccines—and some providers. This is testimony to the huge effort to react to COVID-19. No prevention means for corona virus type-caused infections were considered necessary. (Absent from the list are attempts at nasal vaccines)

their vaccine capabilities, limitations, and risks cannot be ignored in the euphoria of having tamed COVID-19 for a while. To block infection—everyone's target—is quite different from controlling transmission. Incubation time changed from almost seven days for the initial variants (Alpha, Beta, Delta) to less than four days for the Omicron kinds. Moreover, epidemiologists ascertain (using sequencing, of course) that "Most people have been infected with the virus… even if they don't realize it (Wernau 2022). That this conclusion is reached through the use of deterministic software (cf. *Nature*, August 11, 2022) is telling in respect to how misguided the science behind COVID-19 is. If the same energy and the same financial resources had been used for developing therapies a totally different situation would have been created. The Paxlovid euphoria can be understood in the context of many, not necessarily successful, attempts. Indeed, lives were saved and hospitalization was often avoided. However, it is impossible to ignore not only price, but also the side effects, and the still unexplained cases of again coming down with symptoms, weeks after the five-day course of treatment (Lazar 2022, commenting on the remarks of Dr. Eric Feigl-Ding's endorsement of the Pfizer product). The oral antiviral drugs raised high hopes, probably due more to vaccine confusion than to their own merits. Remdesivir, dexamethasone, and the monoclonal antibody treatments are in the same situation.

The billions of taxpayer dollars spent on supporting their development have yet to be justified. As things stand, the decline of antibody responses after vaccination—including the booster–are worrisome and therefore explain why patients hope for more adequate therapies, even though they are all of a reactive nature. If the alibi of virus variants could be accepted, one can wonder why was the medical community not able to see them coming, despite all the efforts at sequencing and measurement?

And why—really *WHY?*—if indeed new vaccines were easy to produce—as broadcasted through all communication means possible—then why haven't they yet been? Everyone agrees: don't get infected! Especially since the long COVID is miserable. "We have no proven therapy for LongCOVID. But there are more than 25 randomized trials in progress" (Ledford 2022b). LongCOVID is by now the third cause of neurological disease affecting Europe and the USA.

It is clear that neither the vaccines, nor the various drugs are the answers. There is nothing wrong, in principle, with a second jab, or even with boosters—and if need be, a third booster. But to claim credit (and glory) for submediocre science and questionable technological progress puts a dent in the credibility of science. Within its own logic: Injecting vaccines that will wane is a stopgap measure. Providing the means for individualized treatment, including individualized vaccines could prove more consequential. And since genetics is supposedly the new frontier in medicine, why not engage the body in healing, instead of continuing the physics and chemistry attack on it? The answer is simple, but not easy to take. Everyone wants a quick fix similar to getting their car to a mechanic, or their iPhone to a shop specialized in cell phones. Neither physicians, nor patients, have the patience and the determination it takes to engage the whole organism in its own healing. Individualized treatment is the only path towards effectively helping the ill, instead of making them even more dependent on chemicals and technology. The digital natives—Generation Z, as well as the Millennials—have an attention span no longer than what it takes to switch from one channel to another. They prefer short term fixes, no matter how high the price—compromised health included. But life, pandemic or not, is not a TV show or a computer game. To maintain it in good health implies a behavior informed by the sense of future. Indeed, sustainability cannot be ignored.

The negative yield of medical dependency means that the investment in fixing—the strategy for dealing with COVID 19 returns less value in relation to what was spent. After the many billions of dollars thrown at the crisis, at the beginning of the fourth year of the pandemic, vaccinated and Paxlovid-treated persons still become infected in great numbers. In a broader perspective, reacting to COVID-19 had a greater effect on the environment.

Nobody openly discusses what yet another headline broadcasts: New COVID Vaccines Need Absurd Amounts of Material and Labor (Schmidt 2021). No doubt, the chemical processes used in the mRNA vaccines take less time than growing attenuated-activity viruses in chicken eggs, or in some other culture. The genetic sequence is eventually turned into the mRNA—a highly unstable substance. It is packed into oily lipid nanoparticles, which also increase absorption through the cells. But there a large variety of reagents is involved, enzymes, glass vials, syringes, sterile containers. Bioreactors, where biochemical reactions take place, are expensive. To produce the substance that prevents degradation of the mRNA is impossible without them. The great investment of public money in the mRNA and the Emergency Authorization made it possible to speed up trials and, after approval, to scale up production. But it also favored them over other vaccination methods. Never mind the billions of doses that were dumped since their expiration date passed.

Epilogue: Doctors Should Lead, not the Medical Establishment

A practicing physician, Buzz Hollander (in *RealClear Science*, August 23, 2021), brought to the public domain what practitioners of medicine who care for the trust of their patients should be aware of:

> *Let's stop pretending the vaccines are 90% effective and breakthrough cases are "uncommon."*
>
> *Let's stop pretending that vaccinated people are far less likely to spread SARS-CoV-2.*
>
> *Let's stop pretending that it's rare for vaccinated people to develop severe COVID-19 or die.*
>
> *Let's stop pretending that prior infection should not influence the decision to vaccinate.*
>
> *Let's stop pretending that the vaccines are a no-brainer for adolescents and children.*
>
> *Let's stop pretending that a third booster is definitely going to help.*
>
> *Let's stop pretending that these vaccines are "kill-shots," cause sterility, spread disease, etc.*
>
> *Let's stop pretending that the vaccines are the only way to reduce the burden of COVID-19.*
>
> *And, finally, let's stop pretending that vaccines alone will bring an end to COVID-19.*

The same physician takes note of the fact that the 36% obesity rate in the USA means a two-to-five-fold increased risk for the obese. The better "vaccine" in terms of reducing the COVID-19 risk is changing behavior. Responsible nutrition and exercise are part of it. This is, in fact, a form of prevention. What people choose to eat and drink in the day-and-age of generalized processed food kills their own guts—as they watch a video on the role of the microbiome. Not only how much people eat, but mainly what they eat and when. In the effort to contain the pandemic, science—from a position of arrogance impossible to accept—ignored medical practitioners. Some of them ended up acting like technicians—pressing buttons, making choices on computer menus describing healing protocols. Quite a number could have contributed to addressing the various aspects of COVID-19. But those who did not align with the official line ended up ridiculed. They discovered, the hard way, that the huge army of experts—epidemiologists, molecular biologists, geneticists, computational medicine experts, etc.—in command of the situation, are captive to a science that did away with learning and experiment.

In the post-pandemic tie, one would like to hear that everyone—young and old, white, black, brown, educated or not yet educated—is determined to change their way of life. Reaction entails more regulation. As a result, medicine became more expensive and less efficient. The overhead of medicine in the USA skyrocketed. The doctor has to charge for three to four times more for administration than for actual care. More regulation effectively blocks learning as the source of anticipatory action. Patients who indulge in checking websites for quick fixes have not become smarter, but rather less patient. Their dominant obsession, from the start, was about getting back to how things used to be. Neither pills, nor vaccines, nor more technology, which the public demands, will make the next breakdown—unavoidable as it is—easier to live with. Little if anything was learned from the COVID-19 experience. Of course, learning cannot be mandated; but circumstances for fostering it, instead of limiting it, should be encouraged. Learning to give up the obsession with questionable progress, which always means more consumption, at any price, would be a start. To pursue alternatives inspired by anticipation action implies education as a condition for freedom informed by awareness. Rethinking medicine, making quality medical care available to all, and rediscovering the uniqueness of each individual are goals no vaccine and no therapy can help society reach, or substitute for. Revolutions are always anticipatory.

A Postscript Dictated by Reality

The Sultanate of Brunei occupies third place in the world in respect to vaccination rates. Hong, Kong, Singapore, and New Zealand stood out due to their drastic measures in fighting COVID-19 and in a high vaccination rate. In these countries, the infection rate was reduced to a minimum—up until the beginning of 2022. After that, at one time Brunei occupied first place in the world for the most infections per capita. For the sake of comparison, if the proportion of those infected out of the total population, the USA would have had about 3.5 million cases per day. In a totally different situation, China was pretty successful in its policies of strict lockdowns and tracing. Many people considered such measures not only contrary to what one expects in a democracy, but also not sustainable. They were right. With Omicron, the city of Shanghai became the poster child for a situation that ran out of control despite the drastic measures taken by the government. Mighty China changed course.

Hong Kong, Singapore, and New Zealand could claim: We did everything science told us to do. But now we are powerless. Neither the vaccines nor the few therapies available are really helping. On the other hand, China could say: We followed the science and also immediately reacted using every method we could think of, but we could not eliminate the *Shanghai crisis* that spread in a situation of unrest that could have led to regime change. Japan, the most masked country in the world and widely vaccinated, faces numbers (of infections, mainly) that were lower during the first three years. Nobody aware of these events can rejoice in the outcome. However, it is clear that the credibility of deterministic science and of reactions, even those at the extreme, became questionable. *Disrupt science* is not a matter of millions of papers published, which fail to show any progress in thinking. It is a matter of life and death at the scale of humankind.

Let's not guess at the outcome of the case court between Moderna and BioNTech (implicating Pfizer, that made the product developed by BioNTech the global success it turned out to be). It was easier to predict the outcome of the newest boostering effort. After all, acquired immunity also explains reduced lethality. When vaccines became available (mostly because those in power were desperate to maintain their positions), the demagoguery of reductionist deterministic science reaches its apex: "We are saved you." Of course, fixing machines should be a 100% success given the excellent record of physics and chemistry in describing change in the non-living. But helping individuals avoid contamination, or helping them in mitigating infectious disease, is a different story. The condition of the immune system and the condition of a computer do not compare. The virus affecting computer programs can be identified and the source of "infection" can be blocked. No virus affecting the human being can be treated lie computer viruses are treated. The pseudo-science of deterministic interventions has to be disrupted.

Given the uniqueness of each individual, to expect that the mass-produced intervention can save everyone is sheer fantasy. Realistically speaking, to even conceive of individualized vaccines is at this moment in time utopian. But efforts in defining, within the heterogeneity of life, commonalities cannot be discarded. It is not unrealistic to design vaccines for particular age groups, for those who have in common some comorbidities, for the young and the very young, or for those under certain types of medical treatment. Moreover, the medical practitioners should have been engaged in defining some typologies, and in reporting on efforts to adapt the vaccine to various kinds of patients.

To green-light new vaccine formulations (such as booster) on antibody data from mice—the classic surrogate for human beings—is to ignore the need to redefine vaccine efficiency. It should be based not on statistical models but on understanding anticipatory processes. Ultimately, the entire health system needs to be redesigned on the foundation of a biology that is not a special or particular case of physics or of chemistry, but rather a body of knowledge specific to the nature of change in the living. As deserving as the scientists who were awarded the 2023 Nobel Prize in Physiology or Medicine are, their contribution was actually to chemistry. They modified the chemistry of synthetic RNA (yielding mRNA) so that the human body would not reject it. Medicine has failed spectacularly since the time it became tethered to physics and chemistry. Vaccinations did not bring an end to COVID-19. The endemic phase is here to stay. Disrupting science, in order to integrate reaction and anticipation, will provide a different perspective in the attempt to deal with dangers to human (but not only human) life. Sustainability must be regained.

References

Bastian H (2021) The mRNA Vaccines Are Extraordinary, but Novavax Is Even Better, The Atlantic, June 24. https://www.theatlantic.com/health/archive/2021/06/novavax-now-best-covid-19-vaccine/619276/ [retrieved August 31, 2021].

Schmidt C (2021) New COVID Vaccines Need Absurd Amounts of Material and Labor, *Scientific American*, January 4. https://www.scientificamerican.com/article/new-covid-vaccines-need-absurd-amounts-of-material-and-labor1/. [retrieved December 17, 2021].

Watson JD, Crick FHC (2013) Molecular Structure of Nucleic Acids: A Structure for Deoxyribose Nucleic Acid, *Nature* 171, 737-738, April 25.

Boivin A, Vendrely R (1947) Sur le rôle possible des deux acides nucléiques dans la cellule vivante, *Experientia*, 3, 32–34.

Owen DR *et al.* (2021) An oral SARS-CoV-2 M^{pro} inhibitor clinical candidate for the treatment of COVID-19, *Science* 374:6575, 1586-1593. November 2. https://doi.org/10.1126/science.abl 4784. [retrieved December 17, 2021].

Brenner S, Jacob F, Meselson M (1961) An Unstable Intermediate Carrying Information from Genes to Ribosomes for Protein Synthesis, Nature 190, pp. 576−581. https://doi.org/10.1038/ 190576a0 [retrieved July 2, 2020].

Callaway E (2021) Covid super-immunity: one of the pandemic's great puzzles, *Nature* 598, 393–394, 14 October.

Conry RM *et al.* (1995) Characterization of a messenger RNA polynucleotide vaccine vector, Cancer Research, 55:7, 1397–1400, April 1. https://pubmed.ncbi.nlm.nih.gov/7882341/ [retrieved December 1, 2021].

Corbett KS *et al.* (2020) SARS-CoV-2 mRNA Vaccine Development Enabled by Prototype Pathogen Preparedness, bioR*x*iv, June 11. https://doi.org/10.1101/2020.06.11.145920v1. [retrieved August 21, 2021].

Desmond A, Offit PA (2021) On the Shoulders of Giants—From Jenner's Cowpox to mRNA Covid Vaccines, New England Journal of Medicine 384, March 25, 1081–1083.

Dolgin E (2023) Trial settles debate over best design for mRNA in COVID vaccines, Nature News, January 12. https://www.nature.com/articles/d41586-023-00042-z [retrieved January 13, 2023].

Gordon Joyce M *et al.* (2021) A SARS-CoV-2 ferritin nanoparticle vaccine elicits protective immune responses in nonhuman primates, *Science Translational Medicine*, December 16. https://doi.org/ 10.1126/scitranslmed.abi5735#con4 [retrieved December 17, 2021].

Hotez PJ, Bottazzi ME (2021) Whole Inactivated Virus and Protein-Based COVID-19 Vaccines, *Annual Review of Medicine* (Online ahead of print). https://pubmed.ncbi.nlm.nih.gov/34637324/ [retrieved October 21, 2021].

Irrgang P *et al.* (2022) Class switch towards non-inflammatory, spike-specific IgG4 antibodies after repeated SARS-CoV-2 mRNA vaccination, *Science Immunology*. December 22 [retrieved December 25, 2022].

Kendall EJ, Bynoe ML, Tyrrell DA (1962) Virus isolations from common colds occurring in a residential school, *British Medical Journal*, July 14, 2:5297, pages 82–86 [retrieved July 22, 2022].

Koff WC, Berkley SF (2021) A universal coronavirus vaccine, Science 371:6531, page 759. February 19. https://doi.org/10.1126/science.abh0447 [retrieved September 1, 2022].

Lazar K (2022) A puzzling phenomenon: Patients report a rebound of COVID-19 symptoms after taking the antiviral Paxlovid, Boston Globe April 21. https://www.bostonglobe.com/2022/04/ 21/metro/puzzling-phenomenon-patients-report-rebound-covid-19-symptoms-after-taking-ant iviral-paxlovid/?p1=StaffPage [retrieved April 22, 2022].

Ledford H (2022a) How common is long COVID? Why studies give different answers, *Nature News*, June 20. https://www.nature.com/articles/d41586-022-01702-2 [retrieved July 1, 2022].

Ledford H (2022b) Long-COVID treatments: why the world is still waiting, *Nature News*, August 9. https://www.nature.com/articles/d41586-022-02140-w [retrieved August 15, 2022].

Malone RW (1989) mRNA transfection of cultured eukaryotic cells and embryos using cationic liposomes, *Focus* 11:4, 61–66

Mesobreanu L (1948) Noui cuceriri ale biochimiei - în jurul visitei Prof. Boivin în România, *Natura* XXCVII:5–6, Mai-Iunie [retrieved December 5, 2021].

Milijkovic MD, Prasad V (2023) PAXLOVID™ – A Regulatory Gamble, The American Journal of Medicine . January 2. https://doi.org/10.1016/j.amjmed.2022.12.017 [retrieved January 6, 2023].

Muller M (2022) Omicron's Mutations Impaired Vaccine Effectiveness, CDC Says, *Bloomberg* August 25. https://www.bloomberg.com/news/articles/2022-08-25/omicron-s-rapid-changes-impaired-vaccine-effectiveness-cdc-says?leadSource=uverify%20wall [retrieved September 1, 2022].

Nadin M (2018b) Rethinking the experiment: a necessary (R)evolution, *AI & Society* 33:4, 467–485. New York/London: Springer

Pizza M, Pecetta S, Rappuoli R (2021) Vaccines 2020: the era of the digital vaccine is here, *Science Translational Medicine*, December 15. https://doi.org/10.1126/scitranslmed.abm3249?download=true [retrieved December 17, 2021].

Taquet M *et al.* (2022) Neurological and psychiatric risk trajectories after SARS-CoV-2 infection: an analysis of 2-year retrospective cohort studies including 1 284 437 patients, *The Lancet Psychiatry*, 9:10, pages 815–827, October.

van Zelm MC (2022) Immune memory to SARS-CoV-2 Omicron BA.1 breakthrough infections: To change the vaccine or not? *Science immunology*, 7:74. June 2. https://doi.org/10.1126/sciimmunol.abq5901 [retrieved September 1, 2022].

Venter JC (2013) *Life at the Speed of Light: From the Double Helix to the Dawn of Digital Life.* New York: Viking

Walker J (2022) Monkeypox Antiviral Began as Bioterrorism Defense, *Wall Street Journal*, August 25.

Wallace-Wells D (2023) 9 Pandemic Narratives We're Getting Wrong, Opinion, *New York Times*, January 4. https://www.nytimes.com/2023/01/04/opinion/covid-pandemic-history.html [retrieved January 6, 2023].

Wernau J (2022) Think You've Never Had Covid-19? Think Again, *Wall Street Journal.* July 25. https://www.wsj.com/articles/think-youve-never-had-covid-19-think-again-11658741403 [retrieved September 1, 2022].

Zweig D (2022) Why Are We Boosting Kids? *The Free Press.* January 27. https://www.thefp.com/p/why-are-we-boosting-kids [retrieved December 26, 2022].

Confluence Versus Convergence or QED (Quid Erat Demonstrandum)

Convergence is a necessary outcome. Rays converging to a focal point is an example. The past unequivocally determines the process.

Contrasting confluence and convergence illustrates the difference between contingent and necessary phenomena. Living processes are confluent. Physical processes are convergent. Machine functioning is the result of convergence: machines are designed to reach a state in which all parts are designed to achieve an integrated function. Given the nature of processes in a machine, they can be described mathematically. All parts of a mechanical clock can be replaced by identical copies. Life processes are contingent in nature. No part of a living process can be replaced without affecting the whole of the organism. CoVID—the crisis of vision—is the result of a convenient but misleading confusion: knowledge reduced to its representation as data. It is a limiting perspective that ultimately undermines the role of knowledge without realization of consequences. In the tradition of scientific discourse, the Sustainability Thesis articulated in the opening of this book is validated through empirical evidence. Anticipatory action is the way to reverse the flipped yield of reactions to breakdowns.

Legacy—Inherently Deterministic

Generations after generations have been indoctrinated with the theology of determinism and reductionism. Once the doctrine became the backbone of science, anything questioning it—including the anticipatory perspective—is deemed blasphemy. Quite tellingly, religion went through a similar process. Incidentally: the doctrine called *dialectic materialism* played the same role in countries where the authority of the communist regime extended to what is right or wrong in science.

As long as religion-based explanations of human concerns (e.g., disease, natural disasters, conflicts within families and tribes and between nations) were accepted, social life was governed more by religion than by knowledge. This changed in the

© The Author(s), under exclusive license to Springer Nature Switzerland AG 2023

M. Nadin, *Disrupt Science*, https://doi.org/10.1007/978-3-031-43957-5_7

seventeenth century with the Cartesian Revolution. It is by no accident that the affirmed causality of phenomena pertinent to the living and the non-living was also adopted as the model for life in society. Many accomplishments, including the emergence and proliferation of social and political institutions, go to the credit of the rationality pursued. However, the consequences of the dogmatic reductionist-determinism view adopted as it became the unwritten foundation of life in society cannot be downplayed. In the absence of awareness of the anticipatory nature of human activity, societies accepted progress—as ill-defined as it is—as a goal, regardless of the price it entails. The Industrial Revolution is probably the epitome of what it means to pursue progress at all price. The breaking point turned out to be reaching the much-glorified scale of digital globality: the promise of more growth on account of a larger market and the illusion of the unlimited resources of the rest of the world: capitalism without borders. The crisis of vision dates to the time when opportunity and risk reached the global scale. The curve representing the Upside/Downside Ratio changed: instead of the ever-higher returns on the investment, more and more endeavors end up as success stolen from the future.

In particular, the equation of energy illustrates what it means to live on borrowed time. Burning fossil fuels is an example. Various "champions" of new technologies propose solutions that instead of protecting the environment in reality undermine its condition. Instead of more accountability for the human contribution to climate change—i.e., instead of requesting reduction of frivolous energy use—they advance alternatives which are nothing but the same old same old: use of fossil fuels, but hidden behind the scenes. The World Bank made public the view shared by many experts: "technologies assumed to populate the clean energy shift…are in fact significantly more material intensive in their composition that current traditional fossil-fuel-based energy supply systems" (Arrobas et al. 2017). Each windmill and solar cell is nothing but fossil fuels in some other form. This does not discourage the demagogues of progress at any price from pushing their cause. Currently, solar and wind facilities need 300 times more land, 300% more copper, and 700% more rare elements than facilities fired by fossil fuels. Feynman famously noticed that the light of flames burning wood is that of the sun rays metabolized in the trees. The electric car—the only one to be accepted in California after 2028—is more of the same self-delusional solution to a simple problem. Zero-emission—a noble target—is an illusion as long as the production of the electric vehicle (EV) itself generates 30,000 pounds of carbon dioxide. The production of internal combustion (IC) cars is less toxic. For this reason, but not only, it is quite obvious that reducing carbon emissions is actually a matter of changed behavior. In short: Drive less! The problem is with the "sprawling death machine," i.e., the car, is how it is used.

The determinism of progress at any price suggests that more (consumption, of course, in order to keep the economy running) is the way. No matter how self-destructive this can be. Anticipatory thinking suggests that the *meaningful* should underlie human actions. The decision to drive, for instance, even when there is no justification for it except convenience and laziness, is where society should start. Making more vehicles available ends up in adding more carbon emissions to the atmosphere, even if the vehicle is so-called electricity powered. Not more, but more meaningful!

The social unrest in the context of the pandemic was not about solar panels, wind mills, or electric cars. The demonstrations revealed a precise aspect of what defines the crisis of vision: *Your past is not our past.* In other words, the deterministic cause-and-effect of a shared past became a test of viability instead of being a stepping-stone towards the future. Statues of the heroes of an epoch of different values were taken down. More such acts will follow. *Presentism* took over: judge the past from the perspective of the present. The past, that the statues symbolized, extends into an unacceptable present, and a possible future that not everyone wishes for. Denouncing racism and lethal police violence is part of a broader scope of society's ills:

- extreme consumerism—the anesthetic to dull the pain of an operation to excise what is left of individual sovereignty;
- the hijacking of new technologies proclaiming to give people more power but actually taking away from them what is left of their privacy.

In a way, *it is all about legacy and how to shake it off*.

The confluence of many movements, some spontaneous, others well organized, some short-lived, others of unexpected duration, resulted in a degree of instability seldom experienced. Of course, there is no future for those whose response to past and present injustice is looting, burning homes and businesses, afflicting pain on whoever stood in their way. But there is future in the determination to honestly address the underlying aspects. Not surprisingly, in the context of the pandemic, it became difficult, if not outright impossible, to distinguish between protest and crime, between right and wrong, between hope abandoned and blind hatred. Blurred vision instead of clarity of goals. The right and the left, those passionately pursuing what they proclaim to be a positive outlook, and those who are dedicated to burning everything to ashes were all affected. Even if those involved did not know what the alternative might be, never mind what it might entail, everyone wanted a new beginning. And no more pandemic.

The virus and the sickness-loaded narration made the failure of science evident. But they are only marginally the reason why COVID-19 took the path it did. A tortuous history of hopelessness and lost trust in one another damages society more than what the virus triggered. A humankind that brought upon itself sickness as the price of questionable prosperity, egotistic to the hilt, surprised itself in an explosion of desperation. The legacy of Cartesian Rationalism, i.e., the narrow perspective it advanced, is so prohibitive that it must make room for a broader and more encompassing view. Many realized that the future is more critical than ever. To focus on legacy is to increase awareness of the toxic presence of that part of the past that reduces causality to the cause (in the past) $\rightarrow$ effect (in the present) sequence, to the exclusion of what the possible future entails. Within the pandemic, it became once again clear that anticipatory action, in particular as prevention, is not a mere option among many, but a necessary dimension of social life.

Civilizations succeed each other along the manifold of continuity and discontinuity. Compatibility between succeeding cultures and negation of the past cannot be separated in describing our civilization. There is much to acknowledge: Science and technology made possible amazing accomplishments. There is also much to be aware of: wars, crimes against the ecology, unfairness, intolerance. Like the newly born to parents they did not choose, individuals cannot pick their past, or their gender. A growing number of young and not so young, willing to have their genetic makeup genetically re-engineered, pushed their revolt against the past to an extreme harmful to themselves. They cannot choose their language, which comes with the biases built into it. They cannot choose, at least for a while, their religion. They cannot choose the economic system within which they will live. Capitalism, more and more subject to various interrogations, might not be their choice. But they want to choose their identity, even against their own genetics. And they want to have some say regarding their future.

In this respect, one question in particular stands out: Can capitalism contribute to sustainability? (Rull 2011). "The pitchforks are coming if we don't reform capitalism" warned the founder of the World Economic Forum (Schwab and Malleret 2020). The demagoguery of the message illustrates the extreme of questioning the economic system to which the model of progress, which got humankind in trouble, is associated with. Of course, the mass extinctions along the record of life (Courtillot and Gaudemer 1996)—five of them preceeding not only capitalism but also the human being (200,000 years in the making (Tattersall and Schwartz 2009)—are not cause-and-effect phenomena. The geological and paleontological record show that each extinction wave was followed by bursts of diversification. That means: more life, in an increased number of expressions. Evolutionary change is neg-entropic. The physics of the universe does not provide guidance in explaining what happened in such mass extinctions, and even less in explaining why they happened. A multitude of phenomena involved in the process are documented: methane producing microbes for example, contributed to the decrease in temperature. Photosynthesis explains the oxygen and rich atmosphere. Lower temperatures affected greenhouse gases. Multicausality is quite obvious. There is a confluence of related and unrelated physio-chemical and biological processes.

Knowledge of the past can assist in understanding the present context. According to some authors, it is sometimes framed as a prelude to another life extinction. But to expect some cycle (and the implicit suggestion of inevitability) is to fall into the trap of determinism. Evidently, it is impossible to avoid thinking in terms other than anthropocentric. But it would definitely help in realizing that the living Earth—since living matter is predominant—is anticipatory: possible futures inform actions at all its levels. It is a holistic process that transcends socioeconomic processes usually invoked by the doomsday demagogues. The rhythm of nature—relatively steady and slow—and that of human activity—faster and faster—are to a large extent out of sync. These are theoretical considerations which might help in approaching the concrete aspect of sustainability as an expression of awareness of the distinction between the physics and chemistry of our world, and the biological.

The expectation of the compatibility of animal husbandry, to go back in time, and industrial farming explains previous pandemics: e.g., the swine flu, the chicken flu, the cow flu. The expectation of compatibility between generations is even more evident in comparing the functioning of machines. Each new generation is expected to be compatible with the previous and with the functioning of human beings. This expectation of compatibility is, ironically, one reason why there are computer viruses. Those causing harm usually break in at the weakest point (obviously in the version of the past). The computer evolved along the legacy model. The designers of the first operating system made decisions that continue to affect generations of users. All the shortcomings at the lowest level (machine language) proliferate from one generation of machines to the next. Artificial languages were designed to "translate" what users want into instructions that machines can understand. Such languages carry with them a logic that is often disconnected from the way humans think and express themselves. And then the inevitable virus—*digital* "poison"—infiltrates. One malicious person or program is all it takes to undermine the efforts of everyone relying on their computers to execute certain tasks. Not painful when writing a simple document, or keeping accounts; quite dangerous when driving a car, or running home security systems; life threatening when monitoring an implant is hacked.

The legacies of slavery, racism, inequality, unfairness, and moral duplicity, more vicious and destructive than the machine legacy, are examples of the shared understanding that the past influences the present, but is not reducible to it. In this respect, Descartes, who courageously advanced rationality as a substitute for the faith-based understanding of reality, is the victim of his own predicament. In time, the Cartesian Revolution, after celebrating Newton's view of natural law and other discoveries as its own victories, locked itself in the golden cage of *machinism*, setting the groundwork for the Industrial Revolution, as well as for the machine-making of things, of human beings, of societies. A new human condition was shaped under the premise of unlimited progress, without ever questioning the long-term consequences of this assumption. In terms of social and political life, it meant a gradual transfer of power and self-determination to the machine called government, and to the operators identified as political parties. The consequences:

- the inadequate and unaffordable legal system—the justice machine;
- the inadequate and unaffordable medical care system—medicine of the human machine
- political cannibalism and the parasitic political class that took over society—the political machine;
- an ever more costly system of education that is unable to address the future of its students, treated more and more as clients—the educational system as machine.

In full awareness of previous breakdowns, in particular the extinction of life, it should be clear that in terms of the pandemic humanity faces a different challenge: a self-inflicted wound. The legacy of determinism and the focus on machines led to a way of life that diverges from the evolutionary path. Once the disease took hold, an exclusively reactive practice, in meeting its many challenges, was furiously pursued. Science submitted, not surprisingly, to the political and to economics. There is no pro-active activity to reference, even though prior to COVID-19 a series of coronavirus-triggered maladies were already on record. In the framework of reaction to the pandemic, the most sophisticated machinery was deployed. In the spirit of the expensive legacy of reductionism, images of the virus (or of virions, the copies generated by the infected subject) were produced. But knowledge about the immune response, never mind the specific antibody dynamic, did not ensue. The WHY? of the multisystem affliction, which continues to present new aspects, is still waiting to be addressed.

The fact that cryo-electron microscopy—a method that earned a Nobel prize for its discoverers—was deployed to see the virus in the lung, gut, kidney cells (Bradley et al. 2020 *op. cit.*) is testimony to how serious the danger COVID-19 is perceived. But it is also a convincing example of how costly legacy, in the sense of following on the beaten path, is. Histopathology and ultrastructural findings of fatal conditions are, without a doubt, a major technological accomplishment. Histopathological features in fatal cases led to the assumption that the SARS-CoV-2 has greater infectivity, but less lethal than previous SARS. No room left for considering acquired immunity as more and more individuals got COVID. In time, the advanced technological exercise was repeated with the mutated virus (Delta, Omicron, etc.). Yet a simple question arises: Machines enable scientists to see the virus, but why do we still not know what to do? Some hurried answers:

> *First*: because what is visualized has nothing to do with the life processes that are affected. These are images of dead matter (frozen at cryogenic level);
>
> *Second*: because the "glasses" through which reality is examined affect what we see. Is this the real virus, or is it the theory behind the visualization method?

The legacy aspect of letting people see the virus is more subtle. Examining this aspect will help explain why the broader issue of the ever-higher costs of reaction summons evaluation of the means and methods deployed. At a cost of over ten million dollars, the advanced microscope and its peripherals are not unlike other technology used in the medical care system: the hammer in search of nails—even where there are none. A plethora of examples, starting with the microscope itself, speak in favor of seeking as much knowledge as possible when humankind is confronted with new questions. Disease turned out to be a state associated often with the invisible. Physics guided the technological effort towards making visible the invisible (remember the X-ray?). It definitely helped physicians to fight the many agents of their patients' undermined health that they themselves were not even aware of. Furthermore, within the machine model it promoted, it delivered what it takes to cut, adjust, replace. The spectacular domain of surgery today extends to neurons and diseased cells. Chemistry focused on alternatives to natural healing. It became another "hammer." Thus, what the body previously took care of via various anticipatory processes (not all related to the immune system) was overridden by chemical intervention. The antiviral drugs developed in response to the pandemic are part of this reductionist path. Healing was replaced by fixing. This legacy includes mechanical ventilators (which made headlines in the COVID-19 context), prostheses for replacing knees and hips, chemotherapy, and spectacular interventions that save lives, or at least diminish suffering.

The social, economic, and political legacy of the Industrial Revolution in medical care made the American system the costliest—but only marginally better than medicine in places where health is a shared concern, not an economy of high profit margins. Each surgery center—a high-profit establishment often funded by physicians themselves—was fully booked until COVID-19 hit. Brick and mortar, and a lot of technology, subject to obsolescence, make for a highly profitable rent capitalism, but not for sustainability, and almost never for healing. Rather, they generate the "patient for life," dependent on more and more fixing. New buildings bear the names of donors, who are not always aware of the fact that their endowments will cost society more than what bought them the illusion of immortality.

Given the nature of legacy, the Post-Industrial is turning everything into a digital process in which scientists can engineer ACE2 (Chan et al. *op. cit.*): i.e., make an artificial substitute of the enzyme to which SARS-CoV-2 attaches itself. It is like a mouse trap (described as "decoy receptors"), to which the virus will bind. No effort is too great when it comes to saving lives. Every venue should be tried. Provided that the effort is not driven by the obsession with technical performance, but by the genuine desire to help patients. Securing new funding along the same path of deterministic reductions of phenomena that are fundamentally non-deterministic can be explained, but not justified. The engineered ACE2 is not made of living cells. Therefore, the expectation that it will be a better receptor for the virus is questionable. It is an aggregate of accomplishments in physics and chemistry. The virulence of SARS-CoV-2 interaction with a living enzyme is the expression of processes undermining the existence of the afflicted. Unless such processes are well understood, chances are minimal that the decoy will prove to be a "high affinity" receptor. To decrease the rate at which the binding of the virus to living cells takes place invites a different way of thinking. From a broader perspective, prevention would avoid even the thought of expensive technological performance.

A High Price-Tag

The order of magnitude of the costs associated with the legacy of deterministic science and technology cannot be reduced as long as the focus remains on the *How* and *What*. To explain this thought: to reveal the structure of proteins by probing a flash-frozen solution with a beam of electrons consumes energy at a level at which the cry for protecting the ecology can no longer be heard. Through the combining of 2D images of molecules and producing a 3D image, researchers in well-funded laboratories generated all those pictures that appear on the covers of establishment science publications. To accelerate electrons to energies of 300 kiloelectron volts entails a pretty large carbon footprint—all for an operation that actually results in limited actionable knowledge. This is yet another example of the diminished return (the inverted yield) on an investment characteristic of the crisis of vision.

Toxic by-products and the use of sulphur hexafluoride gas—which affects the ozone layer—as well as X-rays generated when electrons hit metallic surfaces, were, of course, not considered in the process. They were not reported in the "breakthrough" findings. Many people are involved in servicing such a machine. Employment, for good and absurd reasons, is the obsession of a society focused on consumption. Public money is spent on making more and more machines. This is the Big Science drive. Entrepreneurs, subsidized by public money, justify the high price of their products through more applications—meaningless or meaningful. Legacy extends into business. The owners of such technology, even when it is paid by the public, are, of course, privileged. They get more grants for more of the same. "You own the machine, now you have to use it"—even if this makes no sense. It is a win–win situation.

If statues associated with the legacy of a disputable past are pulled off their pedestals by a generation claiming its right to the future, it is foreseeable that science itself, and the authority it claims, will be questioned. The much-envied Nobel Prize is named after the inventor of dynamite, and funded by the fortune he made with it. Its record is a testimony of celebrating everyone flying the flag of the deterministic vision of its eponymous foundation. The prize in medicine or physiology, where no progress has been made in recent years, actually became a reward for chemistry and physics. There is almost no one to name from among the distinguished laureates who challenged the reductionist-deterministic foundation. The more dynamite is made and used, the more money comes in for a self-celebratory theology as dangerous to humankind as dynamite itself is. When each brick of the buildings embodying the thinking of the past is engraved to the glory of some name (for "legacy" purposes), legacy itself becomes subject to the greatest impetus of questioning authority. The people on the streets affirming the principle of equal opportunity have nothing against science and technology, even when they pay for these and find out that they cannot afford treatment. Or worse: the treatment kills more than it saves life. Opportunity itself being more and more a privilege, they wish for their own slice.

It is obvious that there is science denialism—predating COVID-19 and many other breakdowns (Diethelm and McKee 2009; Hannson 2017). In the larger context of CoVID, a realization seeps in. It is the title of a possible workshop: *Science denialism as grassroots resistance: Digital discourses and social networks* (Hatzidaki et al. 2021):

> …different forms of grassroots science denial amount to highly active and influential collective/mass actions and social movements which resemble conventional/familiar types of anti-authoritarian movement such as class struggle, fight for independence, or mobilization against discrimination/exclusion.

The Road to Hell

Science denialism as a proxy for rebellious stances and activities show how good intentions pave the road to hell. Blindly opposing science is yet another outlet for more general social grievances. The "anti-establishment" aspect and the rejection of legacy explain its strong resonance with grassroots activities all over the world. Science justified, for example, the endless tests and lockdowns (recently described as a "giant experiment that failed," Nocera and McLean 2023). But also in the name of science, Sweden successfully took the opposite direction. "Disrupt science," in its variety of forms of expression, became part of the unwritten agenda of our times, exactly because science as practiced in the service of reductionist-determinism failed.

Those who protested in the streets of the USA, Belarus, Thailand, Germany, France, and in many other places, are facing the legacy of a democracy that is none. A parasitic political class took over. The show of confrontations, which are not about the *demos*—the citizenry—but about privileges, involves the state machinery as well as the machinery of misinformation. When science gets into politics, as an establishment pursuing its own agenda, the mixture is dangerous. If the virus could be alive, it would laugh: nothing protects the virus more than partisanship in exploiting the danger for motives of political and economic gain. Social media, effectively under the supervision of state agencies, became the battleground—until the opposing voices were literally shut up. On Twitter, self-proclaimed experts found a stage for never-ending rants. In their views, no matter how you toss it or turn it, society is doomed if it does not follow their prescriptions. Opposing views were vilified and eventually silenced.

Despite all attempts to extend control over the people's thinking, society is becoming more aware that the legacy of the Cartesian Revolution entails the deplorable state of understanding the pandemic, ecology, and, in particular, sustainability. More people realize that for the sake of the past, the future is sacrificed. If the ecology or the weather were physical or chemical entities, the science of the day would suffice for controlling the "ecology machine," or the "weather production facility." But they are not. They are related to living processes—the evolutionary record of previous extinctions of certain life forms says it all. But ignorance, associated with the impertinence of entitlement, has the upper hand. The iconic young climate change "expert," speaking for Generation Z, flanked by her two dogs, brought it all to the point: Estranged from each other, we need a pet to talk to or run with, while carrying the honorable banner of "Save the planet!" COVID is the outcome of individuals confusing friends on the computer monitor with real people. Estrangement is probably more toxic than the virus that brought with it the requirement of self-quarantine, or the lockdowns that eventually doomed China's containment experiment. Isolation was never a good way to fight disease.

A word generated by the most impressive AI application of natural language processing (the LLM path, for example) costs, in terms of the energy used, as much as what a household uses during a day. But people are asked to use wearables to monitor their isolation. Data Centers, where the data from various monitors converge, consume energy that could help those in real need to take care of themselves. Prevention is almost an offense in a society that accepted to be monitored. Instead of prevention, which monitoring could in principle contribute to, maximizing profits is the objective. At all cost. The bots on the social media, spewing whatever their backers want to inject into the public conversation, are proof of the fact that in automating even human interaction, humanity itself, at the zenith of its performance, is automated away.

A Crisis of Knowledge

COVID-19 is more than yet another pandemic in a long open-ended sequence. In the context of relatively autarchic community life, the extent to which a local viral or bacterial event spread reflected the scale of the pragmatic framework: i.e., what people did for a living and how they interacted. Scale is important. Data associated with the Antonine plague (165–180), or with the Japanese smallpox outbreak (735–737), the Plague of Justinian (541–541), the Black Death (1357–1351), or even the Great Plague of London (1665) support this assessment. Together with measles, typhoid fever, and influenza, smallpox, on record since the third century in Europe, wiped out 90%-95% of the native population in the New World. To characterize such outbreaks as pandemics would stretch the meaning of the word. To ignore the contagion characteristics—what factors influenced each of them—would mean to waste an opportunity to understand what we are facing today on a global scale. The successive waves of flu—Russian, 1889–1890, Spanish, 1918–1919, Asian, 1957–1958, Hong Kong (also known as H3N2), 1968–1970, subsequent corona viruses (swine flu, H1N1, 2009–2010, SARS, 2002–2003, MERS) ongoing since 2015—pertain to the context of progressive globalization. They qualify as pandemics, as does the HIV/AID virus (ongoing since 1981). Therefore, the COVID is dated in reference to that not-so-distant past, more precisely, at the time when the outcome became less important than the price paid for it.

The timeline of economic breakdowns tells a similar story. The scale of the pragmatic framework and the scale of consequences of the disturbed economy are quite similar. The Financial Panic of 33CE, when banks in Rome issued unsecured loans, the crisis preceding the Thirty Years War (1618–1622), the "seventeenth century crisis" (Hobsbawm 1965), the various panics at the end of the eighteenth century, and even the Great Depression (1929–1939) are, in terms of consequences of a scale that corresponds to a stage preceding the quest for globalization following World War II. That is no longer the case with the energy crises (1970, 1973, 1979), with the Savings and Loan banks (failure of more than one-third of them between 1986 and 1995), the new energy crisis (2003–2009), the housing bubble (2003–2011), the financial crisis (Great Recession) of 2007–2009, and the European sovereign debt crisis (2009–2019). The entire world felt the tremors, not evenly, but no area was spared.

In our time, high unemployment and financial insecurity highlight the social dimension of the CoVID. The theology of reductionist-determinism, into which the Cartesian Revolution morphed, had a different outcome in a world of relatively disjoint nations and their respective autonomous economies. Strained race relations, lack of confidence in the authorities, and lack of public safety aggregate, in the sense of a confluence of unrelated "rivers." It is a crisis of many interlinked crises, including that of the medical system. The failure of the political system is apparently a major factor. This is what makes what we called CoVID different from any other previous challenge: It represents an entangled image. It is the confluence of pandemic, economic and financial inadequacy, social unrest, political instability, misguided science, technology that ran amok, inadequate medical care, and inadequate education. Terrorism and the associated fanaticism of the mob cannot be left out of this enumeration. The landscape of moral, economic, social, racial, sexual, religious, scientific, ideological, etc. values—changed. In this landscape, like after an earthquake, work itself, as the pre-requisite for consumption, ceased to be a productive endeavor. Ever fewer people work in a productive capacity in order to make a living. They chase after their own tail; that is, make money for the sake of spending it, frequently relying on credit. Dependencies increase against the background of compromised sustainability. Borrowing more and more against their "expiration date", i.e., beyond being liable, is no longer the exception. Past pandemics and past economic breakdowns were more discrete events, of marginal, if any, correlations among the factors involved. They were easier to contain. This is no longer the case with what is retraced to a virus (one "member" of the Corona virus family already settled in the ecology), but actually reaches deeper, extending well beyond its time stamp. The crisis did not start in December 2019, but rather at the time when the Cartesian underpinning became global. The Upside/Downside Ratio of human activity under the reductionist-deterministic doctrine was turned upside down through digital globalism. This is the aftermath of the previous economic breakdown, at which huge amounts of money were thrown (especially for saving the banks). The fire seemed extinguished. This was also the time of the HIV/AIDS pandemic, and of the conflicts that were captured in the formula "global war on terrorism."

CoVID is a crisis of knowledge. The limiting reductionist-deterministic perspective from which reality is examined pre-empts the possibility to transcend reaction as the only recourse to breakdowns. In the pre-digital global world, reactions corresponded to a reduced scale of risk and containment. In the global digital world, this crisis is manifested in contemporary society in an expensive reactive manner. The record of events shows that there was no attempt to pre-empt the global disaster through a path of long-term pro-active actions.

An observation regarding the role played by science itself needs to be added. Empirical evidence accumulated since the beginnings of scientific inquiry provided proof that life dynamics and the dynamics of the non-living are fundamentally different. But those illustrious scientists, worthy of their renown, who adopted the view of determinism as all-encompassing would not accept the evidence. This is why physics, and all the disciplines that sprang from it, proceeded towards forcing life to submit to the laws of physics and chemistry. Simply put: If the living cannot be explained by physics and chemistry, let's change it. Let's make it behave like the non-living, in particular, like machines. A suggestive analogy: Our god is the only god. If you convert, and accept our god's explanation of everything for everybody, you are fine. Otherwise, we will deny your right to be heard, even to exist. This worked in the pre-global world. Of course, the scientists could not turn people into hydraulic or pneumatic devices—even though they claimed that that's what humans are. Books were written, and peer-reviewed articles published. Scientists of other times could not make them into thermic or electric engines. But they produced mathematical "proofs," as ridiculous as what they were ascertaining. With the advent of the mathematical machine—the computer—a lot more in making humans behave like machines could be achieved. This is currently the global religion, translated for example as

"Human behavioral patterns are *hardwired*." Through the network, individuals—reduced to nodes in a centralized structure—were forced to give up their identity for the convenience of automating their lives. Imagine a world without online appointment calendars, without navigation systems, without maps to the pizzeria around the corner. When analogies, i.e., the metaphors of science, are confused with what they refer to, they become theologies. Science accomplished what religion was not able to: a global human condition, alienated from its natural condition and formative environment.

When the virus spread around the globe, people realized the many shortcomings of a system working towards making them more and more dependent on the power structure in place. Magnified under the lens of unequal access to survival, COVID-19 perpetuated the practice of reaction. Entitlements—as a reaction to real or alleged unfairness—are the bribe into submissiveness. On the streets of American cities—but also in Europe and Asia—alienated young people with a skewed sense of future—expressed their accumulated grievances, sometimes violently. When you don't have future—because the obsession with unsustainable progress (read: more of everything and cheaper) fed by the deterministic-based planners robbed you of it, you try to reclaim at least your present. Re-ascertaining ownership of the past, not to say of ownership of one's identity as a living person and not as an entry in a database, ignited many fires. Mass protests (that include looting, setting fires, killing, etc.) document the multitude of forms of rejecting treatment as anything but the unique person each is. The Cartesian Revolution can claim parenthood of admirable achievements. However, it also owns the more intense crisis indexed to digital globality, and thus the detestable manifestations through which reactions are expressed.

The untangling of the various factors leading to the COVID confluence is quite different from seeking the meanings of a text (literary, philosophic, scientific). It is different from trying to peel away the layers of specialized knowledge that make up the current state of science. The various aspects of the pandemic, as part of the broader crisis, cannot be conceived or assessed as independent of each other. They cannot be reduced to one of its parts separated from the whole. Reductionism already poisoned the well of knowledge to the extent that no effective course of action can be retraced to it.

How Effective is Life?

Science and technology based on the foundation of reductionist determinism is effective to the extent that our current civilization bears the name *Cartesian Revolution* on its "birth certificate." In more ways than can be listed in this context, the human beings who adopted rationality as the premise of their activity benefited from all that this rationality made possible. At the same time, they increasingly lost a sense of vulnerability. The "Tower of Babel" meant to reach to heaven grew and grew until its foundation gave way. More than once. The price was the loss of the one language that unified humanity, replaced by many, i.e., specialized, languages so different that in our time, everyone understands each other less and less. With the algorithmic, the effort of building a tower reaching to heaven restarted, more ambitious, and more seductive. But also ever more in danger of collapsing. The fact that unlimited progress at any and every price is as addictive as the ambition to reach heaven is telling of the price paid each time a new height is reached. Hence, the question: Is being effective within the Cartesian foundation different from being effective within the dynamic of life?

At this juncture, a necessary inference cannot be avoided:

Thesis of life effectiveness: Life is effective on account of reaction (informed by the past), and on account of anticipatory action (informed by possible futures).

In this thesis the conjunction (AND) is the essential part: not one after the other, but together. No reaction, regardless how intense and even how prolonged can substitute for anticipatory action.

To repeat: biological matter is matter endowed with anticipatory capabilities. Therefore, every known living process, which means every form of change, takes place

1. in reaction to the environment (what Jakob von Uexküll called *Umwelt*)
2. in anticipation of possible futures.

They need to be seen in their unity. The reductionist-deterministic description of life, and the associated activities involving living entities, ascertains the

Cartesian Thesis: "Effective" means reducible to the causality of its physical substratum (matter).

It is easy to demonstrate that no matter how successful anticipatory actions can be, they are not a substitute for reaction. But effective in the spirit of the Cartesian Thesis is at most a subset of life effectiveness.

The current crisis of vision, including the handling of the pandemic, documents the limited perspective of the Cartesian Thesis. The crisis is the unavoidable outcome of this limited view. The fact that life always originates in life and extends beyond a certain embodiment (individual organisms) in anticipation of possible future states illustrates the meaning of the inference Considering the *Thesis of life effectiveness* and the *Cartesian Thesis*, we can derive a necessary conclusion. The outcome of science and technology on the foundation of reductionist determinism is, without exception, subject to breakdown. Repair, in whatever form, is reactive, and at a cost similar (or higher) than the making of whatever breakdowns. In contradistinction, life, while not breakdown-free, is capable of self-repair. Life continues into life: from one organism to others. Biological matter is continuously renewed, i.e., recreated. Without a comprehensive view of what life is—part of the continuum—we end up comparing two different realities: the artificial (outcome of living process) and the natural. There is future only in the living. These considerations are highly consequential.

Let's scan the integrated image, keeping in mind that each detail is part of a whole and has meaning in the entire configuration to which it belongs.

- A new virus, with unprecedented symptoms (the World Health Organization—WHO—initially signaling a new type of pneumonia)
- A fast-spreading choke flu against the background of vigorous containment (state-controlled lockdown of a great number of people in Wuhan, and other cities in Hubei, China)
- All kinds of reactions (tracking, social distancing, partial face-masking, improved hygiene, quarantine, etc.)
- Limiting mobility
- Testing
- Treatment of patients
- Applying brakes to the economy
- Attempts at prediction (modeling, in particular)
- Wild market fluctuations
- Infusion of financial support (unemployment benefits, loans, bailout funds, purchase of medical equipment, allocation in hospital capacity, ad hoc research, etc.)
- Social unrest
- Racial conflicts
- Failure of government agencies, and the bankruptcy of the political system.
- Of course, there are some aspects that stand out:
- State of confusion in the medical care system
- Panic in the system in charge of caring for the elderly
- Measures adopted by the educational system not well thought out
- Attempts at redefining work
- Confused strategy of remote activity
- Revolts (from peaceful to violent)
- Revindications—between justified and absurd.

These clusters are part and parcel of the whole that defines the current worldwide crisis (more evident in some parts of the globe than in others).

The Value of Life

The death of a black man in police custody eventually triggered not only protest, but also a macabre spectacle of looting, destruction, and mistreatment of many innocent people. This is part of the same integrated reality for which CoVID stands, although this might not be immediately apparent. But a first take says it all: Similar tragic events, i.e., people killed in confrontation with the police, in a context of almost full employment, led at most to an outcry. With the USA in lockdown for a while, and the rest of the world affected by the economic breakdown, all conditions were created for integration of the pandemic into social unrest, political turmoil, instability, with the prospect of civil conflict. The confluence model adopted in this study has the advantage of opening a broader perspective than the one afforded by the physical model of convergence.

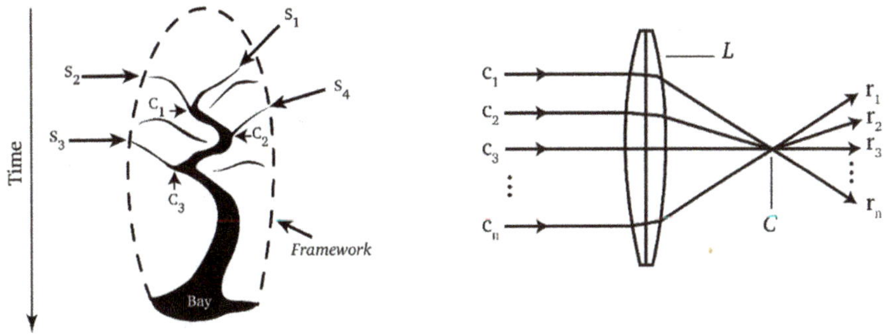

Fig. 1 Confluence and convergence

Confluence describes open-ended dynamics, i.e., contingent changes. The possible future (represented as a Bay) corresponds to metabolized energy. Cycles of life, as they are at various scales (from microorganisms to whales and elephants), represent metabolic processes. Birth takes place on the energy of the host organism. Life is sustained and multiplied through energy from outside the organism. The Bay metaphor points to the variety of such sources of energy enlisted in survival and reproduction. Sources of possible instability (s_1, s_2, … s_i)—think about predators—are loosely connected. *Convergence*—as opposed to confluence—describes necessary dynamics: the model of a necessary end-state determined by the past (represented as causes). The "lens" in the diagram (Fig. 1) is the embodiment of the description of law, such as in optical refraction. Reactions are causally (C) determined by the necessary end-state.

Unrelated events, of divergent ideologies, can come to confluence and furthermore, within a framework, accumulate as a potential source of disruption (Gorard 2020). They are fired by the energy of the conflict among them. The instability of the economic, social or political system ensues as an outcome of qualitative changes. The pandemic is an example of confluence (see Fig. 7 in "Numbers and Meaning") regarding the confluence of societal inequities). For those who act within the deterministic-reductionist model of reality, or try to impose it on society, the pandemic was supposed to be a convergence. In reality, it turned out to have such a variety of expressions (such as the Long COVID open ended list of symptoms) that it looks more and more like a spectrum of disease. With each new mutation, a variety of manifestations are recorded.

A similar commentary regarding the larger COVID crisis is justified at this juncture. The breakpoint came at the time (40-plus years ago) when the benefits of reacting to breakdowns started to fall way below the price that reactions exact. The process leading to this breakpoint can be represented as a confluence of events: breakdowns in politics, the economy, in society and culture. It is not accidental that the HIV/AIDS pandemic was triggered in this context. Likewise, it is not accidental that the war on terror, in the broader framework of digital globalism, led to the invasion of Afghanistan and the war in Iraq—an out-of-scale reaction of small or no return on the huge investment. The globalization of reaction measures, evident in the Ukraine-Russia confrontation, reflects a new scale of risk. Reignited Jew-hatred, with the purpose of exterminating Jews, brought back the miasma of the crematoria. Determinism translated as industrialized killing. A baby was torn from a pregnant woman's womb. Another baby was literally baked in an oven. And many newborns were decapitated. With this in mind, revisiting the historic background of CoVID affords an understanding of the meaning of confluence, in contrast to deterministic convergence.

Cause-and-effect as related to slavery reveals continuity: the obsession with cheap labor, in disregard of those forced to provide it, or enticed to do so. Therefore, to look back at the role of slavery in the establishment of the USA—not a nation, but an economy (Nadin 2013)—is justified as the source of what became the systemic racism deplored, for good reasons, in our days. The confluence in question pertains to the skewed values informing social, political, and economic life in America. It pertains to historic alienation—of the native population, i.e., the Indian tribes, in particular—misguided priorities in welfare, inadequate access to and a low standard of education, and, over time, obsession with military hegemony. The senselessly ended lives of yet other victims of a system that is part and parcel of the deterministic legacy going back to slavery cry for an understanding that transcends the here and now. The colonial past is a contradictory social and economic reality. The misconstrued right to carry arms cancels the demagoguery of the sanctity of life. Absurdity in pure form: protect the unborn for the right to be killed. The time of this contradictory rationality is coming to an end. However, despite progress—or maybe because of the type of progress it forced upon society—it remains a source of deep-reaching unfairness. Not surprisingly quite a number of Americans of all orientations and colors are contemplating the possibility of a revolution to unseat another Revolution, or at least to get rid of its aura. But, once more surprisingly, within the CoVID context, a broad segment of the citizenry is essentially absent from society. They opted for the role of consumer entitled to everything science and technology facilitate.

The establishment of the USA was in reaction to England's actions in the thirteen colonies. Nevertheless, it was also anticipatory in its Declaration of Independence, in that it advocated freedom of opportunity, if not equal access to seize it. Unfortunately, this anticipatory dimension was abandoned in favor of the path towards rapid harnessing of the capitalist profit-making machine. It made the USA the poster picture of inequalities: in economic and political opportunity, access to education, to fairness and justice, to adequate medical care. Too soon did political parties (which George Washington warned against) take over government from the people forming the constituency (initially, only owners of property and of slaves were electable). Instead of anticipatory actions provided for in the Constitution, a progressively reactionary path was promoted by the political class. The USA of the Revolution gave up its own ideals. It brought to the world the most effective, but also most corrupting, economic system. Slavery was replaced by a far more effective way to take possession of the lives of those whose labor generated wealth for the owners of the means of production. Industrial capitalism, followed by rent capitalism, are the sources of such deep-reaching unfairness. Conditions are ripe for a revolution to undo the Revolution of the beginnings.

The new economic and political identity of the world is still dominated by one superpower. The other one—the USSR—imploded, having reached the limits of what determinism, practiced under the protection of ideology, entails. In competition with the rest of the world (the European Union, China, Russia, Iran, the Arab world), the USA had the upper hand because it could define the rules of the game. Economically, the world depends on the USA. It is the largest market, open to a degree that no other market is. The fact that each of its own crises became a crisis for others did not seem to bother her. Under the circumstances of globality, i.e., the COVID confluence, this might change, especially since China, satisfying the desire of Americans for cheap, is rapidly catching up.

It all boils down to how much life is worth—a question going back to slavery, to labor in mines and factories, to the service economy. But also to wars, in which lives are at stake. Or better yet: to value. Indeed, value is the real reference for understanding the meaning of life. There are many estimates of the economic impact of one life lost, or of what it takes for the newly born to become the productive human being that society expects citizens to be. Banks are already speculating on how money can be made in everything imaginable, from market speculation to rebuilding—so much to rebuild after the COVID-19 crisis—to lives lost. These speculative calculations are guided by the deterministic understanding of life as a "factory for living." Automating it, as artificial intelligence proponents promise to do, would maintain the profit-making engine at the price of ignoring individual destinies. Political parties are hanging their destinies on the outcome. Entrepreneurs in everything see the pandemic, as much as the larger crisis, as yet opening more opportunities to make money, regardless of the outcome. Except that labor ceased to be only for survival—the food on the table, the rent, and not much more. Wealth was accumulated, for a sizable minority, at a scale at which it takes effort to spend it before it loses its value. Still, America of wave after wave of immigration, knows also poverty. For others, living on the expense of the rest of the world, became almost a right. It is possible—and became reality—to live without doing anything.

Let us recall: In January 1964—almost 60 years ago—a rather demagogical President (Lyndon Johnson) declared "unconditional war on poverty in America." No less than 28 trillion dollars went into an ever-expanding number of programs intended to help win the so-called war. And what does America contemplate in these days? One headline from among many of the same tenor: "Are bread riots coming to America?" (Cooper et al. 2020 *op. cit.*). The unemployment due to the pandemic forced many people to apply for benefits. In the absence of a safety net, even when special programs are set up in reaction to events, unemployment forced people into precarious situations. Twelve percent of households are regularly short of food; 21% of renters are behind payments; 28% of children experience what it means to grow up in families facing such problems. Hand-outs—for buying elections more than anything else—worked for a long time. But like the "War on Poverty," they too will fail.

There are two outcomes that correspond to the reactive nature of such and similar cause-and-effect measures:

(1) A large segment of the USA population, in particular the African-Americans and Hispanics, are by now less capable of being in control of their existence. This is the result of having their lives conditioned by the system. Seductive giveaways change life patterns—grants for single mothers were of course well intended. But as many commentators and politicians noted (Riley 2015), "The government paid mothers *to* keep fathers out of the home—and paid them well." Politics of the most abject intentions can be destructive. The Institute for Family Studies (IFS) published a convincing analysis: The black American family provides a stark example. From 1890 to 1950, black women had a higher marriage rate than white women. And in 1950, just 9% of black children lived without their father. By 1960, the black marriage rate had declined but remained close to the white marriage rate. In other words, despite open racism and widespread poverty, strong black families used to be the norm.

   But by the mid-1980s, black fatherlessness skyrocketed. Today, only 44% of black children have a father in the home. In unison, the rate of black out-of-wedlock births went from 24.5% in 1964 to 70.7% by 1994, roughly where it stands today.

   Instead of being empowered to partake in opportunities, Blacks were literally re-enslaved or pushed into degrading dependencies (which evolved into entitlements).

(2) Follow the money. If the trillions spent had made it to the hands of those who were supposed to be helped, the vast majority would have attained the condition of freedom and fulfillment they deserve. In reality, public money wound up in the pockets of the entrepreneurs of this yet another American "war." Yes, the USA still nurtures the illusion that wars are the answer, (fighting" the virus is the new giveaway for this way of thinking. Wealth was transferred from the taxpayer to those who own the economy. It is happening under the eyes of those who just went through the ordeal of a pandemic during which the government failed miserably.

The current cry for yet another massive transfer—the 15 trillion dollar "Restitution" movement—will end up the same way, because this is the reductionist determinism implicit in capitalist economy. If you get money (pretty similar to getting replacement surgery, or chemotherapy, or the right chemical for an illness), everything will be fine. Throwing money at a problem is reactive. Providing opportunity is anticipatory. There are many sources alluding to this principle. The best known is probably from Thoreau: "Give a man a fish, and you feed him for a day. Teach a man to fish, and you feed him for a lifetime." Another is from Maimonides: "[…] anticipate charity by preventing poverty, namely, to assist the reduced brother […] by teaching him a trade, or by putting him in the way of business, so that he may earn an honest livelihood and not be forced to the dreadful alternative of holding up his hand for charity."

The difference in the median net worth between Black families ($17,150) and White ($171,000) reflects the difference in opportunity. Reaction, even at a larger scale than ever, will not do better than it has done all along. There is hard work to be invested in engaging those who have been disenfranchised for generations. An anticipatory course of action, focused on acknowledging differences and stimulating uniqueness, is the path to pursue. Take only the accomplishments of Blacks, Hispanics, Indians, Chinese, Koreans, and Vietnamese who manage to live up to their unique gifts. It will take a concerted effort in a society willing to improve, because otherwise, society itself, in its current fractured state, has no future. Alternatively, it will take a new understanding of life and work in a context in which the value of life is no longer reducible to the outcome of work.

The Cost of the Crisis

Evidently, to claim full knowledge about the costs of the crisis throughout the world right now would be as aberrant as all the reactions it prompted. And as useless. Those who have already faced pandemics testified that greater than the loss of one's well-being (as relative as it can be) is fear. Poincaré noted at his time (1905): "The plague was nothing; fear of the plague was much more formidable." COVID-19 might not be of the scale of the plague, but the COVID confluence makes it comparable. Science and technology have facilitated huge advances since the plague, but there is justified fear to live with.

According to a study by the Centers for Disease Control and Prevention (Czeisler 2020), as many as 25% of young Americans considered suicide; over 21% of "essential workers" experience extreme depression. It does not heal. Surviving COVID-19 means a pretty miserable life ahead: the brain of those affected, to name one example, remains impaired for a long time. One physician who survived the disease described it: "The most distressing for me was the horrible odor emanating from my skin, sweat, breath and urine. It was nauseating and horrid…it reminded me of rotting reptile flesh." She also described details of life after the ordeal (Crosby 2020): Months after "becoming infected, I am still fatigued and have dyspepsia with exertion and musculoskeletal pain […] My brain is not working at 100%, and I still find myself with lapses—fear, anxiety, helplessness, hopelessness, and low mood." The "Long COVID," as it is called, is yet another subject that the medical establishment left to the mercy of those who generate data but not knowledge.

In the absence of effective knowledge, or better yet, in the grips of the reductionist-deterministic knowledge underlying reaction, ignorance wins. Let us recall the H3N2 pandemic (the Hong Kong flu), an avian influenza on record in the USA since September 1968. Estimated number of persons affected worldwide: over 60 million; deaths, around one to possibly four million registered (mainly individuals over 65 years of age). No one proclaimed social distancing—an idea first suggested in 2006. Woodstock took place in August 1969. Hugging and kissing (let's stop at these), "making out" to the extreme, even in deep mud was *de rigueur*. Chickenpox parties became popular for young children and continue in our days as a "natural" alternative to the chickenpox vaccine. "Herd immunity" sounds like a new motto. In our days, seeking immunity (with an ID to prove you have it), although nobody knows what it is, how long or short it would last, is a new goal. The headline reads: "Studies Report Rapid Loss of COVID-19 Antibodies"—and automatically the future vaccine is under question. In plain English, "After SARS-CoV-2 infection, people are unlikely to produce long-lasting protective antibodies against this virus" (Heidt 2020). Therefore, the misleading call to vaccinate even those who were sick. Someone declared, "This fact might make it difficult to develop a vaccine that works equally well for all people," (a thought attributed to Dr. Anthony Fauci, at that time Director of the National Institute of Allergy and Infectious Diseases). None other than the co-discoverer of HIV as the cause of AIDS tried to explain, in several interviews, that, given the structure of the protein shielding the virus (a so-called glycan shield), antibodies might not last. This did not make the over 200 companies rushing to produce a highly profitable vaccine happy. Ignorance is expensive. Vaccination, even at its best, is not a convergence process with a 100% rate of success. The much-publicized boosters had only a modest effectiveness—below 30% (Shesthra et al. 2022). Nobody has the science necessary to understand how China's decision to give up on the zero target will affect the rest of the world. Individualized vaccination might not be a realistic goal, but vaccines conceived for certain segments of the population would definitely be more efficient. If the effort of custom designing vaccines could return profits comparable to "one size fits all" vaccines is a question with many aspects. For those arguing for the value of each life, the effort qualifies as justified. But as long as patents are made into profit centers, it is highly unlikely that the alternative will get any attention.

Even less happy than the pharmaceutical companies trying to reap maximum profits are those who are betting on herd immunity, which proved to be short-lived. There are already studies in circulation (Kissler et al. 2020) that predict a "resurgence in contagion…as late as 2024." At the end of December 2022, this prediction was justified: over 200,000 infections on some days. The RA.5 variant of the virus—in Japan (with 95% masking success) caused a wave of infections higher than any previous variant: BQ1, BQ1.1, BA.5, BF.7. The series does not stop. Recurrent wintertime outbreaks are also taken into consideration since the capacity to take care of those stricken is seriously tested by the concomitant seasonal flu and RSV (respiratory synctial virus). These projections are essentially driven by deterministic models—estimates used on time series—or yet by other models, some with an underlying machine learning component (see https://COVID-19-projections.com). To generalize from a few seasons of corona viruses (SARS, MERS, etc.), or from certain regions of the world (mostly temperate climates), ignoring their global nature is already specious. No one has the science needed to understand how China's decision to give up on the zero target will affect the rest of the world. Change human behavior or the possibility of pro-active measures should be the focus. Stricter attempts at isolating the sick, tracing and tracking, limiting interactions etc., are not be palatable in so-called "free societies." In the end, they failed in China. But for a while, they worked in Taiwan, Hong Kong, South Korea, New Zealand. To abdicate to one's own ignorance is a dangerous choice. Of course, the broad COVID crisis poses additional challenges. Self-deception, extreme individualism, indifference, succumbing to the dictatorship of data-driven government policies constitute a confluence of factors.

The deterministic social experiment of COVID-19 will eventually be quantified: number of people infected; number of non-symptomatic cases; total number of people who experienced a life-threatening malady (in a variety of ways); the number of dead; testing costs, hopefully against comparisons of their usefulness, also expressed in numbers; the economic impact, and so forth. The image resulting from such numbers would be incomplete in the absence of acknowledging everything that physicians have faced, and sometimes managed. Heroic in many ways, inadequate because of the bad science they were asked to follow, but tragic—so many of them committed suicide.

Quite a task, with an inherent margin of error that, in the final analysis, undermines the credibility of exclusively quantitative assessments. For the non-living, numerical descriptions (i.e., measurements) are all that is needed to understand their dynamic. For the living, with uniqueness as its concrete expression, what counts is meaning. Some examples: deaths due to COVID-19 or rather on account of comorbidities (e.g., high blood pressure, coronary problems, HIV/AIDS, etc.). Even more specific: clotting turned out to be a major factor in lethality (Toshiaki et al. 2020) and in heart attacks. Predisposition to clotting, even in younger, healthy subjects, is a matter of a person's entire history (Goshua et al. 2020). But this time SARS-CoV-2 is the culprit. Let us also remember that there was a disproportionate number of the elderly among the dead. However, numbers disconnected from context are meaningless. The "machines" for getting the old under supervision (and sometimes abuse)—i.e., the nursing homes in the USA and in those parts of the world that adopted the same "technology"—produced the disproportionate numbers. In Mexico, where nursing homes are rare, this was not the case. The meaning becomes apparent. It is not because people are getting old—this is the unavoidable course of life. Rather, because getting older in nursing homes brings with it risks of all kinds. In Sweden, most of the dead in a country of "Let's do nothing" were aged (probably with underlying comorbidities). One more: Was it COVID-19 or the inadequate healthcare system that explains the unhappy outcome? In countries with adequate medical care (e.g., Taiwan, Germany, Japan, Singapore), the number of those who died is relatively low. The USA, with a population that adds up to only four percent of the world total, had more infections and four times more death per capita than the rest of the world. Medical care quality is not a matter of numbers but of meaningful science made available to physicians in anticipation of new challenges, not in hasty reaction to them.

To understand that the individual is sacrificed for the sake of economic growth (more commerce, in the end), but companies are protected, is to overcome the CoVID component. In a broader view, the individual deserves more protection than institutions and businesses. But within CoVID, the individual is taken for granted, even as a subject to risky experiments. Is working from home a success or only an experiment in evaluating future automated tasks? Return to work or have your work rationalized? Within the pandemic-affected work environment, the security of the data processed remotely counted more than what happens when social aspects of work are cavalierly left out of the equation. Encryption figured high on the agenda. Indeed, those who seek a venue for airing frustrations of all kinds—some more legitimate than others—discovered that you can simultaneously love and hate your dependence on the Internet, your cell phone, games, and car. Studying online, in synchronous of asynchronous mode—raised a major consideration: Is remote education a success story or an experiment that actually unveiled the mediocrity of education driven by all sorts of evaluation criteria (via the accreditation mechanism)? The MOOCs (Massive Open Online Courses), which preceded COVID-19, offer between free and moderately priced classes that certify skill and training. To go into debt for the same at elite or top-tier colleges and universities is not a sound investment. On the streets of many cities, disenfranchised students joined crowds of people voicing concerns over racial, gender, political, and social aspects, and questioning privilege—of course, not their own. One year's tuition in some colleges (or even high schools) can be higher than what two-thirds of the people on Earth earn in a year (and some, during a lifetime). To pay the same for the remote experience of obtaining "canned" knowledge is at least morally questionable. Telemedicine is in this respect often in the same situation. But that is the price—never disclosed as such—of reacting. Anticipatory actions, in preparation for breakdowns that are inevitable, could have made a difference.

It cannot be repeated too often: in all living processes, only data associated with meaning carries actionable information. Reactions can be described in terms of quantitative assessments—which means *data*, i.e., numbers. Laws describing reactions do not reveal meanings, but causes and outcomes connected to the assumptions of reductionist-determinism. They can be in linear or non-linear relation. Let's take the much-discussed magic number of eleven years of life that, presumably, the affected elderly lost. In translation, in the absence of COVID-19, actuarial tables indicate that they would have lived, on average, another eleven years. The meaning, however, is lost since meanings are concretizations, not statistical generalizations. A patient on life support, an elderly person immobilized in a wheel chair, a dialysis patient, experience the meaning of life differently from an active grandmother, a volunteer in a training facility for autistic children, or a physician past retirement age working in a location where medical care is scarce. The price that can be attached to life is not based on averages, but on the meaning of each existence. No life is less precious than another, but definitely of a different significance—to oneself, to family, friends, community, society. To compare the life on a person to the life of an engine (a machine) is not only indicative of the ignorance regarding each, but also disingenuous.

No country has ever tried reactive social experiments at the scale recently practiced in the context of the pandemic. Life was put on hold, and everything made dependent on numbers from unreliable tests, or from the questionable practice of tracking by governments and private enterprises. The number of people who die from the flu, from car accidents, from cancer and cardiovascular disease, from medical treatment (second or third place in the database of causes of death) would justify, if not closing down the economy for good, at least mandated preventive measures. For instance: no more cars that can run faster than 50 miles per hour, since most auto accidents are connected to speeding; no more alcohol (along the line of prohibiting smoking in public places); no more foods, beverages, and behaviors known to lead to cancer. The social experiment in progress showed "How Authoritarians Are Exploiting the COVID-19 Crisis to Grab Power" (Roth 2020), or how "The Coronavirus Is Emboldening Autocrats the World Over" (Diamond 2020), and worse: "Big Pharma's COVID-19 Profiteers. How the race to develop treatments and a vaccine will create a historic windfall for the industry—and everyone else will pay the price" (Taibbi 2020). "Making $1,000 profit every second" (Office for the Coordination of Humanitarian Affairs—OCHA—posting of November 2021) applies not only to the BIG players (Pfizer, BioNTech, Moderna) but also to various other companies active in the COVID-19 business (from cleaning supplies to hospital equipment, from thermometers and testing devices to remote work software, etc.). Shareholders profited as well—i.e., the affluent who "own" the economy. Two-thirds of Congress—72 Senators and 302 Representatives—cashed campaign checks from pharmaceutical companies. This explains their influence on the government. The same holds true for Silicon Valley (all set for more tracking), for retailers (Walmart, CVS, etc.), for weapons-makers. The inadequate government structures all over the world are part and parcel of the crisis of vision. The nefarious role of bureaucracies—the machines that administer the lives of citizens in disregard of their individuality—was documented many times; but they continue to expand.

Neither last nor least, the terrible reality of an unfair system: capitalism as a deterministic machine. Hailed as an expression of freedom and opportunity, once it was hijacked by politics, it morphed into a growing collection of rules and regulations benefiting almost exclusively those who own the economy. Where there is regulation, there is inevitably corruption. The ethics—or lack thereof—of discounting the well-being of future generations inspired Ramsey (1928) in devising *A Mathematical Theory of Saving*. Morality suffers not because of a virus, but rather because of a skewed system of values that prevents anticipatory actions. It qualifies as a crisis of vision because nothing less but the future is at stake. In the age of general unaccountability, crime pays more than prevention. From the over five billion dollars of government relief funds, probably 20% was syphoned away by "entrepreneurs" eager to exploit every weakness of the system. Instead of preventing moral failure, government now invests in investigating crimes made possible by politics.

Social engineering encapsulates the reductionist-deterministic perspective into regulations—and thus into more corruption. Implementation costs money and results in the progressive loss of freedom of choice. Consider only the vaccine promise: the miracle that will save us all, save the economy, bring everything back to a normality that we just discovered was anything but normal. From every corner, demagogues—politicians, scientists, manufacturers, marketing experts—formed a coalition focused on one policy: spend public money for the best mirage available. Responsible scientists—yes, there is such a rare specimen—are still examining the science of the mRNA vaccines and ways to improve them. Nevertheless, the public was forced to pay for an endeavor with uncertain, if not questionable, outcome, at least in the form in which vaccination was promoted and carried out. In the context of government vaccination mandates in most advanced countries, the extremely low rates of vaccination in Africa and in some Asian countries stand in contrast. The "gold rush" got into in high gear, but laws for distributing the bounty are not new: guaranteed high profit, no liability, protectionism in every imaginable form. There is no real incentive to acquire knowledge regarding an improved understanding of what this new method is. The alternative to this situation could be the integrated reactive and anticipatory paths.

Evidently, a pro-active behavior of prevention cannot be achieved retroactively. Also, it cannot be instituted by edict. The motivation for promoting it is straightforward: avoid the price for letting bad things happen. Conditioning individuals to feel and behave like machines and to expect them to pay for their breakdowns undermines their own anticipatory action. "Let things happen. Someone else—yes, the government—will take care of us." What reaction takes away from individuals—an active role in making choices meaningful to each and to those with whom they interact—cannot be re-established through laws and regulations. Or by bribing them to hold on until the storm passes. The goal should be the informed freedom to understand the future as one of choices that individuals and communities make, in anticipation of events, not *only* in reaction to them. But this goal is probably the victim without a grave of the CoVID crisis. It transcends the pandemic in which humankind was forced into, exactly because the future matters less than the immediate.

The value of a life endangered by disease, by social and political unrest, and an insecure future has many dimensions. The unfolding of creativity, including the continuation of the sequence of life, is a matter of individual and social significance. There is more, much more, in interdependencies, particularly in relation to the environment. Everyone's involvement in increasing the knowledge on whose foundation humans become who they are, is a goal that escapes planning or regulation. Beyond this uncensored poetry of the value of human life, and by extension of the value of everything that makes life meaningful, there is the reality of the deterministic conditioning of society. Unwilling to give up measuring and quantifying life in exchange for the benefit of understanding its meaning, society ends up paying an ever-increasing price.

The Price of Life and the Value of Life

The killing machine precedes the Cartesian Revolution by thousands of years. The animal instinct—survival drive in action—eventually turned into the science and technology of wars. Archimedes contributed to this, so did Pliny the Elder, not to mention all scientists since their time who facilitated wars on a larger and larger scale. This is where the numbers attached to the price of a person come from. In modern times, the numbers mean less the price of the bullet, which replaced the cheap stone that smashed the enemy, and more the price of data processing guiding war actions. The remote-controlled drones aimed at a target cost millions—way more than what the assassinated are worth (in terms of the damage they could inflict). Included in the price is the investment in the vast network of data processing supporting the operation, the technology of targeting, and the cost of legal advice supposed to lend immunity to those in charge. Once upon a time, the bounty reward proclaimed an enemy's worth. Today, the expense of ending a life struck by the SARS-CoV-2 virus—because even with antivirals we still do not know how to treat the disease—can be as high as the danger, real or assumed, that such a person poses to others. People protesting conditions affecting their lives and the lives of those they care for are the public expression of the price of life. There is an awareness almost no one cares to make explicit: While the *price* of life increased, the *value* of life is at its lowest point. This is yet another characteristic of the crisis of vision affecting the world.

Instead of developing only more powerful weapons, some countries choose the more dangerous path of biological weapons. "Gain of function" is part of the process of making bio-weapons. This is the ugly face of anticipation at work: a possible future—contamination—that becomes a reality in subduing the enemy. No reason for celebration. The world in which we live is not populated with angels. Yet again, for the budget ledgers, it turns out that the cost of life itself is less important than the change in behavior. The drop in the Gross Domestic Product associated with the COVID-19 social experiment will affect lives in a manner no numbers can describe: the lives of the rich probably less than the lives of the poor, or of those affected by racial, ethnic, gender, or religious discrimination.

Numbers associated to their meaning reveal the place where ignorance, social unrest, political hysteria, and ineffective social policies aggregate. Society does not provide citizens with a clear evaluation of the price of reaction. Even less does society focus on what it would take to learn the value of pro-active measures and to stimulate anticipatory action. Again: Japan is, more by necessity than by choice, an example. In Japan, earthquakes informed a culture of prevention. You cannot stop the earthquake, but you can prepare for it. Japanese citizens wear masks because their culture is one of extreme prudence. Another example: Vietnam—if your ideal is to live in a controlled state. Or, if the numbers are not "cooked," so was China until December 2022, with its extreme containment measures. For the price of freedom, nobody (actually only few) died—or was allowed to die. By extension, the pandemic became everyone's business, not only that of physicians or of the government. The military was deployed, as in war. Each person observed the others.

The value of life is reflected in the shared culture of prevention. It is expensive to build in Japan—housing, bridges, power plants, everything. However, it is much cheaper than what repairs necessitated by building based on a reactive framework would cost. "Cheap" homes in the USA end up costing more in order to compensate for the low quality (in Texas or Louisiana, for example). Moreover, removing one million homes from flood zones—a socially touchy policy, no matter how reasonable—could save one trillion dollars, according to a report from the Federal Emergency Management Agency. Evidently, within a shared culture of prevention—absent in the USA and in quite a number of other countries—a buyout program could be easy to pursue. That it could save lives goes without saying. Moreover, it could provide a sense of stability—exactly what reliance on reaction cannot offer. Could it be abused, just as the money made available to those facing hard times during COVID-19 was stolen? (Fine 2022).

Given the scale at which breakdowns are met by expensive reactive measures, it is not surprising that the value of life became a matter of statistical concern. Designing, producing, training for, and maintaining ejection pods for the crew of the by now outdated B-58 bomber cost 80 million dollars. It saved one to three lives a year—which means that for the US Air Force, the "money valuation of a pilot's life" was around eight to ten million dollars. It is from such considerations (and many more) that the so-called Value of Statistical Life (VSLI, Schelling 1968) was derived—yet another offspring of reductionist-deterministic thinking. The trade-off rate between fatality risk and money was eventually encoded in rules in order to avoid endless bargaining. The ejection pod is in anticipation of what might endanger the life of a pilot on such a bomber, or on very fast attack jets. Interestingly, the price paid for a pro-active measure served as reference for calculating the statistical value. In pricing drugs, which under certain circumstances can help to reduce the number of days a COVID-19 patient must stay in a hospital (assuming recovery), companies considered the money thus saved: number of days times price per day. This is how Redemsivir (and other antiviral drugs), which would cost $15.00 to $20.00 per treatment, ended up with a contract from the government for no less than $3,120.00 per patient. (Taibbi 2020). The fact that it turned out to be ineffective hurts even more than the price charged. The reactive dimension on which the Value of Statistical Life is based comes from the replacement mentality characteristic of considering the living as machines.

Without subscribing to the statistical model, let us play devil's advocate: follow through with the calculations of how social engineering—lockdown, social distancing, telework, telemedicine, and teleteaching—could affect the cost criterion. As reactions, they invite the game of determinism: What's the return on the investment? Keeping the USA economy going—the whole world depends on it. The well-being of those who make it up—investors, workers, the world markets—hangs on it. To spur it on to the ever-increasing numbers qualified as progress (the next generation should be better off than the parents is one particular expression of this understanding), is by now reflected in the magic number that the GDP became. Save one life through social engineering—as you would extend the life of a machine, an airplane, a navy battleship—and you saved million dollars. Take the worst prediction—2.5 million dead due to COVID-19. The cause-and-effect social engineering would claim victory for each percentage point of lower lethality. Reducing the number of dead by only 10% would save more money than the value of the decreased GDP affected by the social engineering practiced (lockdown, businesses closed, telework, etc.). The fact that a great number of the stricken are persons in nursing homes or in jails is irrelevant for statistics. Even though they are consumers, and consumption is 70% of the engine that drives the economy. The effective absence of a social safety net in the USA makes the Value of Statistical Life even more questionable. Of course, a social safety net, such as other countries maintain, is expensive. But as an anticipatory action, a social safety net helps maintain the necessary relation between the well-being of individuals and of the society. Economic activity—only one form of activity among many through which humans creatively identify themselves—plays an important role. The USA has practically no encompassing social program. Americans at first hated the idea of Social Security; likewise, some still hate Medicaid, and have mixed feelings about Medicare. The country accepted a medical care system as a profit-making capitalistic enterprise. In other words, the USA chose the immediacy of success to the detriment of sustainability. No prevention to talk about. Most personal bankruptcies in the USA bear the fingerprint of the medical industrial complex. Once treated, the patient is hostage to the system: medication forever, one condition leading to a cascade of future conditions. It is therefore not surprising that the Value of Statistical Life for an average American is around $7.2 million. The rest of the world gets up to $1.3 million (Sweis 2022).

It is quite a learning experience to examine what an ideologically driven economist called "the economics of not dying" (Krugman 2020). A quote: "A Columbia University study estimated that locking down just a week earlier (the USA went into lockdown in March of 2020) would have saved 36,000 lives by early May." Of course, there is no science to speak of in such pronouncements. In the same vein, an extensive study (Greenstone and Nigam 2020) monetized the benefits of social distancing policies: "the estimated benefits of this mitigation social distancing scenario are roughly 8 trillion." Such assertions have in common the incompetence resulting from *the surrender* to the cause-and-effect scheme. The fact that lockdowns, social distancing, masks, hygiene, vaccination, etc. influence those who practice them as well as those who doubt them is not even open to debate. Disturbing, however, is the lack of understanding the non-deterministic nature of the multitude of processes that resulted in the pandemic. Reductionism is dangerous because, once accepted, it opens the possibility of reactions that prove to be expensive in many ways. One case in point: "The CDC's six-foot distancing rule is the costliest mitigation measure taken during the pandemic." This comes from Gottlieb (2021), former head of the Federal Drug Administration, who argued against diverting finite public health resources to the wrong mission. For example: applying disinfectants to contaminated surfaces. Rivers of useless concoctions turned the skin of those who used them into lifeless dry surfaces. Masking—from being mandated to asking those not masked to have understanding for those still wearing them—is subject to never ending misrepresentations. Medicine by edict was practiced by some of those who should know better—the medical establishment most of all.

A Difficult Choice

The contrast between what society spends on reactive measures and what is invested in prevention is illustrated (for reasons of expressivity) by the image of an excavator bucket and that of a spoonful (Fig. 2). The exaggeration is intentional—to raise awareness more than anything else. The more precise calculation is the following: A heaped teaspoon holds about 0.9 cubic inches. A cubic yard bucket would be 46,656 cubic inches. The order of magnitude—around 100,000!

Let us exemplify this by considering the cost of racting to HIV/AID compared to the amount spent on prevention: "For every HIV infection prevented, an estimated

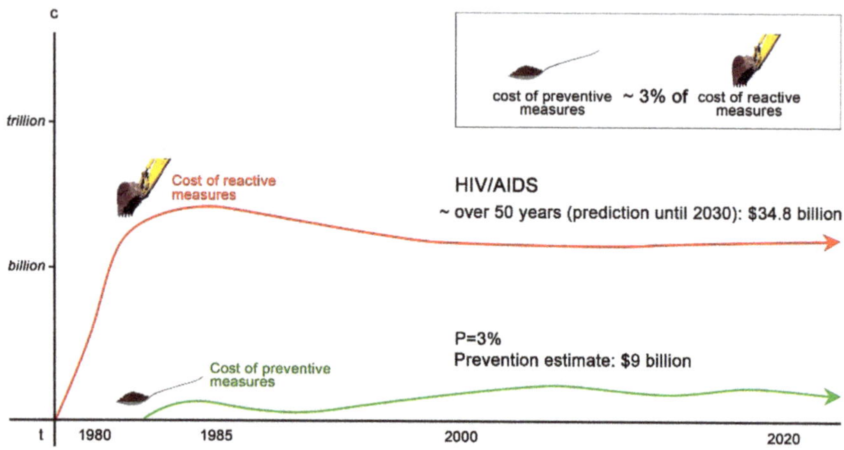

Fig. 2 The excavator bucket versus the spoonful image captures the difference in the scale at which reactive measures were taken and prevention was practiced (care of orphans and vulnerable family members, to preventive sexual practices, abstinence, male circumcision, etc.) in respect HIV/AID. Preventing infections saved not only in the cost of providing treatment

$355,000 is saved in the cost of providing lifetime HIV treatment," says the Surgeon General's report (Beaton 2017).

Comparing the various breakdowns of the last 40 years, the exponential increase of the cost of reactions (some more reasonable than others, but almost all of low effectiveness) stands in stark contrast to what it would cost to act in anticipation. Concretely, society should assess not only the price, but also the benefits of investments in providing:

- a comprehensive social net
- an adequate medical system accessible to all members of society
- programs of opportunity that address inequities in social, racial, gender, and citizenship status
- education anchored in the present and which stimulates shaping the future, as opposed to remaining captive to the reductionist-deterministic (industrial) model that shaped the past.

An exhaustive plan would include, but not be limited to, anticipation-informed city planning, reliance on energy sources that do not damage the environment, and prevention of various ailments through an anticipation-grounded understanding of nourishment, entertainment, recreation, and mobility.

Let us examine the way society handled the breakdowns of the last 40 years (Fig. 3).

Reactions are by necessity short-lived opportunities (usually taken advantage of in an opportunistic manner). Preventive actions are long-term opportunities, that society can scrutinize.

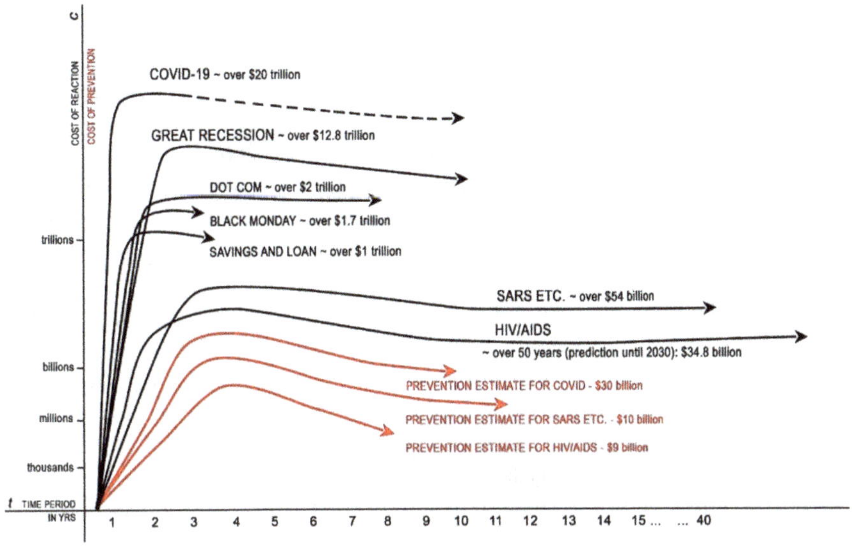

Fig. 3 A "*quod erat demonstrandum*" literally means "what was to be shown." **Sustainability Thesis (ST)**: Reaction is more expensive, by many orders of magnitude, than anticipatory actions; and in the long run, it is unsustainable

If we take the numbers from the Value of a Statistical Life, we can easily infer that over twenty trillion dollars associated with an inadequate knowledge model will fix the dent, but it won't fix society. Reaction precludes anticipatory action. The cost is indicative of the high price and ineffectiveness of reaction. Those countries where even modest anticipatory actions are carried out have a lower percentage of victims and, more important, are better prepared for the inevitable next breakdown (Council of the European Union (2020) Recovery Plan for Europe). It is worth repeating: prevention is by necessity a long-term activity. Reaction is always *to* something, to a well-defined condition, to the "now" of the situation. Therefore, it cannot be generalized. It does not accrue value. The price that the Germans or the French, Singapore or Japan paid for providing across-the-board healthcare, as well as free education, is small compared to the value of life saved. The shackles of the reactive model, still dominant even in such countries despite progress in pro-active policies, prevent an even better outcome. This could qualify as yet another proof that the path of anticipatory action remains a difficult choice in the context of a society conditioned to expect, on the premise of infinite progress, instant gratification—at no matter the cost. A COVID-19 vaccine based on what was in the process of being developed already 20 years ago is an example of opportunity missed.

An Outcome We Could Have Lived Without

For those still trying to understand the breadth of the *Sustainability Thesis*, yet another example: On August 31, 2021, the war in Afghanistan was declared over. It started in reaction to the vicious terrorist attack of September 11, 2001 that cost lives (over 2700) and caused substantial physical, economic and emotional damage. The most massive reaction operation ever carried out included the invasion of Afghanistan and the war in Iraq.[1] Many countries joined the USA in its fight. Focused on terrorism, it costs eight trillion dollars, and over four million lives. It was and continues to be a confrontation between an anticipation-driven multitude of forms of aggression and the large-scale reaction to them. No one can be satisfied with the outcome: Iraq is destabilized. The Taliban is back in power in Afghanistan, controlling the lives of millions of people and threatening those not in line with their fanatic views. Terrorism did not come to an end; but societies keep paying a huge price for combatting it, including the loss of many everyday liberties. Removing shoes when checking in at airports is a nuisance, as are the entire boarding and re-boarding security measures. It is not just an inconvenience. Disguised as a preventive procedure, but rather yet another loss of freedom. Everyone under observation—is a nightmare accepted as the new normal. What is left after the lives of sovereign people in democratic societies are under continuous surveillance is no longer an ideal worth fighting for. As a matter of fact, the anticipatory dimension of life is curtailed. Outsourced to governments, claiming to protect the citizen, prevention is of no interest to the individual.[2]

It would be at least a late expression of remorse if the most advanced science and technology—on whose account so many lost their lives—would be disrupted. As things stand now, awareness of anticipatory processes is no longer optional if the future still has meaning to the human species. The future of life on planet Earth depends on it. Pursuing the same path of reaction will lead only to more failure.

References

Arrobas *et al.* (2017) *The Growing Role of Minerals and Metals for a Low Carbon Future (English).* Washington, DC: World Bank Group. http://documents.worldbank.org/curated/en/207371500386458722/The-Growing-Role-of-Minerals-and-Metals-for-a-Low-Carbon-Future [retrieved August 26, 2022].

[1] The American military presence in over 80 countries is part of reacting to terror instead of preventing it.

[2] The Russian-Ukraine war, like the reaction to terrorism that led to the lost wars in Iraq and Afghanistan, could have been prevented. After the implosion of the Soviet Union, the world should have chosen an anticipatory course of action: engage all the parties of the defunct political system, instead of directly or indirectly instigating conflict among them. In the absence of such actions, wars, some with deep roots in history (such as those between Armenia and Azerbaijan, Osseatia and Georgia, Chechnya and Russia) unavoidably erupt. And people die. As it happened on October 7, 2023, the bloodiest day for Jews after the Holocaust, it could have been prevented. At a much lower price than the justified reaction it triggered.

Beaton T (2017) How Preventive Healthcare Services Reduce Spending for Payers, *Healthpayer Intelligence*. https://healthpayerintelligence.com/news/how-preventive-healthcare-services-reduce-spending-for-payers [retrieved December 3, 2021].

Bradley BT *et al.* (2020) Histopathology and ultrastructural findings of fatal COVID-19 infections in Washington State: a case series, *The Lancet,* 396:10247, 320–332, August 1.

Cooper KW *et al.* (2020) COVID-19 and the chemical senses: supporting players take center stage, *Neuron*. https://doi.org/10.1016/j.neuron.202006.032July 1 [retrieved August 15, 2020].

Council of the European Union (2020) *Recovery Plan for Europe*. https://ec.europa.eu/commission/presscorner/detail/en/ip_20_940 [retrieved August 14, 2020].

Diamond L (2020) The Coronavirus Is Emboldening Autocrats the World Over, Foreign Affairs, June 13. https://www.foreignaffairs.com/articles/world/2020-06-13/democracy-versus-pandemic [retrieved June 14, 2020].

Courtillot V, Gaudemer Y (1996) Effects of mass extinction on biodiversity, *Nature*, 381, pp. 146–148.

Crosby SS (2020) My COVID-19, *Annals of Internal Medicine*, 11 August https://www.acpjournals.org/doi/full/https://doi.org/10.7326/M20-5126?journalCode=aim [Retrieved August 15, 2020].

Diethelm P, McKee M (2009) Denialism: What is it and how Should Scientists Respond? *European Journal of Public Health* 19:1), 2–4.

Fine G (2022) fighting fraud, waste, and abuse—COVID 19, Brookings Institute. February 11. https://www.brookings.edu/blog/fixgov/2022/02/11/fighting-fraud-waste-and-abuse-covid-19-pandemic-relief-expenditures/ [retrieved September 22, 2022].

Gorard J (2020) Some Relativistic and Gravitational Properties of the Wolfram Model. https://www.wolframcloud.com/obj/wolframphysics/Documents/some-relativistic-and-gravitational-properties-of-the-wolfram-model.pdf. [retrieved September 18, 2020].

Goshua G *et al.* (2020) Endotheliopathy in COVID-19-associated coagulopathy: evidence from a single-centre, cross-sectional study, *The Lancet Hematology*, June 30 https://www.thelancet.com/journals/lanhae/article/PIIS2352-3026(20)30216-7/fulltext [retrieved August 10, 2020].

Gottlieb S (2021) Uncontrolled Spread: Why COVID-19 Crushed Us and How We Can Defeat the Next Pandemic. New York: Harper.

Greenstone M, Nigam V (2020) Does Social Distancing Matter? Becker Friedman Institute Working Paper, March 25. https://bfi.uchicago.edu/working-paper/2020-26/ [retrieved December 1, 2021].

Hannson SO (2017) Science Denialism as a Form of Pseudoscience, *Studies in History and Philosophy of Science* 63, 39–47.

Hatzidaki O, Saridakis IE, Karampelas P (2021) Organizers of Workshop *Science denialism as grassroots resistance: Digital discourses and social networks*. Planned for August-September 2021. See: https://www.academia.edu/44167676/Call_for_Papers_Science_denialism_as_grassroots_resistance_Digital_discourses_and_social_networks.

Heidt A (2020) Studies Report Rapid Loss of COVID-19 Antibodies, The Scientist, June 19. https://www.the-scientist.com/news-opinion/studies-report-rapid-loss-of-covid-19-antibodies-67650 [retrieved June 19].

Hobsbawm E (1965) The General Crisis of the European Economy in the 17th Century. In: (Aston T, Ed.) *Crisis in Europe 1560–1660: Essays from Past and Present*, pp. 5–58. London: Routledge & Kegan Paul.

Kissler SM *et al.* (2020) Projecting the transmission dynamics of SARS-CoV-2 through the post-pandemic period, *Science* 368:6493, 860–868. May 22.

Krugman P (2020) On the Economics of Dying, *New York Times*, May 28. https://www.nytimes.com/2020/05/28/opinion/coronavirus-economy-death.html [retrieved December 1, 2021].

Nadin M (2013) *Are You Stupid?* Heidelberg: Synchron.

Nocera J, McLean B (2023) COVID Lockdowns Were a Giant Experiment. It Was a Failure, A key lesson of the pandemic. New York, October 30. https://nymag.com/intelligencer/article/covid-lockdowns-big-fail-joe-nocera-bethany-mclean-book-excerpt.html [retrieved November 1, 2023]

Ramsey FP (1928) A Mathematical Theory of Saving, *Economic Journal* 38:152, 543–559.

Riley JL (2015) Still Right on the Black Family After All These Years. The warnings that Daniel Patrick Moynihan sounded 50 years ago have come true. Will liberals ever forgive him? *Manhattan Institute.* February 11. https://www.manhattan-institute.org/html/still-right-black-family-after-all-these-years-5197.html [retrieved September 22, 2022].

Roth K (2020) How Authoritarians Are Exploiting the COVID-19 Crisis to Grab Power, NYR Daily, March 31. https://www.nybooks.com/daily/2020/03/31/how-authoritarians-are-exploiting-the-covid-19-crisis-to-grab-power/ [retrieved April 15, 2020].

Rull V (2011) Sustainability, capitalism and evolution, EMBO Reports 12, pp. 103–106. https://doi.org/10.1038/embor.2010.211 [retrieved September 2, 2022].

Schelling, TC (1968) The life you save may be your own, *Problems in Public Expenditure Analysis*, ed. S. B. Chase, Jr., Washington, DC: The Brookings Institution. pp. 127–162. https://www.brookings.edu/blog/fixgov/2022/02/11/fighting-fraud-waste-and-abuse-covid-19-pandemic-relief-expenditures/ [retrieved July 3, 2022].

Schwab K, Malleret T (2020) *Covid-19: The Great Reset*, Cologne/Geneva: Forum Publishing.

Shesthra NK *et al.* (2022) Effectiveness of the Coronavirus Disease 2019 (COVID-19) Bivalent Vaccine, medR$_{X}$iv. https://doi.org/10.1101/2022.12.17.22283625v1 [retrieved December 16, 2022]

Sweis NJ (2022) Revisiting the value of statistical life: an international approach during Covid-19, *Risk Management*, 24, pp. 259–272. April 19. https://doi.org/10.1057/s41283-022-00094-x [retrieved September 15, 2022].

Taibbi M (2020) Big Pharma's Covid-19 Profiteers. How the race to develop treatments and a vaccine will create a historic windfall for the industry—and everyone else will pay the price, *Rolling Stone*, August 13. {Retrieved August 14, 2020].

Tattersall I, Schwartz JH (2009) Evolution of the Genus Homo, Annual review of Earth and Planetary sciences, 37, pp. 67–92. May 30. https://doi.org/10.1146/annurev.earth.031208.100202 [retrieved September 19, 2022].

Toshiaki I *et al.* (2020) Coagulopathy of Coronavirus Disease 2019, *Critical Care Medicine* May 27. https://pubmed.ncbi.nlm.nih.gov/32467443/ [retrieved August 10, 2020].

Prometheus and Epimetheus—An Epilogue

Prometheus and his twin brother Epimetheus—mentioned in the writings of Plato (*Protagoras*) and Hesiod (*Theogony* and *Works and Days*)—are characters from a time closer to mythology than to science. Prometheus is associated with foresight (that's what his name means), his brother with hindsight (for which the name stands). Some refer to forethought and afterthought, respectively. Zeus asked the twin Titans to populate the world. Epimetheus made the animals and endowed each species with whatever was available to him: wings for flight, claws to predators, bodies for speed, stealth, cunning, and so on. By the time Prometheus made man, nothing was left for the object that he had molded out of clay in the form of the gods. To give them life, he stole fire from Zeus. He taught humans, who ended up sacrificing animals to the gods, to keep the best—"chops and roasts"—for themselves. The audacity of telling humans to leave "scraps and entrails and fat" for the gods on Olympus brought trouble upon him. Nailed to a mountain in the Caucas, an eagle preys on his innards. The liver is healed continuously—he remains alive to experience the torture. The punishment is so descriptive in its details, that the meaning of his having empowered humans through the gift of fire is ignored. Tragedy is more powerful than inspiration. Zeus punished Prometheus but also wanted to bring woe upon mankind. He gave Hepheastus the assignment to make a woman of clay—Pandora—into which the four winds would breathe life. She was intended for Epimetheus, the prototype of all those who act in afterthought, i.e., in reaction. Pandora brought a jar with her, which—surprise!—her husband could not keep from opening. Out rushed Old Age, Labor, Sickness, Insanity, Vice, Passion, etc. But inevitably, Hope also: illusion cannot be left out—but what a curse for humankind to live by.

If there were no Epimetheus at hand, Zeus would have chosen Descartes, some centuries later, to open Pandora's jar. The method that Descartes eventually affirmed was the reaction-based answer to everything that needs a rational explanation. Recall one of his many afterthought (i.e., cause-and-effect) explanations: envy (this was not in the jar) "forces up the yellow bile from the lower part of the liver, and the black bile that comes from the spleen, which diffuses itself from the heart to

© The Author(s), under exclusive license to Springer Nature Switzerland AG 2023

M. Nadin, *Disrupt Science*, https://doi.org/10.1007/978-3-031-43957-5_8

the arteries." Of course, there is no bile in the spleen, the example illustrates the scientific method based on his reductionist-determinism. This mechanistic view of physiology—amended by discoveries made since his time, extends to our days. In spirit—but omitting Cartesian doubt—it is probably at the level of the molecular and genetics explanations of how and why COVID-19 victims succumb. But let's return to Prometheus.

The gift of fire is actually the hyperbole of life as an expression of anticipation. Actions associated with fire illustrate how the possible future affects the present. Fire brings light into the dark and heat into shelter; it renders raw food into easily edible, good-tasting cooked meals. It prevents disease by killing bacteria, parasites, microbes, and more. The culture of fire is one of discovering the future in both its desired aspects—the flame of passion (leading to sexuality and offspring), deferred gratification (as in cooked food vs. raw meat and vegetables)—and in its less than desired consequences: devastating fires, killings (the FIRE! command), smoke intoxication. Good or bad—and sometimes good and bad—it is, in short, emblematic of non-determinism. A spark might lead to devastating flames—or to nothing. The dynamics of fire escapes predictability. It can easily get out of control. And often, all that is left to fight fire is fire.

Epimetheus was into repairing, fixing what broke down in the past. Associating cause and effect was his take—hindsight, afterthought—on causality. Prometheus and the gift of fire can be associated with an anticipatory view—foresight/forethought, as his actions, not only his name, disclose. Vaccinations, for example, work like fighting fire with fire, exposing the organism to a controlled form of a disease or causing the disease for the purpose of engaging the organism to fight it. Epimetheus is rather the force necessary to confront foreseeable (but not prevented) breakdowns after they get out of control.

The crisis of vision—CoVID—brought the disaffected population to the streets, and with them the passions that led to violence and destruction. In the aggregate crisis into which COVID-19 morphed, Epimetheus would do what we see happening when ignorance takes over: He would ask for more gifts from Zeus. (In our days he would ask the government, i.e., the taxpayers.) Would Prometheus try to save the world by setting it on fire? Fixing or starting anew?

Contrasting mythological characters is different from contrasting fundamental views of reality. If Descartes had described his view of how science deals with reality (Discourses 1637) at the time when Zeus asked the twin Titans to populate the world, only Epimetheus would have appeared in the story. There is no room for Prometheus in Descartes' perspective, as there is no room for understanding the fundamental distinction between the living and the non-living. The unity of the reactive (which describes the non-living) and the anticipatory (which describes how life unfolds) can benefit from the story's expressive power, but not from its logic.

Indeed, in taking note of the Western view of the world, one cannot avoid the realization that the Industrial Revolution rode successfully into the world scene on Descartes' mechanical horse. It is with him, and those who think like him, that machine theology (Nadin 2018) became the new religion. Truth be told, Descartes saw only animals as God-made machines; he did allow that humans possessed a soul. But this was forgotten and the human, who invented machines, began to be considered a machine. Manufacturing—making things by hand with the aid of tools—was superseded by the factory model in which machines are used to produce more and more cheaply. For those who are addicted to consumption, the fact the human is turned into a part of the larger machine each factory is, does not enter into their thinking process. The machine model shaped architecture: Le Corbusier's *machine à vivre*, (1923, here called *machine à habiter*) is a vivid description of the machine-based view of shelter. Hospitals, military camps, schools, and universities became machines for processing the sick, for training the military, for molding the young in the spirit of determinism. Before the Cartesian Revolution, the Catholic Church took care of people needing medical care, of the elderly, of children (mostly orphans), and also of future fighters. It made sure that everyone was exposed to its doctrines. Cartesianism, undisputable in affirming the rationality of science, emerged from the fertile ground of religion. But once established as the new doctrine, its proponents ended up practicing it as a theology. With the Industrial Revolution, the mechanistic pattern of automating everything that had to be made—shoes, canned soup, sausage, clothing, and practically all there is to consume shaped a new human condition. Every living being that had to be taken care of—the young, the sick, the elderly, criminals—ended up in forms of dependency similar to those of cogs in a machine. The underlying causality of profit-making, characteristic of capitalism, is the foundation of industrial (and post-industrial) society. There is no profit-making in being pro-active.

In the final analysis, the result is a usurpation of individual freedom of choice, replaced by fully determined cycles easier to maintain and keep up to date than the creative effort of individuals. Anticipation, on account of which the living ascertains individuality in creative activities (from reproduction to skilled work, to art and literature, to dedication to the well-being of others) is progressively eliminated. True, anticipation cannot be automated. Neither can behavior, unless it is disconnected from its motivation and purpose divorced from life.

Machine theology embodying the belief in determinism—the "mother of all causes"—and social Darwinism—based on a distorted understanding of evolution as the outcome of competition—fused into capitalism. Reaction to breakdowns instead of actions to prevent them: the subject is as old as the human awareness of change. Reaction comes with costs attached to it: fixing, for all practical purposes, is of the scale of effort and investment needed to make what you now must fix. Pro-active actions, from improved maintenance to facilitating alternatives, saves money (and sometimes lives)—provided that those who benefit from infrastructure-based services are motivated to support them. The reactive path entails a price tag exceeding what it cost to build the infrastructure in place. It is accepted because it takes less time than preventive maintenance, and distributes funds to more beneficiaries, especially to those who profit from large government projects. Of course, it goes without saying that performing the job correctly in the first place is an anticipatory action *par excellence*. But it does not pay in an economy of preprogrammed obsolescence. To throw away is still cheaper than to make it last and adapt.

In a society shaped by the theology of the machine, the consequence is that solidarity is overridden by exacerbated individualism or self-centeredness. The expectation of higher prosperity and security is all that matters. That it comes at a cost—in particular, at the cost of the future—is left for other people to worry about. The human being thus conditioned becomes part of the larger machine, free of ethics, of course. Therefore, unable to realize that what is deemed as unlimited progress is unsustainable. Above and beyond the immediacy of social events that brought people to the streets all over the world, let's once again remember the industrial model of farming: chickens/poultry, cows/dairy, beef, pigs, fish farming, antibiotics; dependence of intensive agriculture on powerful chemical fertilizers and pesticides; the morphing of medical care into automated maintenance procedures of the living machines that humans are being turned into. Society driven by the insatiable instinct for ever-more at the lowest price created many ecological niches for more and invasively dangerous viruses. To be alive might mean to face risk, but definitely not to increase it to the level of undermining it.

All this can seem too far away from the choke flu known as COVID-19, or from the larger crisis going by CoVID, a crisis of knowledge and values. Why reaction to everything that the SARS-CoV-2 brought about will cost way more than anticipatory action does not seem self-evident. The profound social and economic disparities of the larger crisis are reflected in the statistics of those who became contaminated, those who died, those who suffered most (in regard to unemployment, to access to medical care), but also those who, for better or worse, decided "Enough is enough!" Whether this applies to opportunity, police brutality, political demagoguery, or any form of real or perceived discrimination remains open. A generation without a future reclaims the right to one.

There is no reason to reject determinism, on account of which civilization reached its current stage of well-being. But there are profound reasons to question the price of progress, and, even more, to argue in favor of a more comprehensive view of reality that includes the anticipatory perspective. Exclusive reliance on physics—de facto the god of science—and on chemistry leads to a situation in which every breakdown becomes an economic opportunity. And thus a source of profit. At the expense of lives lost, destinies affected, and deeper dependence on the science and technology that failed, progress is reported. In the long run, this changes the human condition. Machine-like performance is expected. Therefore the economy forces the human being into non-human behavior, for eventual replacement by machines. For an observer from another planet, CoVID is relevant in regard to the inadequacy of reactive understanding. It is embodied in a variety of disciplines and practices vying for recognition as science worthy of funding. The various models of the pandemic, as well as the models of economic breakdowns, social unrest, political turmoil, and election results, compete for being as irrelevant as possible. But even more, society is forced to find out, as the imaginary observer from afar might realize, how primitive the Cartesian cause-and-effect understanding of causality is. This is the most important revelation—inviting change from the automatic acceptance of the positivist Cartesian foundation of science to a critical examination of it.

In the firm embrace of the Cartesian view, science—in particular, what is defined as medical science—failed miserably. Physicians, trained to see the world through the limited perspective of determinism, ended up on the front line, with only protective gear (sometimes questionable), but no valid knowledge for understanding a pandemic that broke from all patterns on record. Medical scientists, comfortably ensconced in their academic institutes or research facilities, delivered after-the-fact minutiae to consolidate their tenure. If anything, physicians were at least trying the repurposing of treatments. All of this while laypersons, from various professions and interest groups, argued whether one drug or another, or anything else—vaccines, in particular—deserved the public attention they received.

It was the moment of truth for the pharma industry ("Big Pharma")—the most profitable lobby in the world—competing for every imaginable path to higher profits. The health industry, in whose hands medicine became a "nail-and-hammer" profession, pushed testing. This means: "Measure for us, produce the data"—because each individual in a database is worth its weight in the gold of our worries and concerns. The fact that measuring never does away with a problem, providing at most coordinates for where society is at some moment in time, never made it into the public domain. Curves were to be flattened by more testing, no matter how faulty or misleading some tests were. In reality, obsession with data proved to be about tracking—a ubiquitous attempt at limiting freedom. Who is infected, who might get infected through exposure to infected persons, who is infected but not symptomatic, and on and on is only a small part of the generalized tracking of everyone, instituted subsequent to the 9/11 terror attacks. Before COVID-19, tracking—which means robbing individuals of their identity with the aim of monetizing it—prompted social activists to revolt and denounce it. *The Age of Surveillance Capitalism* (Zuboff 2019), as it is labeled, is actually the broad stroke expression of deeper dependencies reflecting the loss of individual sovereignty. It is the individual monetized over the entire lifespan—cradle to grave. With COVID-19—to stay with us as do HIV/AIDS, the flu, and SARS—such tracking will settle in as much and as acceptable as tracking a citizen's water and electricity consumption, automobile use and miles driven, not to say, where, when, why—the last tidbits of one's identity.

The pharma industry and the medical science promoters forced physicians to make abstractions of their patients. They are supposed to focus on disease as some kind of engine malfunction in need of chemicals to restore proper functioning—or in need of spare replacement parts. No wonder the streets filled with demonstrators, and social media sucked in, broadcasting the performance, confused people's minds even more. The sense of alienation ran counter to the promises of never-ending progress anchored in the past they started questioning. Those promises, made decades ago, are the best evidence of the failure of determinism. Do this—give money, pass laws, legalize aberrant behaviors—and the result will be positive: opportunity, equality, general well-being. None of this came about. None can.

Dependence on deterministic means and methods increased since COVID-19. And this holds true for the confluence crisis of pandemic, economy, social and political conflicts. Determinism replaces self-determination, transforming democracy—the will of the people—into the right to abandon responsibility for the satisfaction of a questionable prosperity at the price of future generations and a compromised ecology.

This rather broad view, of less than unlimited optimism, is probably more appropriate in a philosophic context. It does not yet fully account for why the anticipatory view is practically deemed as anti-scientific by a science fully captive to the theology of the machine and its underlying reduction of causality to reaction only. Misunderstanding of the living as nothing other than an embodiment of electrons and protons, of atoms, molecules, genetic components, and of whatever else physics can deal with (in particular, in computational form) does away with how life is actually expressed in the living: purpose, choice, value. In other words: meaning. This particular reductionism—which totally ignores meaning—is extremely consequential: the next breakdown, of a larger scale, is closer than any previous breakdowns. Unless the liberated Prometheus inspires not only the understanding of the importance of foresight, but also the revolution in anticipatory thinking and acting. *Prometheus Unbound* (the title of Aeschylus's play, now lost) could mean liberating humankind from subservience to idolatry. Hope is always anticipatory. Revolutions are grounded in hope.

References

Descartes R (1637) *Discourse on the Method of Rightly Conducting One's Reason and of Seeking Truth in the Sciences* (Discours de la Méthode Pour bien conduire sa raison, et chercher la vérité dans les sciences). Leiden: Jan Maire

Le Corbusier (1923) *Vers une architecture* (Towards an Architecture) (Trans. Etchells F, London: J. Rodker, 1931)

Nadin M (2018) Rethinking the experiment: a necessary (R)evolution, *AI & Society* 33:4, 467–485. New York/London: Springer

Zuboff S (2019) The Age of Surveillance Capitalism: The Fight for a Human Future at the New Frontier of Power. New York: Public Affairs/Hachette

Afterword

No book has yet to leap from an author's mind onto the printed page, to be embraced by a well-designed cover, such as that created by Baruch Gorkin. Each book has its own story.

On March 27, 2020, Floriana Jucan, who publishes Q magazine in Romania, provoked me: "Would you like you write something about this virus? About what is waiting for us tomorrow?

Instead of responding to the questions, I wrote an essay, which became the Epilogue of this book: "Reaction to Crises: Prometheus and Epimetheus." After that, the book wanted to be written. At the end of 2020, the text was ready. Several publishing houses rushed to read it. From among all of them, Spandugino Publishers—more precisely, Lavinia Spandonide—accepted the text and generously offered to have it translated into the Romanian. The book, written in English, was to have its world premiere in Romanian. It was not the final version. The fact that the text, in its first rendition, appeared in Romania, where I was born, honors me. Maybe those who are experiencing a new way of life after dictatorship will find through it their opportunity to challenge the current model of science. Hope originates in anticipation.

Writing this book is an attempt to share what I discovered and learned. Almost one hundred medical practitioners from many countries taught me the humility of a dedication to life that I would like to feel in respect to everyone. Many doctors have lost their lives trying to save the lives of others—relying, unfortunately, on a science that betrays them. Therefore, the high rate of doctor suicides cannot be ignored.

Valuable insights were provided through either personal communication or through their writing and commentaries on the subject: Dr. Anders Tegnell (Sweden), Chen Shih-chung (Taiwan), Christian Drosten (Germany), Jean-François Delfraissy (France), Lisa Cosimi, M.D. (USA) Todd Pollack, M.D. (USA), Vo Thi Tuyet Nhung (Vietnam), Víðir Reynisson (Iceland), Þórólfur Guðnason (Iceland), Alma Möller (Iceland), Mohammed Jobayer Chisti (Bangladesh), Ronni Gamzu (Israel). The untiring Eric Topol used Twitter (with ca. 500,000 "subscribers" to his postings) to highlight various medical aspects. I don't care for his activism, but I am grateful

© The Editor(s) (if applicable) and The Author(s), under exclusive license to Springer Nature Switzerland AG 2023

M. Nadin, *Disrupt Science*, https://doi.org/10.1007/978-3-031-43957-5

for his daily reports (which became a Substack title). Whoever writes so much, especially under daily pressure, is bound to make mistakes. He made many! Eran Segal (Weizmann Institute) remained the cool head of data analysis, from which we can all learn. Peter Hotez (*yasher koach*), a voice of competence in a deluge of strident misrepresentations, found time to advise me in matters of vaccinology (and more). It is too bad that he also took the path of activism.

Shlomo Ber Estrin triggered an inescapable call to clarifying why a new science perspective—"Second Cartesian Revolution" described in the book—became critical. Marcos Novak, faithful to his Greek heritage, also inspired the beginning of the adventure. Asma Naz, one of the best PhD candidates I had the privilege of advising, accepted to think with me about intuitive ways to explain the idea expressed in the Sustainability Thesis. Anne Balsamo, former Dean of the ATEC School at the University of Texas at Dallas, supported the intense effort, fully aware of the fact that during a pandemic, when there are so many things to take care of, writing a book would not qualify as essential from a University Administration viewpoint focused on enrollment numbers and donations. Last but not least, Susan Evans, Karsten Lücke, Baruch Gorkin, Maryam Ashkaboosi: for assisting in the visual thinking involved in visualizing realizing the intricate nature of human interactions in a context of simultaneous crises. Through their patient reviews, … and …. Helped me to clarify the message. Many others (asking for anonymity) deserve credit for serving as sounding boards, and often bringing me back from the abstraction of thinking to the concreteness of life. COVID-19 stole many lives from me, and from us as a community; but it gave me a chance to confront crises (my own included) from a perspective of hope. This world will not become a better place if we do not become better human beings.

Bibliography

Altmann D (2022) Vaccines are no match for long Covid. Treating it is science's next great challenge, The Guardian, April 16. https://www.theguardian.com/commentisfree/2022/apr/16/vaccines-long-covid-science [retrieved April 20, 2022].

Ballhous R (2020) As Covid Hit, Washington Officials Traded Stocks with Exquisite Timing, *Wall Street Journal.* https://www.wsj.com/articles/covid-washington-officials-stocks-trading-markets-stimulus-11666192404 [retrieved December 21, 2022].

Baum A et al. (2020) REGN-COV2 antibodies prevent and treat SARS-CoV-2 infection in rhesus macaques and hamsters, *Science*, October 9, 370:6520, pp. 1110–1115. https://www.science.org/doi/10.1126/science.abe2402 [retrieved March 3, 2022].

Begley CG, Ellis LM (2012) Drug development: Raise standards for preclinical cancer research, *Nature* 483:7391, pp. 531–533. March 28. https://www.nature.com/articles/483531a [retrieved March 3, 2022].

Button KS, Ioannidis JP, Mokrysz C et al. (2013) Power failure: why small sample size undermines the reliability of neuroscience. *Nature Reviews Neuroscience*, 14, pp. 365–376. https://doi.org/10.1038/nrn3475 [retrieved March 15, 2022].

Cavaillon J-M (2019) Andre Boivin: A pioneer in endotoxin research and an amazing visionary during the birth of molecular biology (Review Article), *Innate Immunity*, 25:3, pp. 1–7.

Curtis A (2011) All Watched Over by Machines of Loving Grace (TV series) Part 2, "The Use and Abuse of Vegetational Concepts." London: BBC Two.

Dawkins R (1976) *The Selfish Gene.* Cambridge: Oxford University Press.

Dolgin E (2021) The tangled history of mRNA vaccines, *Nature*, 14 September, 22 October).

Dumas-Mallet E et al. (2017) Low statistical power in biomedical science: a review of three human research domains, Royal Society Open Science, 4:2, 160254, February 1. https://doi.org/10.1098/rsos.160254. [retrieved March 15, 2022].

Esland van SL, O'Hare R (2020) Coronavirus pandemic could have caused 40 million deaths if left unchecked *Imperial College of London NEWS* 26 March. https://www.imperial.ac.uk/news/196496/coronavirus-pandemic-could-have-caused-40/ [retrieved August 15, 2020].

European Lung Foundation (2020) COVID-19 patients suffer long-term lung and heart damage; Can improve over time. https://newsabode.com/covid-19-patients/ [retrieved September 7, 2020].

Fahrenthold D (2022) Prosecutors Struggle to Catch Up to a Tidal Wave of Pandemic Fraud *New York Times*, August 16. https://www.nytimes.com/2022/08/16/business/economy/covid-pandemic-fraud.html [retrieved August 18, 2022].

Freedman L, Cockburn IM, Simcoe T (2015) The Economics of Reproducibility in Preclinical Research, PLoS Biology 13:6:e1002165, June. https://www.researchgate.net/publication/278046305_The_Economics_of_Reproducibility_in_Preclinical_Research [retrieved March 3, 2022].

© The Editor(s) (if applicable) and The Author(s), under exclusive license to Springer Nature Switzerland AG 2023
M. Nadin, *Disrupt Science*, https://doi.org/10.1007/978-3-031-43957-5

Funk RJ, Owen-Smith J (2017) A dynamic network measure of technological change, Management Science 63, pp. 791–817.

Hamilton WD (1963) The evolution of altruistic behavior, *The American Naturalist*, 97:896, September-October, 354–356. Chicago: The University of Chicago Press Journals.

Ioannides JPA et al. (2015) Meta-research: Evaluation and Improvement of Research Methods and Practices, PLoS Biology, October, 13:10, e1002264. https://www.ncbi.nlm.nih.gov/pmc/articles/PMC4592065/ [retrieved March 4, 2022].

Illich I (1974) Medical Nemesis, *The Lancet*, 303:7863, pages 918–921. May 11. https://www.thelancet.com/journals/lancet/article/PIIS0140-6736(74)90361-4/fulltext [retrieved September 4, 2022].

Iwasaki A et al. (2020) Why and How Vaccines Work, *Cell* 183:2, October 15, pp. 290–295.

Khan M et al. (2021) Visualizing in deceased COVID-19 patients how SARS-CoV-2 attacks the respiratory and olfactory mucosae but spares the olfactory bulb, *Cell*, 184:5932. November 24. https://doi.org/10.1016/j.cell.2021.10.027 [retrieved January 2, 2022].

Lancet (2022) Estimating excess mortality due to the covid-19 pandemic: a systematic analysis of COVID-19-related mortality, 2020–2021, March 10. https://www.thelancet.com/journals/lancet/article/PIIS0140-6736(21)02796-3/fulltext [retrieved March 13, 2022].

Ledford H (2020) Coronavirus reinfections: three questions scientists are asking, *Nature* 585:7824. pp. 168–169, September 4.

Lepore J (2020) How the Simulmatics Corporation Invented the Future, *The New Yorker*, August 3&10, 18–24.

Lewis D (2021) Superspreading drives the COVID pandemic—and could help to tame it, Nature 590, 544–546, February 23. https://www.nature.com/articles/d41586-021-00460-x [retrieved March 13, 2021].

Malone RW, Felgner PL, Verma IM (1989) *Proceedings of the National Academy of Science*, 86, 6077–6081, August 1. https://www.pnas.org/content/86/16/6077 [retrieved December 2, 2021].

Moore K (2013) *Disrupting Science: Social Movements, American Scientists, and the Politics of the Military, 1945–1975*. Princeton: Princeton University Press.

Niarchos N (2021) The Dark Side of Congo's Cobalt Rush, *The New Yorker*, May 31.

Prinz F, Schlange T, Asadullah K (2011) Believe it or not: how much can we rely on published data on potential drug targets? *Nature Reviews Drug Discovery*, 10, 712, September. https://doi.org/10.1038/nrd3439-c1 [retrieved March 3, 2022].

Regalado A (2022) How Pfizer made an effective anti-covid pill, MIT Technology Review. February 4. https://www.technologyreview.com/2022/02/04/1044714/pfizer-covid-pill-paxlovid-pandemic [retrieved April 20, 2022].

Ritchie H et al. (2021) Coronavirus (COVID-19) Testing, Our world data https://ourworldindata.org/coronavirus-testing (retrieved March 13 2021). See also: Hasell J et al. A cross-country database of COVID-19 testing. *Sci Data* 7, 345 (2020). https://doi.org/10.1038/s41597-020-00688-8

Seneff S, Nigh G (2021) Worse Than the Disease? Reviewing Some Possible Unintended Consequences of the mRNA Vaccines Against COVID-19, International Journal of Vaccine theory, Practice, and Research, 2:1, May 10, pp. 38–79.

Siddhartha V, Dhanak M, Frankenfeld J (2020) Visualizing droplet dispersal for face shields and masks with exhalation valves, *Physics of Fluids* 32:9. September 1. https://doi.org/10.1063/5.0022968 [retrieved September 1, 2020].

Sterling P, Eyer J (1988) Allostasis: A new paradigm to explain arousal pathology. In: Fisher S, Reason JT (Eds.) *Handbook of life stress, cognition, and health*. Chichester, NY: Wiley.

Sudharsanan N et al. (2020) The Contribution of the Age Distribution of Cases to COVID-19 Case Fatality Across Countries, Annals of Internal Medicine, July 22. https://www.acpjournals.org/doi/10.7326/m20-2973 [retrieved July 22, 2022].

Thao TTN et al. (2020) Rapid reconstruction of SARS-CoV-2 using a synthetic genome platform, (Accelerated Article Preview, April), Nature, 582, 561–565, May 4. https://www.nature.com/articles/s41586-020-2294-9 [retrieved August 10, 2020].

UN Environment Programme (2020) The next wave—How to beat future pandemics. July 6. https://www.unenvironment.org/news-and-stories/video/next-wave-how-beat-future-pandemics [retrieved July 7, 2020].

Windelband, 'Geschichte und Naturwissenschaft' Präludien Band 2 (Tübingen: Möhr, 1924) pp. 136–160.

Printed in the USA
CPSIA information can be obtained
at www.ICGtesting.com
LVHW051113291123
764947LV00028B/225

9 783031 439568